Analytical System Dynamics

Brian C. Fabien

Analytical System Dynamics

Modeling and Simulation

Springer

Brian C. Fabien
University of Washington
Mechanical Engineering Department
Seattle, WA 98195

ISBN: 978-0-387-85604-9 e-ISBN: 978-0-387-85605-6
DOI: 10.1007/978-0-387-85605-6

Library of Congress Control Number: 2008935329

Printed on acid-free paper

springer.com

To Mary, Maurice and Zoë

Preface

This book combines results from *Analytical Mechanics* and *System Dynamics* to develop an approach to modeling constrained multidiscipline dynamic systems. This combination yields a modeling technique based on the energy method of Lagrange, which in turn, results in a set of differential-algebraic equations that are suitable for numerical integration. Using the modeling approach presented in this book are able to model, and simulate, systems as diverse as a six link closed-loop mechanism or a transistor power amplifier.

The material in this text is used to teach dynamic systems modeling and simulation to seniors and first-year graduate students in engineering. In a ten week course (4 hours per week) we cover most of the matcrial in this book. The basic prerequisites for the class are undergraduate level classes in; (i) physics (specifically, Newtonian mechanics and basic circuit analysis); (ii) ordinary differential equations; and (iii) linear algebra.

A brief summary of the chapters that make up the book are as follows.

Chapter 1 This chapter introduces Paynter's unified system variables; effort, flow, displacement and momentum. Here, we show how these fundamental system variables are used to relate power and energy, within, and between, the various engineering disciplines, i.e., mechanical, electrical, fluid and thermal systems.

Chapter 2 The modeling technique developed in this book requires that we determine analytical expressions for the kinetic coenergy, potential energy and the dissipation function. In the case of mechanical systems this requires that we determine expressions the position and velocity of various points on the system. In the case of electrical, fluid and thermal networks this requires that we determine the relationship between the flow variables of the system. This chapter develops an approach to the kinematic analysis of both planar and spatial mechanical systems. In addition, we consider the structural properties of network systems that give rise to constraint relationships between the flow variables.

Chapter 3 Lagrange's equation of motion is derived in this chapter by using a differential/variational form of the first law of thermodynamics. Here, via

numerous examples, we also present a systematic approach to deriving the equations of motion for multidiscipline dynamic systems.
Chapter 4 In this chapter Lagrange's equation motion is extended to accommodate systems with displacement, flow, effort and dynamic constraints. This modeling approach leads to the Lagrangian differential-algebraic equations (LDAEs) of motion.
Chapter 5 This chapter presents numerical techniques for the solution of differential equations, and differential-algebraic equations. Here, we give particular emphasis to the explicit and implicit Runge-Kutta methods. Also, we consider some of the subtleties involved is solving differential-algebraic equations. For example, we address the problem of computing consistent initial conditions, as well as the problem reducing the differentiation index of the system in order to obtain accurate solutions.
Chapter 6 The concepts of equilibrium, and stability, in the sense of Lyapunov, are introduced in this chapter. These ideas are used to analyze the behavior of some simple dynamic systems. This chapter also presents simulation results for various the models developed in the book. Here, we use two programs to demonstrate the efficacy of the modeling technique developed here. The first program, `ldaetrans`, translates an input file describing the model into the LDAEs. This is accomplished via symbolic differentiation of the system energy terms and constraints. The program `ldaetrans` also generates MATLAB/Octave files that can be used to integrate the equations of motion. A second program presented in this chapter is `ride`. This is a MATLAB/Octave implementation of an implicit Runge-Kutta method that is used to integrate the Lagrangian differential-algebraic equations. In addition, some of the example problems in this chapter are used to introduce concepts from linear and nonlinear feedback control.
Acknowledgment A number of my graduate students have contributed to the development of the material presented in this book. They include; Richard A. Layton, Jonathan Alberts and Jae Suk. In addition, the research that has culminated in this book was supported by the National Science Foundation under Grant MSS-9350467. This support is gratefully acknowledged.

Brian C. Fabien
Seattle, Washington
August, 2008

Contents

Chapter 1
A Unified System Representation

In this text we will develop an analytical approach to modeling multidiscipline dynamic systems. This modeling technique relies on examining the change in energy of the dynamic system. Here, the dynamic systems considered may contain components from various engineering disciplines. In particular, we may have mechanical, electrical, fluid and/or thermal components present in our dynamic system model. In classical approaches to modeling multidiscipline dynamic systems each of these engineering disciplines use a different set of variables to define the energy in the system. Here however, it will be shown that, in fact, the variables used to describe energy state in each of these disciplines all share common relationships.

Section 1.1 presents particular set of variables that can be used in each discipline to define the system energy. In Section 1.2 the components of the system are classified based on how they manipulate the system energy. In particular the system is divided into (1) energy storage components (ideal inductors and capacitors), (2) energy dissipative components (ideal resistors), (3) energy transforming components (constraint elements) and (4) energy sources. It will be shown that the energy storage components can be described in terms of the kinetic energy and potential energy. The dissipative components will be described in terms of a dissipation function. The transforming components give rise to constraints between the system variables. Finally, the energy sources will specify the energy input to the system.

1.1 Fundamental System Variables

In each engineering discipline (i.e., mechanical, electrical, fluid and thermal) the state of the system is defined using a pair of *kinematic* variables and a pair of *kinetic* variables. The kinematic variables are called *displacement*, $q(t)$, and *flow*, $f(t)$. The kinetic variables are called *momentum*, $p(t)$, and *effort*, $e(t)$. In these definitions t denotes the time. Also, $q(t)$, $f(t)$, $p(t)$ and $e(t)$ are

B. Fabien, *Analytical System Dynamics: Modeling and Simulation*,
DOI 10.1007/978-0-387-85605-6_1,

each vectors of dimension n, with n being the number of coordinates used to describe the system. Thus, associated with the i-th displacement $q_i(t)$, there is a corresponding flow, $f_i(t)$, momentum, $p_i(t)$, and effort, $e_i(t)$, variable for $i = 1, \ldots n$.

The quadruple $\{q(t), f(t), p(t), e(t)\}$ are called the *fundamental system variables* because they appear in all the engineering disciplines considered here. The familiar engineering terms for these fundamental variables are listed in Table 1.1. The table shows the names used for displacement, flow, effort and momentum in various engineering disciplines. We note that (i) mechanical systems are divided into components that are in translation and components that are in rotation, (ii) the fluid subsystem includes compressible and incompressible flow, and (iii) there is no momentum variable defined for thermal systems. The systems modeling approach, developed in this text, will account for the coupling that may exists between all of the disciplines defined in Table 1.1.

1.1.1 Kinematic variables

The kinematic variables (displacement $q(t)$ and flow $f(t)$) are related to the geometric or spatial properties of the system. These kinematic variables, listed in Table 1.1, share a differential (integral) relationship. Specifically, we have that

$$f(t) = \frac{dq(t)}{dt} \quad (a) \quad \text{and} \quad q(t) = \int f(t)\, dt \quad (b). \tag{1.1}$$

In each of the engineering disciplines these relationships can be illustrated as follows.

- Mechanical Translation:

  For an object undergoing pure translation the displacement variable, $q(t)$, corresponds to linear displacement, and the flow variable, $f(t)$, corresponds to the linear velocity. Consider an object moving in a straight line with position given by $x(t)$, then the velocity of the object is $v(t) = dx(t)/dt$. Conversely, if the velocity, $v(t)$, is a known function of time we can compute the displacement using $x(t) = \int v(t)\, dt$.

- Mechanical Rotation:

  For an object undergoing pure rotation the displacement variable, $q(t)$, corresponds to angular displacement, and the flow variable, $f(t)$, corresponds to the angular velocity. If we consider an object with angular position given by $\theta(t)$, then the angular velocity of the object is $\omega(t) = d\theta(t)/dt$.

Table 1.1 Fundamental Multidiscipline System Variables

	Effort e	Flow f	Displacement q	Momentum p
Mechanical Translation	force, F	velocity, v	position, x	linear momentum, p
Mechanical Rotation	torque, τ	angular velocity, ω	angle θ	angular momentum, H
Electrical	voltage, v	current, i	charge, q	flux linkage, λ
Fluid	pressure, P	volumetric flow rate, Q	volume, V	pressure momentum, Γ
Thermal	temperature, T	entropy flow rate, $\dot{S}$	entropy, S	(none)

If the angular velocity, $\omega(t)$, is a known function of time we can compute the angular displacement using $\theta(t) = \int \omega(t)\, dt$.

- Electrical:

  For electrical systems the displacement variable, $q(t)$, corresponds to the charge, and the flow variable, $f(t)$, corresponds to the current. Thus, if $q(t)$ is the charge in a conductor, then the current (the rate of charge flow) is given by $i(t) = dq(t)/dt$. If the current, $i(t)$, is known we can compute the charge accumulated using $q(t) = \int i(t)\, dt$.

- Fluid:

  In fluid systems the displacement variable, $q(t)$, corresponds to the volume, and the flow variable, $f(t)$, corresponds to the volume flow rate. Thus, if $V(t)$ is the volume of fluid in the system, then the volume flow rate is given by $Q(t) = dV(t)/dt$. If the volume flow rate, $Q(t)$, is known we can compute the volume accumulated using $V(t) = \int Q(t)\, dt$.

- Thermal:

  For thermal systems the displacement variable, $q(t)$, corresponds to the entropy, and the flow variable, $f(t)$, corresponds to the entropy flow rate. Thus, if $S(t)$ is the entropy in the system, then the entropy flow rate is given by $\dot{S}(t) = dS(t)/dt$. If the entropy flow rate, $\dot{S}(t)$, is known we can compute the entropy using $S(t) = \int \dot{S}(t)\, dt$.

1.1.2 Kinetic variables

The kinetic variables (momentum $p(t)$ and effort $e(t)$) are related to the inertial properties of the system. The kinetic variables also share a differential (integral) relationship. In particular,

$$e(t) = \frac{dp(t)}{dt} \quad (a) \quad \text{and} \quad p(t) = \int e(t) dt \quad (b). \tag{1.2}$$

These relationships can be verified for each of the engineering disciplines as follows.

- Mechanical Translation:

  For an object in pure translation the effort variable, $e(t)$, corresponds to the force, and the momentum variable, $p(t)$, corresponds to the linear momentum. Thus, if $F(t)$ is the net force acting on an object and $p(t)$ is its linear momentum then equation (1.2a) implies that

$$F(t) = \frac{dp(t)}{dt},$$

which is recognized as Newton's second law of motion.

- Mechanical Rotation:

  For an object undergoing pure rotation the effort variable, $e(t)$ corresponds to the torque, and the momentum variable, $p(t)$ corresponds to the angular momentum. Thus, if $\tau(t)$ is the net torque acting on an object and $H(t)$ is its angular momentum then equation (1.2a) implies that

$$\tau(t) = \frac{dH(t)}{dt},$$

  which is called Euler's equation of motion.

- Electrical:

  In electrical systems the effort variable, $e(t)$, corresponds to the voltage, and the momentum variable, $p(t)$, corresponds to the flux linkage. Let $v(t)$ denote the voltage drop across the terminals of a circuit, and $\lambda(t)$ the flux linkage, then using equation (1.2a) we obtain

$$v(t) = \frac{d\lambda(t)}{dt},$$

  which is called Faraday's law.

- Fluid:

  In fluid systems the effort variable, $e(t)$, corresponds to the pressure, and the momentum variable, $p(t)$, corresponds to the pressure momentum. Thus, if $P(t)$ is the net pressure drop across the fluid system, and $\Gamma(t)$ is the pressure momentum then, equation (1.2a) implies that

$$P(t) = \frac{d\Gamma(t)}{dt}.$$

  Note that this expression can be developed by applying Newton's second law to a element of fluid flowing in a pipe. (See Example 1.3 below.)

1.1.3 Work, power and energy

Suppose an effort $e(t)$ is applied to an element of the system, then the *increment in work* done by the effort in displacing the element an amount $dq(t)$ is defined as

$$\check{d}\mathcal{W}(t) = e(t)dq(t), \tag{1.3}$$

where $\check{d}$ is used to emphasize the fact that $\check{d}\mathcal{W}$ is not an exact differential of work, but instead an infinitesimal quantity (Meirovitch (1970), p. 14). Also, it is understood that $e(t)$ is the effort in the direction of the displacement $q(t)$.

Using equations (1.1) and (1.2) the increment in work can be written as

$$\begin{aligned} \check{d}\mathcal{W}(t) &= e(t)dq(t) \\ &= \frac{dp(t)}{dt}dq(t) \\ &= \frac{dq(t)}{dt}dp(t) \\ &= f(t)dp(t). \end{aligned} \tag{1.4}$$

Thus, $\check{d}\mathcal{W}(t)$ can also be described as the increment in work done by the flow $f(t)$ in changing the momentum an amount $dp(t)$.

Power is the rate at which work is performed and is given by

$$\mathcal{P}(t) = \frac{\check{d}\mathcal{W}(t)}{dt} = e(t)\frac{dq(t)}{dt} = f(t)\frac{dp(t)}{dt} = e(t)f(t). \tag{1.5}$$

Expressions for the work and power in each of the engineering disciplines are listed in Table (1.2).

Discipline	Work		Power
	$e\,dq$	$f\,dp$	ef
Mechanical Translation	$F\,dx$	$v\,dp$	Fv
Mechanical Rotation	$\tau\,d\theta$	$\omega\,dH$	$\tau\omega$
Electrical	$v\,dq$	$i\,d\lambda$	vi
Fluid	$P\,dV$	$Q\,d\Gamma$	PQ
Thermal	$T\,dS$	–	$T\dot{S}$

Table 1.2 Work and Power

Energy is the capacity to do work, and is defined as is the time integral of the power (Nelkon, (1973)). Thus, using equations (1.1), (1.2) and (1.5) we obtain

$$\mathcal{E}(t) = \int e(t)f(t)\,dt = \int_{C_q} e(t)dq(t) = \int_{C_p} f(t)dp(t). \tag{1.6}$$

Here, the last two integrals are along the displacement path C_q and the momentum path C_p, respectively.

In the Problems section the reader is asked to specify units for q, f, e and p, such that work, power and energy are consistently defined across all the engineering disciplines.

1.2 System Components

The modeling technique developed in this text divides the dynamic system into discrete components. These components are categorized according to how they manipulate the energy in the system. In particular, the system components are classified as one of the following:

- Energy storage components which are represented by ideal *inductors*, and the ideal *capacitors*.
- Energy dissipation components which are represented by ideal *resistors*.
- Energy transforming components which are represented by *constraint* elements.
- Energy sources which provide energy to the system.

These system components will be consider next, with examples and basic results for each of the engineering disciplines.

1.2.1 Ideal inductors

Ideal inductors are energy storage components whose behavior is determined by expressions that relate the momentum and flow variables. Here, these relationships are written as *constitutive* equations of the form $p = \Phi_I(f)$, where p denotes the momentum and $\Phi_I(f)$ is a continuous function of the flow, f. The constitutive equation need not be linear but, it is assumed that these relations are such that $\Phi_I(0) = 0$, and that $\Phi_I(f)$ is invertible. Thus, it is possible to find an expression for the flow in terms of the momentum, specifically, $f = \Phi_I^{-1}(p)$.

Examples of ideal inductors and their corresponding constitutive relationship are as follows.

- Mechanical Translation:

For mechanical systems in pure translation, the point mass represents the ideal inductor, and for velocities much less than the speed of light the constitutive equation is given by

$$p = mv,$$

where p is the linear momentum, m is the mass and v is the velocity. Note that the point mass or particle only occupies a point in space, i.e., it does not have any extent.

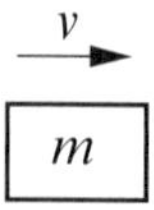

Point mass

- Mechanical Rotation:

For mechanical systems in pure rotation, the mass moment of inertia represents the ideal inductor. In this case the constitutive equation is given by

$$H = I\omega,$$

where H is the angular momentum, I, is the mass moment of inertia of the rotor, and ω is the angular velocity.

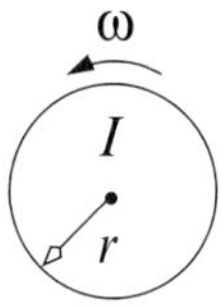

Rotor

Example 1.1.

As an example consider a uniform thin disk of mass m, and radius r. Then, the moment of inertia and the angular momentum about the center of mass of the disk are given by

$$I = \frac{mr^2}{2} \quad \text{and} \quad H = \frac{\omega mr^2}{2},$$

respectively.

- Electrical:

In electrical systems the inductor represents the ideal inductor, and the constitutive equation is given by

$$\lambda = Li,$$

where λ is the flux linkage, L is the inductance and i is the current.

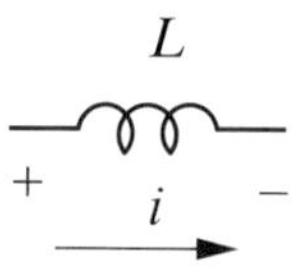

Inductor

Example 1.2.

An example of an electrical inductor is a coil formed by warping a conductor N times around a material with permeability μ. In which case the inductance is given by

$$L = \frac{\pi \mu d^2 N^2}{4l},$$

where d is the diameter of the coil and l is the length of the coil. Hence, the constitutive equation is $\lambda = \pi\mu d^2 N^2 i/4l$ (Smith, (1976)).

- Fluid:

In fluid systems the fluid inertia represents the ideal inductor, and the constitutive equation is given by

$$\Gamma = I_f Q,$$

where Γ is the pressure momentum, I_f is the fluid inertia and Q is the volume flow rate.

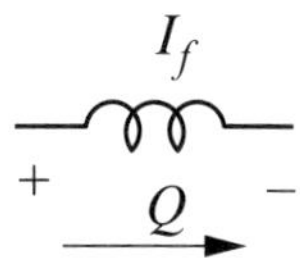

Fluid inertia

Example 1.3.

Consider fluid flowing in a pipe of length L, and uniform cross-sectional area A. Let P denote the difference in pressure at the ends of the pipe, Q the volume flow rate of the fluid, and ρ the density of the fluid in the pipe.

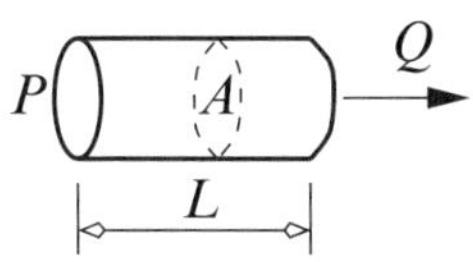

Then applying Newton's second law to the fluid in the pipe gives

$$\begin{aligned} \text{Force} &= \frac{d}{dt}(\text{mass} \times \text{velocity}) \\ PA &= \frac{d}{dt}(\rho AL \times (Q/A)) \\ P &= \frac{\rho L}{A}\frac{dQ}{dt}. \end{aligned}$$

Since $P = d\Gamma(t)/dt$ it can be deduced that the fluid inertia is given by $I_f = \rho L/A$, and the constitutive equation is $\Gamma = \rho LQ/A$.

Kinetic energy and kinetic coenergy

The energy stored by ideal inductors is called the *kinetic energy*. Using the constitutive equation $f = \Phi_I^{-1}(p)$ and equation (1.6) the kinetic energy is defined as

$$\mathcal{E} = \int_{C_p} f\,dp = \int_0^p \Phi_I^{-1}(p)\,dp = T(p). \tag{1.7}$$

By graphing the constitutive equation for the ideal inductor we can obtain an area representation of the kinetic energy. Figure 1.1 plots the constitutive equation for a typical nonlinear inductor and shows the area that determines the kinetic energy. This figure also shows the complement of the kinetic energy, i.e., the *kinetic coenergy* which is defined as

$$T^*(f) = \int_0^f \Phi_I(f)\,df. \tag{1.8}$$

From the definitions of $T(p)$ and $T^*(f)$ it can be seen that at any operating point (p, f) along the constitutive relation $\Phi_I(f)$ the following condition holds,

$$pf = T(p) + T^*(f). \tag{1.9}$$

That is the product of the momentum and flow is equal to the sum of the

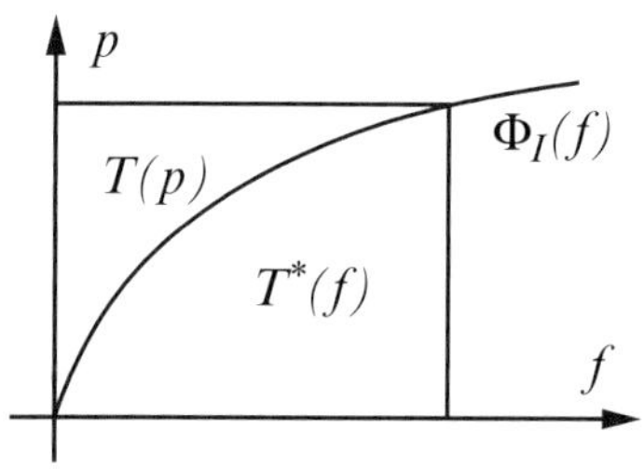

Fig. 1.1 Kinetic energy and kinetic coenergy

kinetic energy and the kinetic coenergy. It should be noted that the kinetic energy, $T(p)$, is independent of the flow f, and the kinetic coenergy, $T^*(f)$, is independent of the momentum p. Moreover, both $T(p)$ and $T^*(f)$ are scalar functions.

If the constitutive relation $\Phi_I(f)$ is linear then the kinetic energy and the kinetic coenergy are equal. In particular, consider the case where $p = \Phi_I(f) = If$, where I is a constant inductance. Then

$$T(p) = \int_0^p \Phi_I^{-1}(p)\,dp = \int_0^p (p/I)\,dp = p^2/2I,$$

$$T^*(f) = \int_0^f \Phi_I(f)\,df = \int_0^f If\,df = If^2/2 = p^2/2I.$$

The kinetic energy and the kinetic coenergy for the linear inductors described above are presented in Table 1.3.

	Translation	Rotation	Electrical	Fluid
Energy, $T(p)$	$p^2/2m$	$H^2/2I$	$\lambda^2/2L$	$\Gamma^2/2I_f$
Coenergy, $T^*(f)$	$mv^2/2$	$I\omega^2/2$	$Li^2/2$	$I_fQ^2/2$

Table 1.3 Kinetic energy and coenergy stored by linear inductors

Finally, note that in its most general form the constitutive equation for ideal inductors has the representation $p = \Phi_I(q, f, t)$ which leads a kinetic energy of the form $T = T(q, p, t)$, and a kinetic coenergy of the form $T^* = T^*(q, f, t)$. Examples of such relationships will be explored in the following chapters (see also Problem 5).

Work and energy

Let $q(t_0)$ be the displacement of the system at time t_0, and $q(t_1)$ be the displacement of the system at time t_1, with $t_1 > t_0$. Then from equation (1.4) it can be seen that the total work done by an effort in carrying the system from displacement $q(t_0)$ to displacement $q(t_1)$ is

$$\mathcal{W}_{q(t_0)\to q(t_1)} = \int_{q(t_0)}^{q(t_1)} e(t)\,dq(t) = \int_{p(t_0)}^{p(t_1)} f(t)\,dp(t) = T(p(t_1)) - T(p(t_0)). \tag{1.10}$$

Here, $p(t_0)$ is the momentum at time t_0, and $p(t_1)$ is the momentum at time t_1. Thus, the work done by an effort in displacing the system from $q(t_0)$ to $q(t_1)$ is equal to the change in kinetic energy.

1.2.2 Ideal capacitors

Ideal capacitors are system components that store energy. The behavior of the capacitor element is described by constitutive equations that relate the displacement and the effort. That is $e_C = \Phi_C(q)$, where e_C is the effort applied *to* the capacitor, q is the displacement, and $\Phi_C(q)$ is a continuous, invertible function that satisfies $\Phi_C(0) = 0$. The effort the capacitor applies to the other elements in the system is $e = -e_C$. Since the constitutive equation is invertible it can be used to obtain the displacement as a function of the effort, i.e., $q = \Phi_C^{-1}(e_C)$.

Some examples of ideal capacitors in the various engineering disciplines are as follows.

- Mechanical Translation:

Springs represent the capacitor elements in translating mechanical systems. For a linear spring, the relationship between the applied force and the deflection of the spring is given by Hooke's law, i.e.,

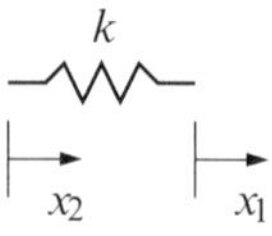

Mechanical spring

$$f = k(x_2 - x_1) = kx,$$

where f is the force applied to the spring, k is the spring stiffness, and $x = x_2 - x_1$ is the net deflection of the spring.

Example 1.4.

Consider the axial deflection of a rod with uniform cross-sectional area A, length L and modulus of elasticity E.

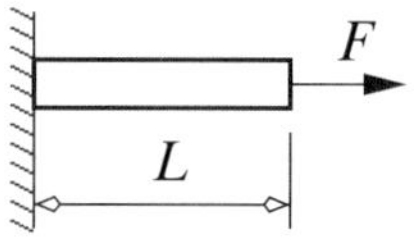

Then the spring stiffness is given by

$$k = \frac{EA}{L},$$

and the constitutive equation is $F = EAx/L$.

- Mechanical Rotation:

Torsional springs are capacitor elements for mechanical systems in pure rotation. The linear torsional spring behaves according to the relationship

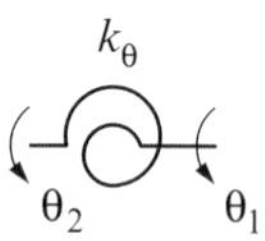

Torsional spring

$$\tau = k_\theta(\theta_2 - \theta_1) = k_\theta \theta,$$

where τ is the torque applied to the spring, $\theta = \theta_2 - \theta_1$ is the net angular deflection of the spring, and k_θ is the torsional spring stiffness.

Example 1.5.

Consider a rod with uniform cross-sectional area A, length L and shear modulus of elasticity G.

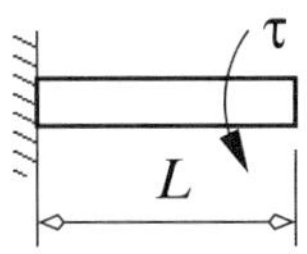

Then, the torsional spring stiffness of the rod is given by

$$k_\theta = \frac{GA}{L},$$

and the constitutive equation is $\tau = GA\theta/L$.

- Electrical:

A linear electrical capacitor behaves according to the equation

$$v = q/C,$$

where v is the voltage applied across the terminals of the capacitor, q is the accumulated charge, and C is the capacitance.

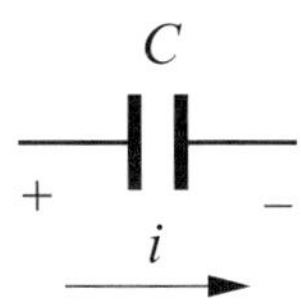

Electrical capacitor

Example 1.6.

Consider an electrical capacitor constructed from two plates of area A and separated by a distance d.

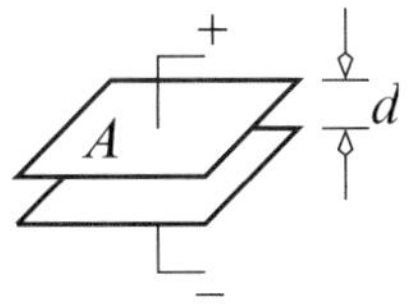

Then, the capacitance is

$$C = \frac{\epsilon A}{d},$$

where ϵ is the dielectric constant for the material in the air gap that separated the two plates. In this case the constitutive equation is $v = qd/(\epsilon A)$.

- Fluid:

  Tanks are the capacitor elements in fluid systems. Linear fluid capacitors satisfy the equation

$$P = V/C_f,$$

  where P is the pressure, V is the volume of the fluid and C_f is the fluid capacitance.

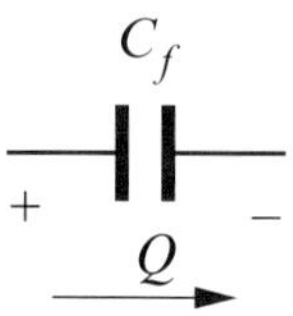

Fluid capacitance

Example 1.7.

Consider a circular cylindrical tank with cross sectional area A, fluid density ρ and fluid height h.

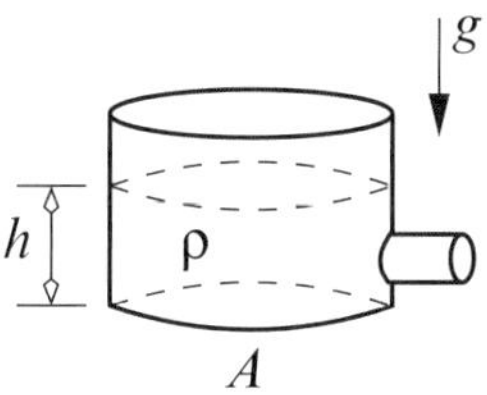

Then the pressure at the bottom of the tank is given by $P = \rho g h$ where g is the acceleration due to gravity. The height of the fluid in the tank can be written in terms of the volume, i.e., $h = V/A$. Thus, the pressure at the bottom of the tank is

$$P = \frac{\rho g V}{A}.$$

This implies that the fluid capacitance is given by $C_f = A/\rho g$.

Potential energy and potential coenergy

The energy stored by ideal capacitors is called the *potential energy*, and is defined as

$$\mathcal{E} = \int_{C_q} e_C \, dq = -\int_0^q e \, dq = \int_0^q \Phi_C(q) \, dq = V(q), \tag{1.11}$$

where we have used the constitutive relationship $e = -e_C = -\Phi_C(q)$.

An area representation of the potential energy is shown in Figure 1.2, where the complement *potential coenergy*, $V^*(e_C)$, is also illustrated. The potential coenergy is defined as

$$V^*(e_C) = \int_0^e \Phi_C^{-1}(e_C)\, de. \tag{1.12}$$

Thus, at any point (e_C, q) along the curve defined by $\Phi_C(q)$ the following condition holds,

$$e_C q = V(q) + V^*(e_C). \tag{1.13}$$

That is, the product of the effort and displacement is equal to the sum of the potential energy and the potential coenergy.

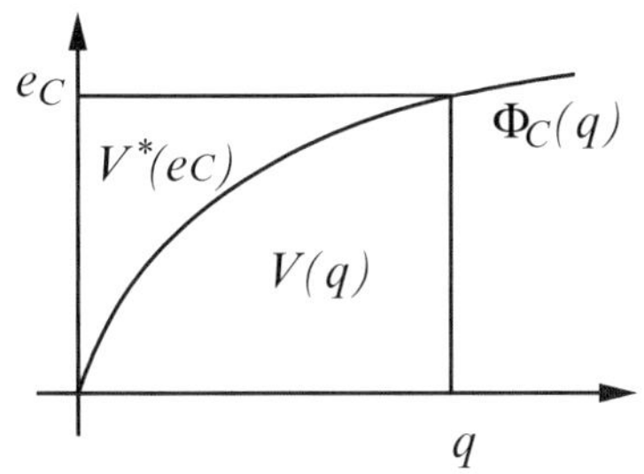

Fig. 1.2 Potential energy and potential coenergy

The potential energy, $V(q)$, and potential coenergy, $V^*(e_C)$ are both scalar functions, with $V(q)$ being independent of the effort, e_C, and $V^*(e_C)$ being independent of the displacement, q.

If the constitutive relation $\Phi_C(q)$ is linear then the potential energy is equal to the potential coenergy. In particular, consider the case where $e_C = -e = \Phi_C(q) = q/C$, where C is a constant capacitance, then

$$V(q) = \int_0^q \Phi_C(q)\, dq = \int_0^q q/C\, dq = q^2/2C,$$
$$V^*(e_C) = \int_0^{e_C} \Phi_C^{-1}(e_C)\, de_C = \int_0^{e_C} Ce_C\, de_C = C{e_C}^2/2 = q^2/2C.$$

The potential energy and the potential coenergy for the linear capacitor elements discussed above are listed in Table 1.4.

	Translation	Rotation	Electrical	Fluid
Energy, $V(q)$	$kx^2/2$	$k_\theta\theta^2/2$	$q^2/2C$	$V^2/2C_f$
Coenergy, $V^*(e_C)$	$F^2/2k$	$\tau^2/2k_\theta$	$Cv^2/2$	$C_fP^2/2$

Table 1.4 Energy stored by linear capacitors

Work and energy

Suppose the only effort applied to the system is due to an ideal capacitor. Then from (1.11) the effort can be expressed in terms of the scalar potential $V(q)$, i.e., $e = -dV/dq$. Such efforts are called *conservative efforts*. Let $q(t_0)$ be the displacement of the system at time t_0, and $q(t_1)$ be the displacement of the system at time t_1, with $t_1 > t_0$. Then from equation (1.4) it can be seen that the total work done by an effort in carrying the system from displacement $q(t_0)$ to displacement $q(t_1)$ is

$$\mathcal{W}_{q(t_0)\to q(t_1)} = \int_{q(t_0)}^{q(t_1)} e(t)\, dq(t) = V(q(t_0)) - V(q(t_1)). \tag{1.14}$$

Thus, the work done by an effort in displacing the system from $q(t_0)$ to $q(t_1)$ is equal to the change in potential energy. However, (1.10) shows that the work done by any effort is also equal to the change in kinetic energy. Combining (1.10) and (1.14) we get

$$\begin{aligned}\mathcal{W}_{q(t_0)\to q(t_1)} &= T(p(t_1)) - T(p(t_0))\\ &= V(q(t_0)) - V(q(t_1)).\end{aligned}$$

Thus,

$$T(p(t_1)) + V(q(t_1)) = T(p(t_0)) + V(q(t_0)), \tag{1.15}$$

which is the *Principle of Conservation of Energy*. That is, if all the efforts acting on the system are conservative, then the total energy (kinetic energy plus the potential energy) is a constant.

Gravity

The force due to gravity can be written as a potential energy function. For example, consider the system shown on the right. Here the acceleration due to gravity acts in the $-y$ direction. The force due to gravity that acts on the mass, m, is $F_{\text{gravity}} = -mg$, as shown.

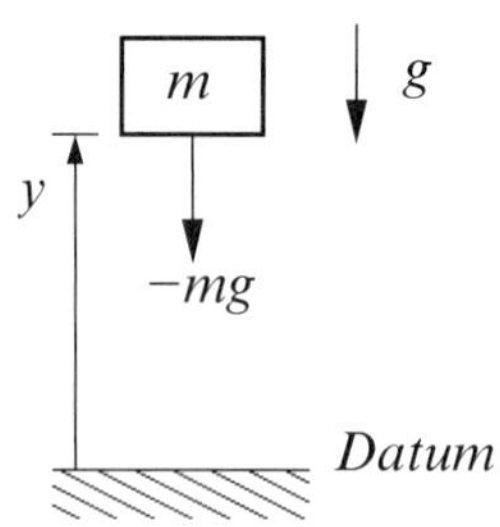

The vertical displacement, measured from the reference or datum is given by y. Therefore, the work done by the force due to gravity in displacing the mass from a height y to the datum 0, is

$$\begin{aligned}\mathcal{W}_{y\to 0} &= \int_y^0 e(t)\,dq(t)\\ &= \int_y^0 -mg\,dy\\ &= mgy\\ &= V(y) - V(0),\end{aligned}$$

where $V(y)$ is the potential energy at y and $V(0)$ is the potential energy at the datum. If we assume that $V(0) = 0$, i.e., the potential energy at the datum is zero, then the work done by the force due to gravity is $V(y) = mgy$. In addition, it can be seen that

$$F_{\text{gravity}} = -\frac{dV}{dy} = -mg.$$

Hence, the force due to gravity can be determined from the scalar potential energy function. We also note that the increment in work done by the gravity force in displacing the object a distance dy is

$$\check{d}\mathcal{W} = F_{\text{gravity}}\,dy = -mg\,dy.$$

1.2.3 Ideal resistors

Ideal resistors are system components that dissipate energy. The behavior of the resistor element is described by a constitutive equations that relate the applied effort and the flow. Specifically, the constitutive equation is given by $e_R = \Phi_R(f)$, where e_R is the effort applied to the resistor, f is the flow, and $\Phi_R(f)$ is a continuous, invertible function of f that satisfies $\Phi_R(0) = 0$. The effort applied to the other elements of the system by the resistor is given by $e = -e_R$.

Some examples of resistors in the various engineering disciplines are as follows.

- Mechanical Translation:

For mechanical systems in pure translation dampers represent ideal resistors. Linear dampers behave according to the law

$$f = b(v_2 - v_1) = bv,$$

where f is the force applied to the damper, b is the damping coefficient, and $v = v_2 - v_1$ is the relative velocity of the terminals of the damper.

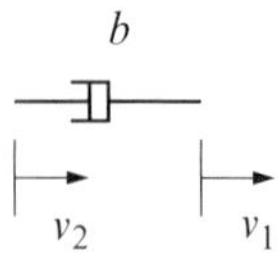

Damper

- Mechanical Rotation:

For mechanical systems in pure translation, torsional dampers represent ideal resistors. Linear torsional dampers satisfy the equation

$$\tau = b_\omega(\omega_2 - \omega_1) = b_\omega \omega,$$

where τ is the torque applied to the damper, b_ω is the torsional damping coefficient, and $\omega = \omega_2 - \omega_1$ is the relative angular velocity of the terminals of the damper.

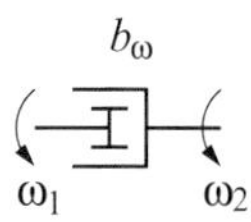

Torsional damper

- Electrical:

Electrical resistors represent ideal resistors. Linear electrical resistors have a constitutive relationship

$$v = Ri,$$

where v is the voltage across the terminals of the resistor, R is the resistance, and i is the current through the resistor.

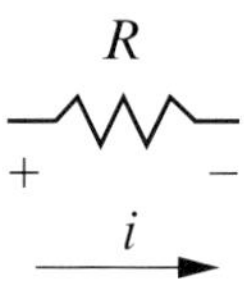

Resistor

- Fluid:

The fluid resistors represent ideal resistors in fluid systems. Linear fluid resistors satisfy the constitutive relationship

$$P = R_f Q,$$

where P is the pressure across the terminals of the resistor, R_f is the fluid resistance, and Q is the volume flow rate through the fluid resistor.

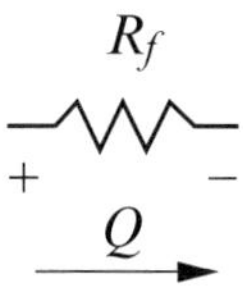

Fluid resistor

Content and cocontent

The effort applied by the resistor is related to a scalar function called the *content* which is defined as

$$D(f) = \int_0^f e_R\,df = -\int_0^f e\,df = \int_0^f \Phi_R(f)\,df, \tag{1.16}$$

where we have used the constitutive relationship $e = -e_R = -\Phi_R(f)$. Hence, $e = -dD(f)/df$. Note that the content $D(f)$ is also called the *Rayleigh dissipation function.*

The *cocontent* is defined as

$$D^*(e_R) = \int_0^{e_R} \Phi_R^{-1}(e_R)\,de, \tag{1.17}$$

where $\Phi_R^{-1}(e_R) = f$ is the inverse of the function $\Phi_R(f) = e_R$. An area representation of the content and cocontent for a nonlinear resistor is shown in Fig. 1.3. Thus, at any point (e_R, f) along the curve $\Phi_R(f)$ the following condition must hold,

$$e_R f = D(f) + D^*(e_R). \tag{1.18}$$

The content and the cocontent are both scalar functions, with $D(f)$ independent of the effort, e_R, and $D^*(e_R)$ independent of the flow, f.

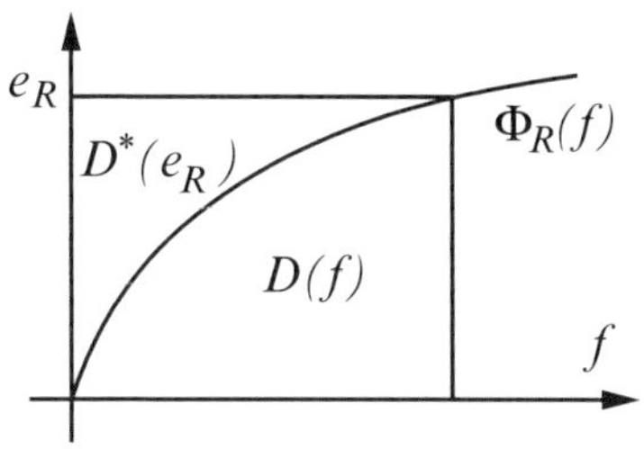

Fig. 1.3 Content and cocontent

If the constitutive relationship $\Phi_R(f)$ is linear the content and cocontent will be equal. In particular, consider the case where $e_R = -e = \Phi_R(f) = Rf$, where R is a constant resistance, then

$$D(f) = \int_0^f \Phi_R(f)\,df = \int_0^f Rf\,df = Rf^2/2,$$

$$D^*(e_R) = \int_0^{e_R} \Phi_R^{-1}(e_R)\,de_R = \int_0^{e_R} (e_R/R)\,de_R = {e_R}^2/2R = Rf^2/2.$$

The content and cocontent for the linear resistors described above are listed in Table 1.5.

	Translation	Rotation	Electrical	Fluid
Content, $D(f)$	$bv^2/2$	$b_\omega\omega^2/2$	$Ri^2/2$	$R_fQ^2/2$
Cocontent, $D^*(e_R)$	$F^2/2b$	$\tau^2/2b_\theta$	$v^2/2R$	$P^2/2R_f$

Table 1.5 Energy dissipated by linear resistors

1.2.4 Constraint elements

Constraint elements define kinematic constraint relationships among the system variables. These constraint elements can be described as *transformers* or *transducers*.

- Transformers
  A transformer transfers energy between the subsystems in the dynamic system model. These idealized elements do not store, dissipate or generate energy, and they behave in such a way that the net power into the device is zero. In the case of transformers the energy transfer takes place within the same engineering discipline. These elements give rise to *displacement constraints* or *flow constraints* that do no work on the system.
- Transducers
  A transducer are similar to a transformer however, the energy transfer takes place between different engineering disciplines. These elements give rise to displacement or flow constraints that do no work on the system.

Specific examples of transformers and transducers are given below.

Transformers

- Mechanical Translation:

The lever shown below represents a transformer for mechanical systems in translation. Here, F_1 is the force input to left hand side of the lever, and x_1 is the corresponding displacement. Similarly, F_2 is the force input to right hand side of the lever, and x_2 is the corresponding displacement. The lever makes angle θ with the horizontal axis, as shown.

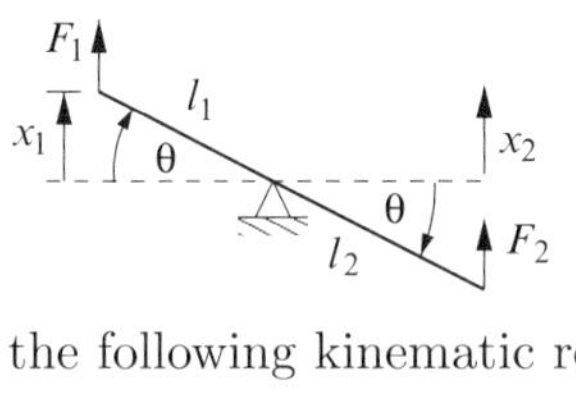

For small displacements the following kinematic relationship holds,

$$x_1 = l_1\theta \ \rightarrow \theta = x_1/l_1,$$
$$x_2 = -l_2\theta \ \rightarrow x_2 = -\frac{l_2}{l_1}x_1.$$

In terms of the velocities these relations become

$$\frac{l_2}{l_1}v_1 + v_2 = 0,$$

where $v_1 = dx_1/dt$ and $v_2 = dx_2/dt$. This equation represents a flow constraint the lever must satisfy.
Now, summing moments about the pivot (counterclockwise positive) gives,

$$-F_1 l_1 + F_2 l_2 = 0,$$

which is a constraint on the effort variables of the device. As a result, the power balance yields

$$\begin{aligned}\text{Power input } + \text{Power output} &= F_1 v_1 + F_2 v_2 \\ &= (F_1 - \frac{l_2}{l_1}F_2)v_1 = 0.\end{aligned}$$

Hence, no energy is stored or dissipated in the device.

- Mechanical Rotation:

The simple gear train shown below represents a transformer for mechanical systems in rotation. The torque input to the left hand gear is τ_1, and ω_1 is the corresponding angular velocity. Similarly, the torque input to the right hand gear is τ_2, and ω_2 is the corresponding angular velocity.

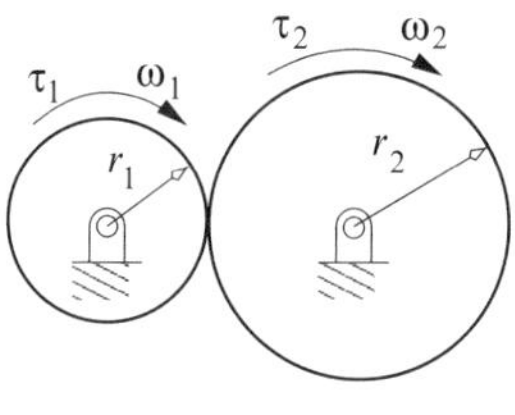

Since there is no slipping or backlash the velocity at the point of contact is $r_1\omega_1 = -r_2\omega_2$. Thus, from the kinematics we have

$$\frac{r_1}{r_2}\omega_1 + \omega_2 = 0,$$

which is a flow constraint the simple gear train must satisfy. By summing the moments about the center of each gear, it can be seen that the force at the point of contact satisfies

$$\frac{\tau_1}{r_1} = \frac{\tau_2}{r_2},$$

which is an effort constraint for the simple gear train. As a result, the power balance satisfies

$$\begin{aligned}\text{Power input } + \text{Power output} &= \tau_1\omega_1 + \tau_2\omega_2 \\ &= (\tau_1 - \frac{r_1}{r_2}\tau_2)\omega_1 = 0.\end{aligned}$$

- Electrical:

A schematic of an ideal electrical transformer is shown below. This diagram shows two coils that are magnetically coupled via an iron core. The coil on the left (the primary coil) has N_1 turns, applied voltage v_1 and current i_1. The current in the primary coil creates a magnetic flux $\phi = N_1 i_1$ that induces a voltage v_2 in the coil on the right (the secondary coil). Here, the secondary coil has N_2 turns and current i_2. Dots are placed near one of the terminals in each coil of the transformer to indicate whether the magnetic flux produced by the coils add or subtract. By convention, if current enters both dotted terminals the coils will produce magnetic fluxes that add (Hambley, (1997), p. 686).

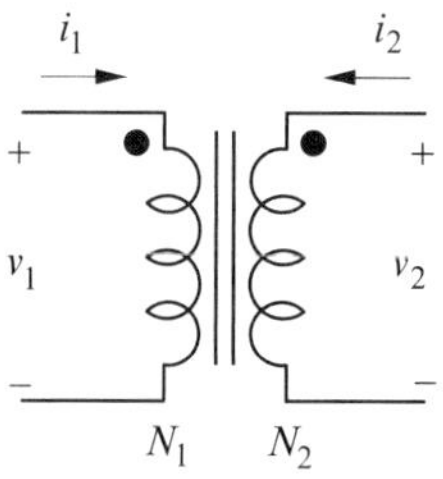

It is assumed that there is no flux leakage and the inductance of the coils can be neglected. Then, a magnetomotive force balance gives,

$$N_1 i_1 + N_2 i_2 = 0,$$

which is the flow constraint for the device. Also, from Faraday's law we have

$$\begin{aligned}v_1 &= \frac{d\lambda_1}{dt} = \frac{d(N_1\phi)}{dt} = N_1\frac{d\phi}{dt}, \\ v_2 &= \frac{d\lambda_2}{dt} = \frac{d(N_2\phi)}{dt} = N_2\frac{d\phi}{dt}, \\ v_2 &= \frac{N_2}{N_1}v_1,\end{aligned}$$

which is the effort constraint that the device must satisfy. The power balance gives

$$\text{Power input } + \text{Power output} = v_1 i_1 + v_2 i_2$$

$$= (v_1 - \frac{N_1}{N_2} v_2) i_1 = 0.$$

- Fluid:

The diagram below shows a fluid transformer. On the left hand side of the device the pressure is P_1, the volume flow rate is Q_1 and the area of the piston is A_1. On the right hand side of the device the pressure is P_2, the volume flow rate is Q_2 and the area of the piston is A_2.

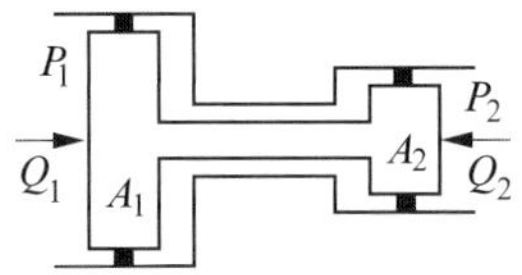

Using these definitions the velocity of the piston is $Q_1/A_1 = -Q_2/A_2$, or

$$\frac{A_2}{A_1} Q_1 + Q_2 = 0,$$

which is the flow constraint that the device must satisfy. Moreover, if the piston is assumed to be massless, the net force acting on the piston is

$$P_1 A_1 - P_2 A_2 = 0,$$

which is a constraint on the effort variables for the device. Finally, the power balance gives

$$\begin{aligned} \text{Power input } + \text{Power output} &= P_1 Q_1 + P_2 Q_2 \\ &= (P_1 - \frac{A_1}{A_2} P_2) Q_1 = 0. \end{aligned}$$

Transducers

Some examples of transducer elements are described next. Recall that transducers are system components that transfer energy from one engineering discipline to another.

- Mechanical Transducer:

A rack and pinion system shown below is an example of a mechanical transducer. This device couples a gear (in rotation) with a rack (in translation). The torque associated with the gear is τ, and ω is the corresponding angular velocity. The rack has a force F and velocity v.

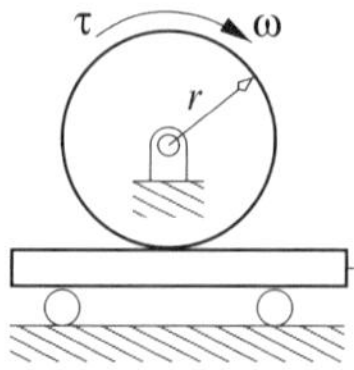

If there is no slipping or backlash the system satisfies the flow constraint

$$v + r\omega = 0.$$

Moreover, if the system is inertialess, summing the moments about the center of the gear gives

$$Fr - \tau = 0,$$

which is the effort constraint that must be satisfied. The power balance for the system gives

$$\begin{aligned}\text{Power input } + \text{Power output} &= Fv + \tau\omega \\ &= (F - \frac{1}{r}\tau)v = 0.\end{aligned}$$

- Fluid-mechanical Transducer:

The hydraulic press shown below is an example of a fluid-mechanical transducer. In this device an incompressible fluid with volume flow rate Q and pressure P acts on a massless piston with area A. The piston has an applied force F and velocity v in the direction shown.

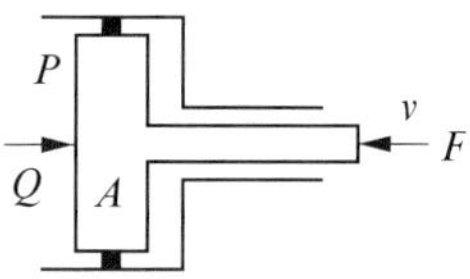

It is assumed that the piston does not deform hence, we have

$$v + Q/A = 0,$$

which is the flow constraint that the device must satisfy. Since the piston is massless, the net force acting on the piston is

$$-F + PA = 0,$$

which is the effort constraint that the device must satisfy. The power balance for the system gives

$$\begin{aligned}\text{Power input } + \text{Power output} &= Fv + PQ \\ &= (F - AP)v = 0.\end{aligned}$$

1.2.5 Source elements

Power is input to the system via components called *sources*. These sources occur in two forms *effort sources*, $e_s(t)$, and *flow sources*, $f_s(t)$. Moreover, the sources can be *ideal sources* or *regulated sources* as described below.

Ideal sources

An *ideal effort source* will provide the specified effort to the system irrespective of the associated flow required by the system. An *ideal flow source* will provide the specified flow to the system irrespective of the associated effort required by the system. Note that the power available to these ideal sources is infinite. An example of an ideal effort source is the voltage supplied by a battery. An example of an ideal flow source is the volume flow rate provided by a pump. These ideal sources are convenient to use in modeling dynamic systems. However, it should be noted that they do not exists in reality, since actual energy sources can not provide infinite power.

The symbolic representations of the ideal sources used in this text are shown below.

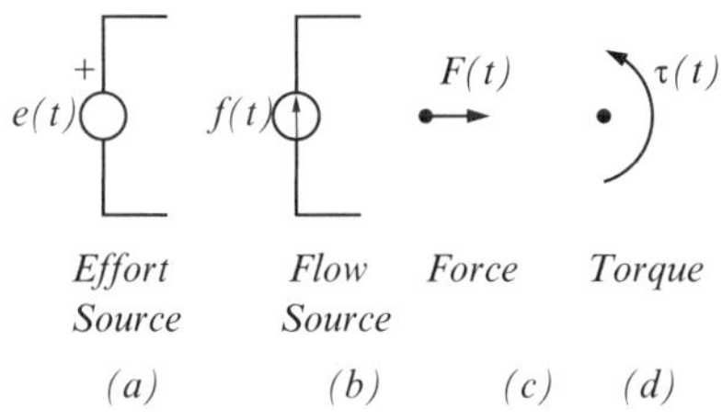

Here, the + sign on the effort source, (a), indicates the positive polarity of the source. We will use (a) to represent ideal voltage and pressure sources. The ideal flow source is shown in (b), where the arrow indicates the direction of positive flow. An applied force is shown in (c), and an applied torque is shown in (d). In both cases the arrows indicate the direction of positive effort.

Regulated sources

At times it will prove expedient to model certain system elements as controlled or regulated sources. Typically, these devices provide one of the following;

1. An effort source that is a function of a flow variable, i.e, a flow regulated effort source.
2. A flow source that is a function of an effort variable, i.e., an effort regulated flow source.
3. An effort source that is a function of an effort variable, i.e., an effort regulated effort source.
4. A flow source that is a function of a flow variable, i.e., a flow regulated flow source.

Examples of these regulated sources are given below.

- Coulomb friction

The Coulomb friction force model is often used to describe the force of interaction between objects. Consider an object moving on a rough surface with velocity v, and let N be the normal reaction force of the surface acting on the object.

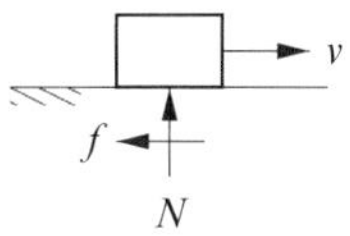

Then the friction force acting on the object can be modeled as follows;

$$\begin{array}{ll} |f| \leq |\mu_s N|, & v = 0, \\ f \;\; = -\mu_k N v/|v|, & v \neq 0. \end{array} \tag{a}$$

Thus, when the system is in static equilibrium, i.e., $v = 0$, the upper bound on the friction force is $|\mu_s N|$ where, μ_s is the coefficient of static friction. The magnitude and direction of the static friction force is determined by the equations of equilibrium.

If there is sliding between the objects, i.e., $v \neq 0$, then the sliding friction force has a constant magnitude, $\mu_k N$, and acts opposite to the direction of motion. The constant μ_k is called the coefficient of kinetic friction, and $\mu_k \leq \mu_s$. Hence, the Coulomb friction force can be considered to be a flow regulated effort source. In this text we call the equations described in (a) *effort constraints*. These effort constraint equations provide a relationship between the effort variables and the flow (or displacement) variables in the system.

- Diode

A diode can be modeled as an effort regulated flow source. In particular, let v_d represent the voltage (effort) across the terminals of the diode, and i be the current (flow) through the diode.

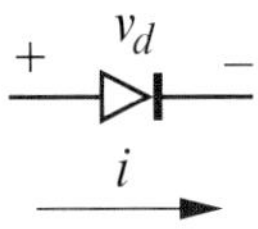

A diode essentially acts as a switch where $i > 0$ if $v_d > 0$, and $i = 0$ if $v_d \leq 0$. An approximate model for the behavior of the diode is given by the effort constraint equation

$$i - I_s(e^{\alpha v_d} - 1) = 0,$$

where $I_s > 0$ and $\alpha > 0$ are constants that are determined by the material properties of the diode. The variable I_s is called the reverse saturation current and has a typical value of 1×10^{-12} amp. The variable α is the inverse of the thermal voltage, and at room temperature it has a typical value of 40 1/volt. Hence, the diode can be modeled as an effort regulated flow source.

We also note that a fluid flow check valve can be modeled as a diode type device. In this case the fluid flows through the check valve (i.e., diode) if the pressure drop across the valve is positive. As a result, the fluid flow through the check valve can be approximated by the effort constraint equation

$$Q - Q_s(e^{\alpha P_d} - 1) = 0,$$

where $Q_s > 0$ and $\alpha > 0$ are constants, and P_d is the pressure across the valve.

- Bipolar Junction Transistor

A bipolar junction transistor can also be modeled as an effort regulated flow source. The figure below shows the schematic of a NPN transistor and a PNP transistor. These devices have three terminals, a base (B), a collector (C) and an emitter (E).

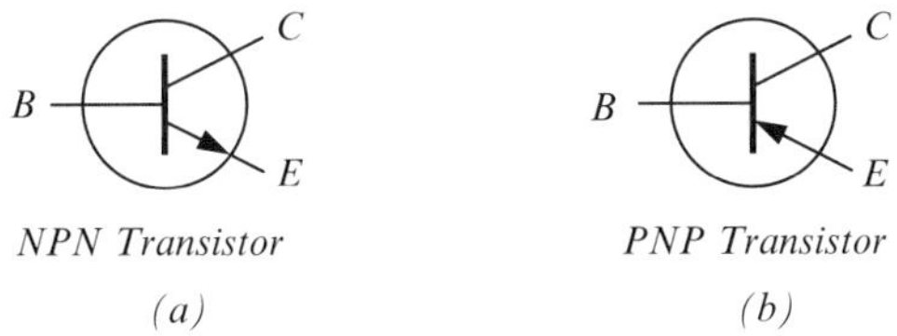

The input-output behavior for these devices can be described using the Ebers-Moll transistor model (see Jensen and McNamee, 1976, pp. 781). In the case of the NPN transistor the Ebers-Moll model yields the equivalent circuit shown below. (The Ebbers-Moll model for a PNP transistor is obtained reversing the current flow directions in the NPN model.)

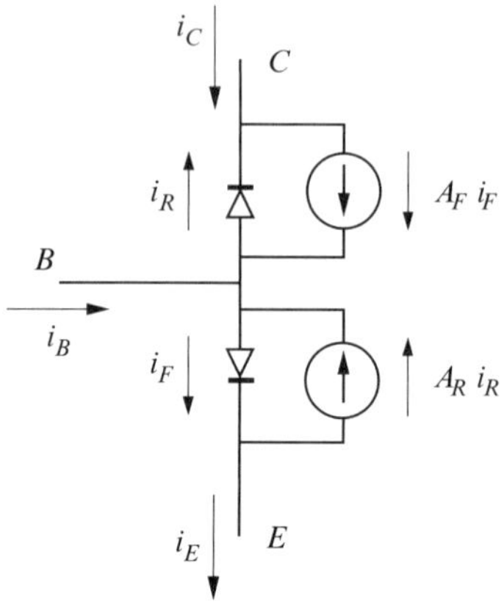

Ebers–Moll NPN Transistor Model

In this model the base current, i_B, the emitter current, i_E, and the collector current, i_C, are defined in terms of the diode currents i_F and i_R as follows.

$$
\begin{aligned}
i_B &= i_E - i_C \\
i_E &= i_F - A_R i_R \\
i_C &= -i_R + A_F i_F \\
i_F &= I_{ES}(e^{\alpha v_{BE}} - 1) \\
i_R &= I_{CS}(e^{\alpha v_{BC}} - 1)
\end{aligned}
$$

where A_R, A_F, I_{ES}, I_{CS} and α are positive constants that depend on the material properties of the transistor, v_{BE} is the base-emitter voltage, and v_{BC} is the base-collector voltage.
It can be shown that $A_R I_{CS} = A_F I_{ES} = I_S$ (see Möschwitzer, 1991). Hence, the behavior of the transistor can be described by the effort constraint equations

$$
\begin{bmatrix} i_E \\ i_C \\ i_B \end{bmatrix} - \begin{bmatrix} I_{ES} & -I_S \\ I_S & -I_{CS} \\ I_{ES} - I_S & I_{CS} - I_S \end{bmatrix} \begin{bmatrix} e^{\alpha v_{BE}} - 1 \\ e^{\alpha v_{BC}} - 1 \end{bmatrix} = \begin{bmatrix} 0 \\ 0 \\ 0 \end{bmatrix}. \tag{1.19}
$$

Using these equations we can assert that the transistor currents are a function of the voltages v_{BE} and v_{BC}. Moreover, these equations imply that the NPN transistor can be modeled using the equivalent circuit model shown below.

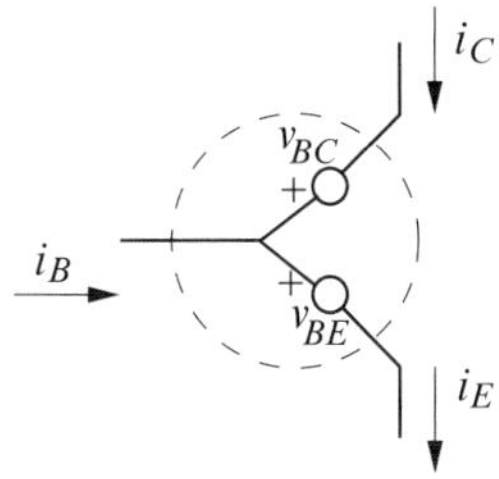

Equivalent circuit model

In this model the currents i_B, i_C and i_E are functions of the voltages v_{BE} and v_{BC}, which are considered to be effort sources.
The operating modes of the NPN transistor are characterized by the voltages v_{BE} and v_{BC}. In particular,

- if $v_{BE} > 0$ and $v_{BC} < 0$ the transistor is said to be in *active normal* mode.
- if $v_{BE} > 0$ and $v_{BC} > 0$ the transistor is said to be in *saturation* mode.
- if $v_{BE} < 0$ and $v_{BC} > 0$ the transistor is said to be in *active inverse* mode.
- if $v_{BE} < 0$ and $v_{BC} < 0$ the transistor is said to be in *reverse* mode.

The circuits considered in this text are constructed so the transistor operates in active normal mode. In which case if $v_{BC} \ll 0$ then, the term $e^{\alpha v_{BC}}$ in the Ebers-Moll equation is negligible. We can therefore rewrite the transistor effort constraint equations as

$$\begin{aligned} i_E &= I_{ES}(e^{\alpha v_{BE}} - 1) \\ i_C &= A_F I_{ES}(e^{\alpha v_{BE}} - 1) \\ i_B &= (1 - A_F) I_{ES}(e^{\alpha v_{BE}} - 1). \end{aligned} \tag{1.20}$$

If we define

$$\beta = \frac{A_F}{1 - A_F},$$

it can be seen that

$$i_C = A_F i_E = \beta i_B.$$

Typically, A_F ranges from 0.999 to 0.980, and saturation current $I_{ES} = 1 \times 10^{-12}$ amp. As a result β ranges from 1000 to 50, approximately. Hence, the last equation indicates that, in normal active mode, the collector and emitter currents are almost equal. Also, the collector current is a significant amplification of the base current.

- Operational Amplifier

The figure (a) below shows the schematic of an operational amplifier. This integrated circuit has input terminals 1 and 2, and output terminal 3. The operational amplifier can be considered to be a effort regulated effort source. The input-output behavior can be described using the diagram (b).

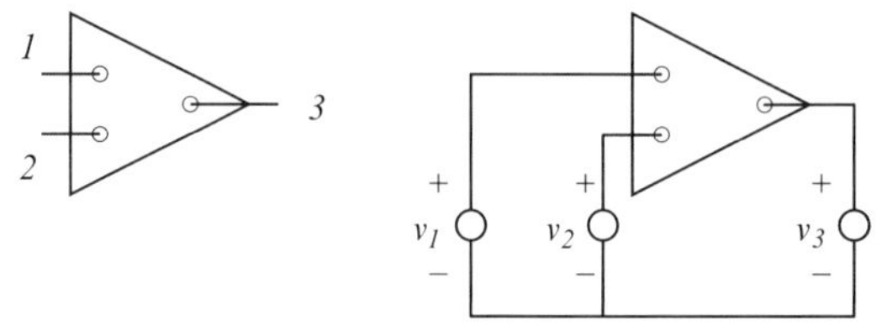

(a) (b)

Specifically, the output voltage of the operational amplifier, v_3, is given by

$$v_3 = k_g(v_2 - v_1),$$

where v_1 is the voltage input to terminal 1, v_2 is the voltage input to terminal 2, and k_g is the gain for the amplifier. The gain for typical operational amplifiers is in the interval $10^5 \leq k_g \leq 10^7$. In addition, the device has very high input resistance ($\approx 10^{10}$ ohm), and the current input to the device is negligible, i.e., $i_1 \approx 0$, and $i_2 \approx 0$.

- DC motor/generator

The schematic of a direct current (DC) motor/generator is shown below.

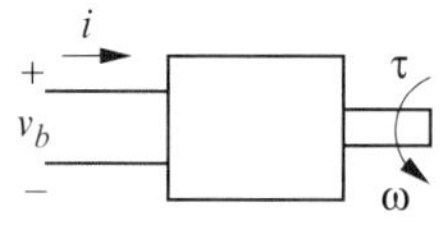

The essential electromechanical coupling in a DC motor/generator is given by

$$\tau = K_t i,$$
$$v_b = K_v \omega,$$

where τ is the torque due to the current i, and K_t is the torque constant. The voltage v_b (back electromotive force (emf)) is the result of the angular velocity ω, and K_v is the back emf constant. Thus, both the motor torque τ and the generator voltage v_b can be viewed are flow regulated effort sources. Note that this model neglects the inductance and resistance of the motor coils, as well as the inertia, damping and flexibility of the rotor.

1.2.6 Paynter's diagram and system models

The relationships between the fundamental system variables can be summarized using Paynter's diagram shown in Fig. 1.4. The diagram is a graph

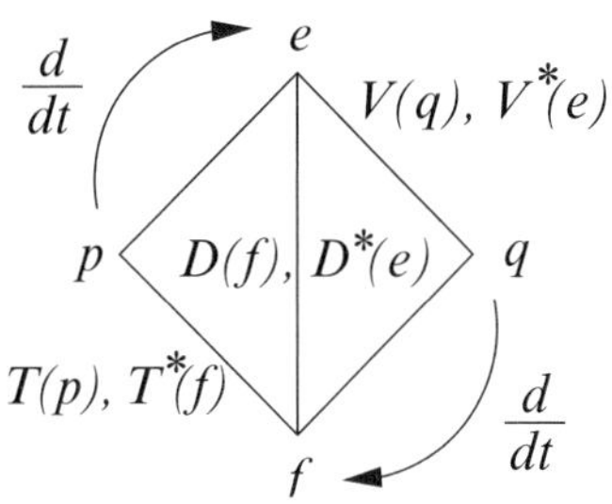

Fig. 1.4 Paynter's Diagram

with the effort e, flow f, displacement q and momentum p at its vertexes. The relationships between the fundamental variables are illustrated on the line segments joining the variables. The line segment joining the kinematic variables q and f shows the differential relationship between these variables. Similarly, the line segment joining the kinetic variables p and e shows the differential relationship between these variables. The energy stored by ideal inductors is shown on the line segment joining the momentum (p) and flow (f) variables. The energy stored by ideal capacitors is shown on the line segment joining the effort (e) and displacement (q) variables. The energy dissipated by ideal resistors is shown on the line segment joining the effort (e) and flow (f) variables. This diagram is useful in making concrete the analogy that exists between the different engineering disciplines. We also note that, given any pair of the fundamental system variables, we can use the relationships in Paynter's diagram to determine the unknown pair of fundamental variables. For example, given the momentum p and displacement q, we can determine the effort and flow as $e = \frac{dp}{dt}$ and $f = \frac{dq}{dt}$, respectively.

In the development of a unified approach to modeling multidiscipline systems we have a choice as to which pair of the fundamental variables will be treated as independent. Our approach, which is based on Lagrangian mechanics, uses the displacement and flow variables (q, f) as the variables that are used to described the equations of motion. In the bond graph approach to system modeling the effort and flow (e, f) are used as the independent pair of fundamental variables (see Karnopp, Margolis, and Rosenberg (1990)). The linear graph approach to system modeling also uses the effort and flow (e, f) as the independent pair of fundamental variables (see Rowell and Wormley (1991)). A Hamiltonian approach to multidiscipline system modeling uses the displacement and momentum (q, p) as the independent fundamental variables (see Layton and Fabien (1997)). The other variable pair combinations, i.e., (f, p), (e, p) and (e, q) are not often used to model multidiscipline systems.

References

1. S. H. Crandall, D. C. Karnopp, E. F. Kurtz, and D. C. Pridmore-Brown, *Dynamics of Mechanical and Electromechanical Systems*, McGraw-Hill, 1968.
2. A. R. Hambley, *Electrical Engineering: Principles & Applications*, Prentice-Hall, Inc., 1997.
3. E. Umez-Eronini, *System Dynamics and Control*, PWS Publishing, 1999.
4. R. W. Jensen and L. R. McNamee, *Handbook of Circuit Analysis Languages and Techniques*, Prentice-Hall, 1976.
5. D. C. Karnopp, D. L. Margolis, and R. C. Rosenberg, *System Dynamics: A Unified Approach*, 2nd ed. Wiley & Sons, 1990.
6. R. A. Layton and B. C. Fabien, "Modeling and simulation of physical systems IV: An introduction to Hamiltonian DAEs," *IASTED International Conference, Applied Modeling and Simulation*, 264–268, 1997.
7. A. G. J. MacFarlane, *Dynamical System Models*, Harrap & Co., 1970.
8. L. Meirovitch, *Methods of Analytical Dynamics*, McGraw-Hill, 1970.
9. A. Möschwitzer, *Semiconductor devices, circuits, and systems*, Oxford University Press, 1991.
10. M. Nelkon, *Principles of Physics*, 1973.
11. H. F. Olson, *Dynamical Analogies*, Van Norstrand, 1943.
12. H. M. Paynter, *Analysis and Design of Engineering Systems*, M.I.T. Press, Cambridge, MA, 1961.
13. R. M. Rosenberg, *Analytical Dynamics of Discrete Systems*, Plenum Press, 1977.
14. D. Rowell and D. N. Wormley, *System Dynamics: An Introduction*, Department of Mechanical Engineering, MIT, 1991.
15. J. L. Shearer, A. T. Murphy, and H. H. Richardson, *Introduction to System Dynamics*, Addison-Wesley, 1967.
16. J. L. Shearer and B. T. Kulakowski, *Dynamic Modeling and Control of Engineering Systems*, Macmillan, 1990.
17. R. J. Smith, *Circuits devices and systems*, John Wiley and Sons, 1976.
18. P. E. Wellstead, *Introduction to Physical System Modelling*, Academic Press, 1979.
19. D. C. White, and H. H. Woodson, *Electromechanical Energy Conversion*, John Wiley & Sons, 1959.
20. J. H. Williams, Jr., *Fundamentals of Applied Dynamics*, John Wiley & Sons, 1996.

Problems

1. Determine a consistent set of English and SI units for the variables listed in Table 1.1. Hence, show that the units of work, power and energy (Table 1.2) are equal in each discipline.
2. Verify the results given in Table 1.3, i.e., show that kinetic energy and the kinetic coenergy are equal for a point mass, a rotor, a linear inductor and a linear fluid inertia. Also, assign consistent SI units to all the variable in these expressions.
3. Verify the results given in Table 1.4, i.e., show that potential energy and the potential coenergy are equal for a linear spring, a linear torsional spring, a linear capacitor and a linear fluid capacitor. Also, assign consistent SI units to all the variable in these expressions.
4. Verify the results given in Table 1.5, i.e., show that content (Rayleigh dissipation function) and the cocontent are equal for a linear damper, a linear torsional damper, a linear resistor and a linear fluid resistor. Also, assign consistent SI units to all the variable in these expressions.
5. Consider a particle of mass m moving in a plane as shown below. Let the velocity of the particle be $\bar{v} = \dot{x}\hat{i} + \dot{y}\hat{j}$, and the linear momentum be $\bar{p} = m\bar{v}$.

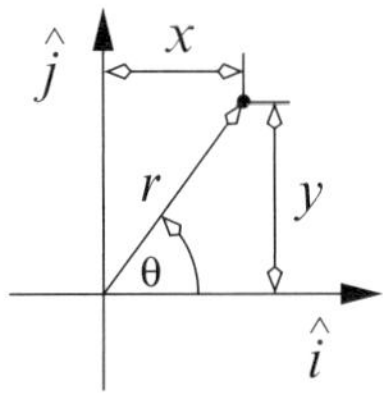

   a. Show that the kinetic coenergy is $T^*(\bar{v}) = \frac{1}{2}m\bar{v} \cdot \bar{v} = \frac{1}{2}m(\dot{x}^2 + \dot{y}^2)$.
   b. Perform the coordinate transformation $x = r\cos\theta$, $y = r\sin\theta$, and show that $T^*(\bar{v}) = \frac{1}{2}m(\dot{r}^2 + r^2\dot{\theta}^2)$. (Note that T^* is a function of the displacement and flow in this formulation.)
   c. Suppose further that $\theta = \frac{1}{2}\alpha t^2 + \omega t + \theta_0$, where α, ω and θ_0 are constants, and t is the time. Show that in general T^* can be a function of displacement, flow and time.

6. The flux linkage for an electrical inductor is experimentally determined to be $\lambda = ki^{\frac{2}{3}}$, where k is a constant, and i is the current. Determine expressions for the kinetic energy and kinetic coenergy for this device.
7. According to the special theory of relativity the mass of particle is given as

$$m(t) = \frac{m_0}{\sqrt{1 - (v(t)/c)^2}},$$

where, the constant m_0 is rest mass, $v(t)$ is the velocity of the particle, t is the time, and the constant c is the speed of light.

a. Show the the kinetic energy is $T = (m(t) - m_0)c^2$.
b. Show that if $v \ll c$, $T \approx m_0 v(t)^2/2$.

8. A block of mass m is constrained to move on an incline. The incline is rough and has a slope with angle α. A constant force, F, is applied to the block at an angle θ.

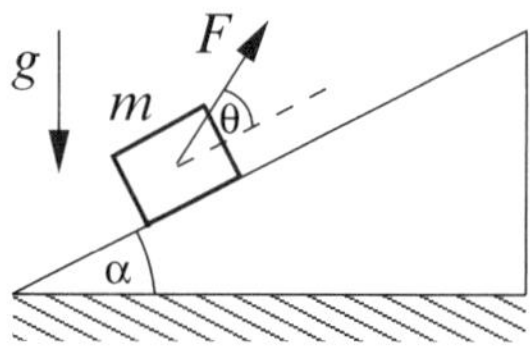

The coefficient of friction is μ, and the acceleration due to gravity acts downward as shown. If the block starts from rest, find an expression for its velocity after it has moved a distance, s, up the incline. (Use the work energy principle (1.10).)
9. The fluid in the conical tank shown here has density ρ. The pressure at the bottom of the tank is P, and the height of the fluid in the tank is h.

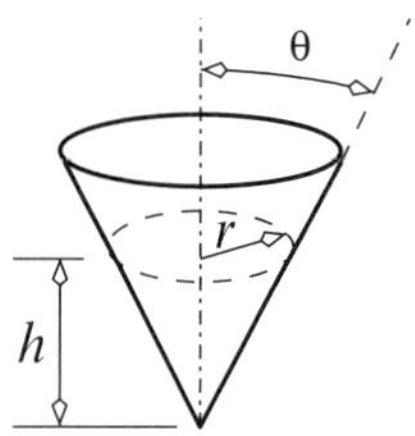

a. Find the constitutive equation that relates the pressure P to the volume V of fluid in the tank.
b. Find an expression for the potential energy due to the volume of fluid stored in the tank.

10. The constitutive equation for a nonlinear spring is $f = kx^3$ where, f is the applied force, k is a constant, and x is the net deflection of the spring. Determine the potential energy and the potential coenergy.
11. A block of mass m fall a distance, h, onto an unstretched spring. The spring has stiffness, k, and the acceleration due to gravity, g, acts downward as shown.

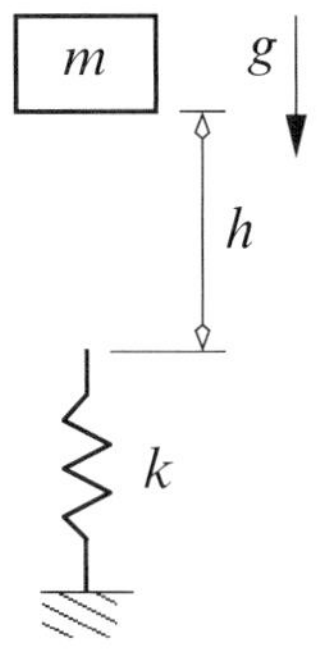

If the block is released from rest: (a) what is the velocity the block just before it hits the spring, and (b) what is the maximum deflection of the spring?

12. Explain how you would modify the model of an ideal electrical transformer to include the effect of the inductance and resistance of the coils.
13. Explain why ideal effort and flow sources do not exist. Illustrate the behavior you would expect from an actual effort source as the systems flow demand increases. Use a typical electrochemical battery as an example. (Plot the effort versus flow for an ideal efforts source and an actual effort source.)
14. The circuit shown here satisfies the equations

$$R\dot{q} = v - v_d,$$
$$\dot{q} = I_s(e^{\alpha v_d} - 1),$$

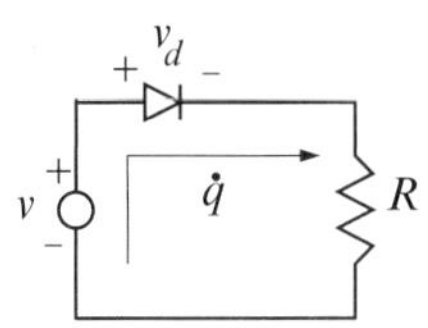

where $R = 10$ ohm is the resistance, v is the voltage source, v_d is the diode voltage, I_s is the diode saturation current, and $\alpha = 40$ volts^{-1} is the diode junction constant. Obtain numerical solutions to these equations for $0 \leq v \leq 10$, and plot $\dot{q}$ versus v_d for the following saturation currents; $I_s = 10^{-12}, 10^{-14}, 10^{-16}$.
15. Sketch Paynter's diagram for each of the energy disciplines. Label the vertexes with the appropriate variable names.
16. For the systems shown below, write expressions for; (i) the kinetic energy and the kinetic coenergy, (ii) the potential energy and the potential coenergy, and (iii) the content and the cocontent. For the systems with applied efforts write expressions for the work done on the system by these efforts. (See Section 3.1.1.)

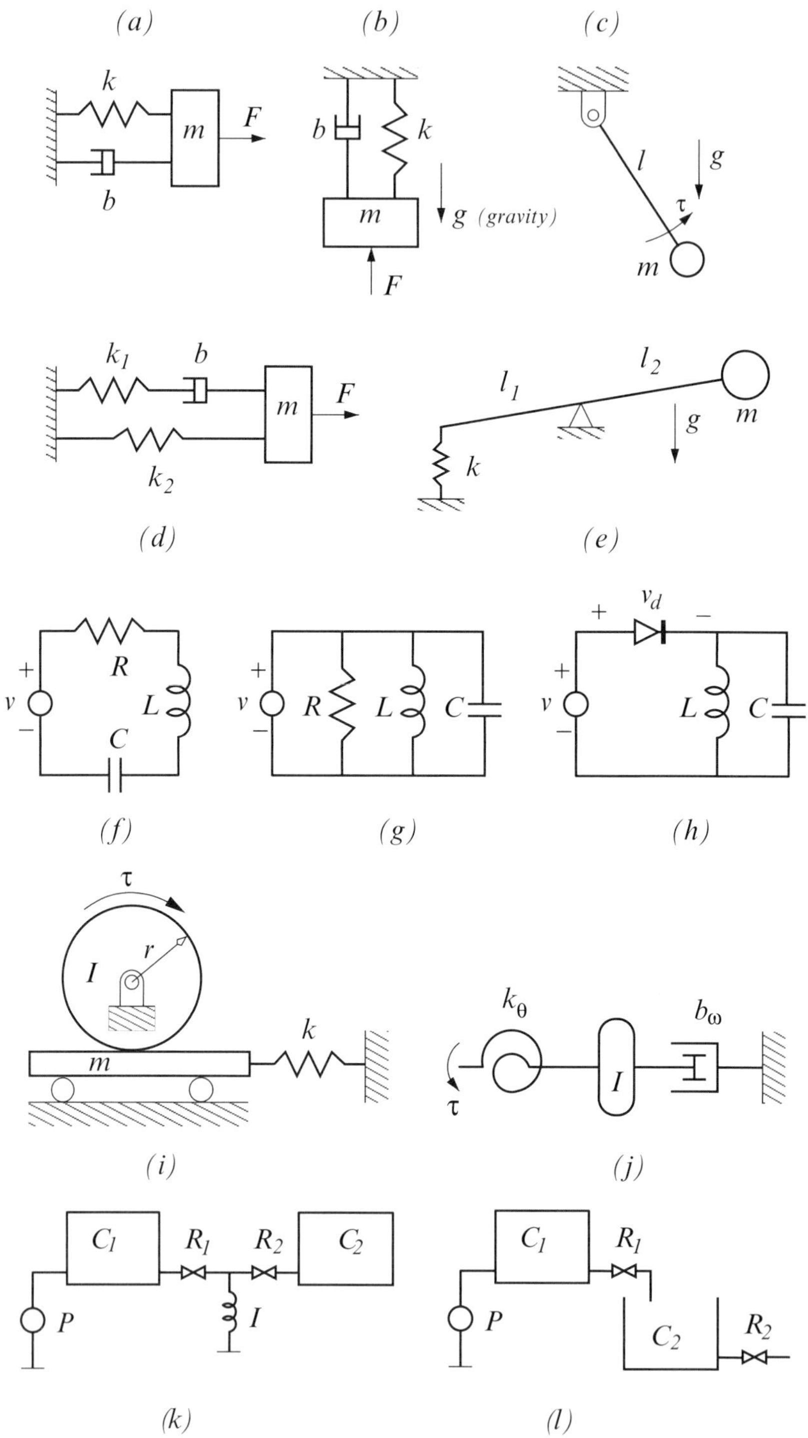
(a)
(b)
(c)
k
m
F
b
b
k
m
g (gravity)
F
l
g
τ
m
k_1
b
m
F
k_2
l_1
l_2
m
g
k
(d)
(e)
R
L
C
v
+
−
R
L
C
v
+
−
v_d
+
−
L
C
v
+
−
(f)
(g)
(h)
τ
r
I
m
k
k_θ
b_ω
I
τ
(i)
(j)
C_1
R_1
R_2
C_2
P
I
C_1
R_1
P
C_2
R_2
(k)
(l)

Chapter 2
Kinematics

In analytical mechanics *kinematics* is called the study of the geometry of motion. For mechanical systems kinematic analysis involves the determination of the position, velocity and acceleration of points in the system. Moreover, this analysis takes place without regard for the forces that give rise to the motion. In this text we will extend these concepts to multidiscipline systems using the unified framework presented in Chapter 1. In particular, this chapter develops techniques to help determine the displacement and flow associated with the components of the system. Section 2.1 considers the kinematics of mechanical systems where the inertia elements are treated as point masses and rigid bodies. Section 2.2 considers the mobility and kinematics of mechanisms. Finally, Section 2.3 presents the kinematic analysis of networks used to model electrical, and fluid systems. Using the results developed in this chapter we will be able to write analytical expressions for the kinetic coenergy, the potential energy and the dissipation function of multidiscipline systems, as well as analytical expressions for the displacement and flow constraints that may exist in the system.

2.1 Mechanical Systems

2.1.1 A point moving in a fixed frame

We begin the kinematic analysis of mechanical systems by considering the motion of a point that moves relative to a fixed reference frame. The point in question could represent a point mass or a point on a rigid body. In either case the objective of the kinematic analysis is to find the position and velocity (i.e., displacement and flow) of the point with respect to some fixed reference.

B. Fabien, *Analytical System Dynamics: Modeling and Simulation*,
DOI 10.1007/978-0-387-85605-6_2,

Coordinate systems

For mechanical systems the coordinate systems most often utilized are rectangular, cylindrical and spherical coordinates.

Rectangular coordinates

Consider a point P that is free to move in space, as shown in Fig. 2.1. In Fig. 2.1a the fixed reference frame is established by the rectangular coordinate system x-y-z. The origin of the system is at Q, and the axes x, y, and z are orthogonal, with unit vectors $\hat{\imath}$, $\hat{\jmath}$, and $\hat{k}$, respectively. Here, we will designate the coordinate system x-y-z as reference frame 0 (zero). In which case, the displacement of the point P relative to point Q as seen from frame 0 is given by

$$^0\bar{r}_{QP} = x\,\hat{\imath} + y\,\hat{\jmath} + z\,\hat{k}.$$

This rather verbose notation will prove beneficial when we consider the kinematics of systems that involve multiple reference frames.

The velocity of P is given by

$$^0\dot{\bar{r}}_{QP} = \dot{x}\,\hat{\imath} + \dot{y}\,\hat{\jmath} + \dot{z}\,\hat{k},$$

where, the superscript $(\dot{\bullet}) = \frac{d}{dt}(\bullet)$. Hence, $\dot{x}$, $\dot{y}$, and $\dot{z}$, are the velocities in the x, y, and z directions, respectively.

Cylindrical coordinates

The displacement of P can also be established using a cylindrical coordinate system as shown in Fig. 2.1b. The cylindrical coordinate system is defined using the unit vectors $\hat{e}_r$, $\hat{e}_\theta$, and $\hat{e}_z$. The vector $\hat{e}_r$ is directed from the point Q to P'. Here, P' is the projection of the point P onto the x-y plane. The vector $\hat{e}_z$ is parallel to the z-axis, and the vector $\hat{e}_\theta$ is the cross product of $\hat{k}$ and $\hat{e}_r$ using the right-hand rule. The distance from Q to P' is ρ, and the angle from the x-axis to the line QP' is θ. In this case the displacement of P is given by

$$^0\bar{r}_{QP} = \rho\,\hat{e}_r + \zeta\,\hat{e}_z.$$

Note that in the cylindrical coordinate system the unit vectors $\hat{e}_r$ and $\hat{e}_\theta$ change direction as P moves.

The displacements in rectangular coordinates are related to the displacements in cylindrical coordinates via the *transformation* equations

$$\begin{aligned} x &= \rho\cos\theta, \\ y &= \rho\sin\theta, \\ z &= \zeta, \end{aligned}$$

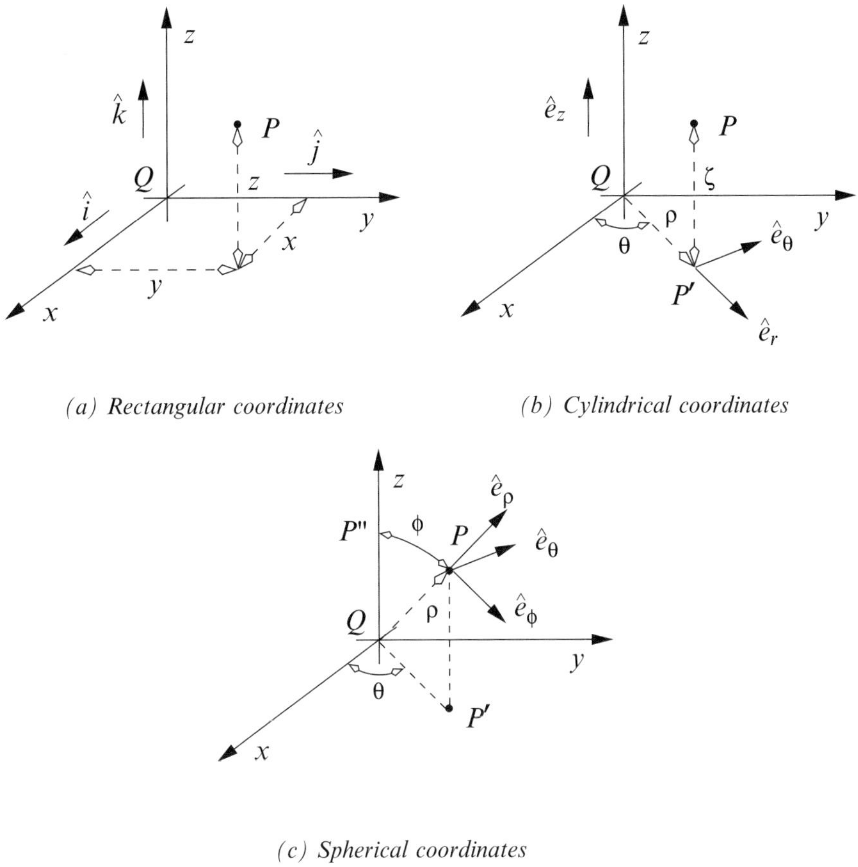

Fig. 2.1 Coordinate systems

where $0 \leq \theta \leq 2\pi$. Also, the unit vectors in the cylindrical coordinate system are related to the unit vectors in the rectangular coordinate systems by

$$\begin{aligned} \hat{e}_r &= \cos\theta\,\hat{\imath} + \sin\theta\,\hat{\jmath}, \\ \hat{e}_\theta &= -\sin\theta\,\hat{\imath} + \cos\theta\,\hat{\jmath}, \\ \hat{e}_z &= \hat{k}. \end{aligned}$$

The velocity of P in cylindrical coordinates is

$$\begin{aligned} {}^0\dot{\bar{r}}_{QP} &= \frac{d}{dt}(\rho\,\hat{e}_r) + \frac{d}{dt}(\zeta\,\hat{e}_z) \\ &= \dot{\rho}\,\hat{e}_r + \rho\frac{d}{dt}\,\hat{e}_r + \dot{\zeta}\,\hat{e}_z \\ &= \dot{\rho}\,\hat{e}_r + \rho\dot{\theta}\,\hat{e}_\theta + \dot{\zeta}\,\hat{e}_z, \end{aligned}$$

where $\dot{\rho}$ is the radial velocity, $\dot{\theta}$ is the angular velocity, and $\dot{\zeta}$ is the velocity in the $\hat{e}_z$ direction.

Spherical coordinates

The position of P can also be established using the spherical coordinates shown in Fig. 2.1c. The displacement variables associated with this system are ρ, θ, and ϕ. Here, ρ is the distance from Q to P. The variable θ measures the angle from the x-axis to the line QP' where, P' is the projection of P onto the x-y plane. The variable ϕ measures the angle from the z-axis to the line QP. Note that the angles θ and ϕ are measured positive in the counterclockwise direction.

The unit vector $\hat{e}_\rho$ is directed along the line from Q to P. The unit vector $\hat{e}_\phi$ is tangent to the arc $P''P$ at point P. The unit vector $\hat{e}_\theta$ is cross product of $\hat{e}_\rho$ and $\hat{e}_\phi$ using the right-hand rule.

The rectangular coordinates and the spherical coordinates are related via the equations

$$\begin{aligned} x &= \rho \sin\phi \cos\theta, \\ y &= \rho \sin\phi \sin\theta, \\ z &= \rho \cos\phi, \end{aligned}$$

where $0 \leq \theta \leq 2\pi$, and $0 \leq \phi \leq 2\pi$. The relationships between the unit vectors in the spherical coordinate system and the unit vectors in the rectangular coordinate system are

$$\begin{aligned} \hat{e}_\rho &= \sin\phi \cos\theta\, \hat{\imath} + \sin\phi \sin\theta\, \hat{\jmath} + \cos\phi\, \hat{k}, \\ \hat{e}_\theta &= -\sin\theta\, \hat{\imath} + \cos\theta\, \hat{\jmath}, \\ \hat{e}_\phi &= \cos\phi \cos\theta\, \hat{\imath} + \cos\phi \sin\theta\, \hat{\jmath} - \sin\phi\, \hat{k}. \end{aligned}$$

The displacement of P is spherical coordinates is

$${}^0\bar{r}_{QP} = \rho\, \hat{e}_\rho,$$

and the velocity of P is

$$\begin{aligned} {}^0\dot{\bar{r}}_{QP} &= \frac{d}{dt}(\rho\, \hat{e}_\rho) = \dot{\rho}\, \hat{e}_\rho + \rho \frac{d}{dt}\, \hat{e}_\rho, \\ &= \dot{\rho}\, \hat{e}_\rho + \rho \frac{d}{dt}(\sin\phi \cos\theta\, \hat{\imath} + \sin\phi \sin\theta\, \hat{\jmath} + \cos\phi\, \hat{k}), \\ &= \dot{\rho}\, \hat{e}_\rho + \rho\dot{\theta} \sin\phi\, \hat{e}_\theta + \rho\dot{\phi}\, \hat{e}_\phi, \end{aligned}$$

where $\dot{\rho}$ is the radial velocity, $\dot{\theta}$ is the rate of change in the angle θ, and $\dot{\phi}$ is the rate of change in the angle ϕ.

Example 2.1.

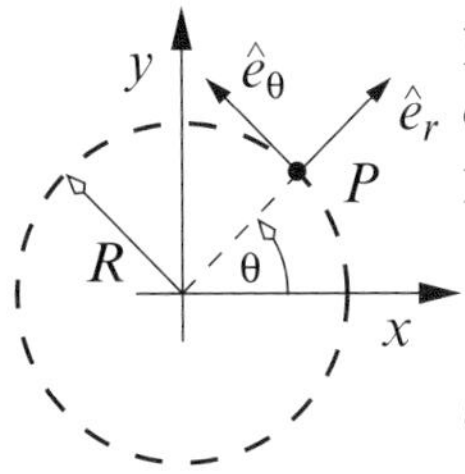

Consider a point P that moves in a circular path as shown here. The radius R is fixed, but the angle θ is a function of the time t. Using rectangular coordinates (x-y-z) the position of P is given by

$$^0\bar{r}_{QP} = R\cos\theta\,\hat{\imath} + R\sin\theta\,\hat{\jmath},$$

and the velocity is given by

$$^0\bar{v}_{QP} = \frac{d}{dt}\,^0\bar{r}_{QP} = R\dot{\theta}(-\sin\theta\,\hat{\imath} + \cos\theta\,\hat{\jmath}).$$

Using cylindrical coordinates the position and velocity of P are

$$^0\bar{r}_{QP} = R\,\hat{e}_r$$
$$^0\bar{v}_{QP} = R\dot{\theta}\,\hat{e}_\theta$$

In either the rectangular coordinate system or the cylindrical coordinate system it can be seen that the velocity is tangent to the path of motion. Moreover, if we define the angular velocity vector as $\bar{\omega} = \dot{\theta}\,\hat{k}$. Then the velocity of P can be written as the cross product

$$^0\bar{v}_{QP} = \bar{\omega} \times {}^0\bar{r}_{QP}.$$

2.1.2 A point moving in a translating frame

This section considers the kinematics of a point that is moving in a reference frame that can only translate relative to a fixed reference frame.

Displacement analysis

Let x-y-z be a fixed rectangular coordinate system with origin Q (see Fig. 2.2). The mutually orthogonal unit vectors $\hat{\imath}$, $\hat{\jmath}$, and $\hat{k}$, are in the x, y, and z directions, respectively. Henceforth, the x-y-z system will be called the fixed reference frame, and will be designated as frame 0. Let x_1-y_1-z_1 be a reference frame with origin Q_1. The mutually orthogonal unit vectors $\hat{\imath}_1$, $\hat{\jmath}_1$, and $\hat{k}_1$, are in the x_1, y_1, and z_1 directions, respectively. The analysis presented in this section assumes that the x_1-y_1-z_1 frame can only undergo translation with respect to the x-y-z frame. That is, the unit vectors $\hat{\imath}_1$, $\hat{\jmath}_1$, and $\hat{k}_1$ do not change direction relative to the fixed frame. However, the position of the point Q_1 can vary. The x_1-y_1-z_1 system will be called the moving frame, and designated as frame 1. Finally, the point P can undergo arbitrary displacements with respect to the x_1-y_1-z_1 frame.

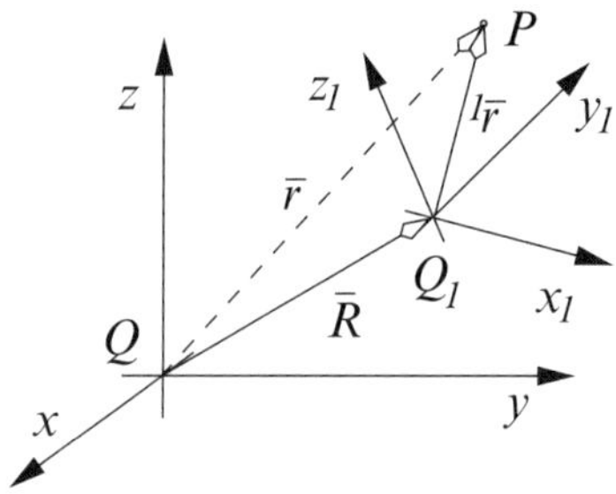

Fig. 2.2 A point moving in a moving frame

Let ${}^0\bar{R}_{QQ_1} = X\,\hat{\imath} + Y\,\hat{\jmath} + Z\,\hat{k}$ be the displacement vector from Q to Q_1 with respect to frame 0. Let ${}^1\bar{r}_{Q_1P} = x_1\,\hat{\imath}_1 + y_1\,\hat{\jmath}_1 + z_1\,\hat{k}_1$ be the displacement vector from Q_1 to P with respect to frame 1. Let ${}^0\bar{r}_{QP} = x\,\hat{\imath} + y\,\hat{\jmath} + z\,\hat{k}$ be the displacement vector from Q to P with respect to frame 0. Then from vector algebra we have

$$ {}^0\bar{r}_{QP} = {}^0\bar{R}_{QQ_1} + {}^1\bar{r}_{Q_1P}. \tag{2.1}$$

That is,

$$x\,\hat{\imath} + y\,\hat{\jmath} + z\,\hat{k} = X\,\hat{\imath} + Y\,\hat{\jmath} + Z\,\hat{k} + x_1\,\hat{\imath}_1 + y_1\,\hat{\jmath}_1 + z_1\,\hat{k}_1.$$

Using the property of the dot product the components of the vector ${}^0\bar{r}_{QP}$ can be determined as follows;

$$\begin{aligned} x &= {}^0\bar{r}_{QP} \cdot \hat{\imath} = X + x_1\,\hat{\imath}_1 \cdot \hat{\imath} + y_1\,\hat{\jmath}_1 \cdot \hat{\imath} + z_1\,\hat{k}_1 \cdot \hat{\imath}, \\ y &= {}^0\bar{r}_{QP} \cdot \hat{\jmath} = Y + x_1\,\hat{\imath}_1 \cdot \hat{\jmath} + y_1\,\hat{\jmath}_1 \cdot \hat{\jmath} + z_1\,\hat{k}_1 \cdot \hat{\jmath}, \\ z &= {}^0\bar{r}_{QP} \cdot \hat{k} = Z + x_1\,\hat{\imath}_1 \cdot \hat{k} + y_1\,\hat{\jmath}_1 \cdot \hat{k} + z_1\,\hat{k}_1 \cdot \hat{k}. \end{aligned}$$

With this result the displacement equation (2.1) can be written in a compact form using matrix notation. Specifically, let ${}^0r_{QP} = [x\ y\ z]^T$, ${}^0R_{QQ_1} = [X\ Y\ Z]^T$, ${}^1r_{Q_1P} = [x_1,\ y_1,\ z_1]^T$, and

$${}^0A_1 = \begin{bmatrix} \hat{\imath}_1 \cdot \hat{\imath} & \hat{\jmath}_1 \cdot \hat{\imath} & \hat{k}_1 \cdot \hat{\imath} \\ \hat{\imath}_1 \cdot \hat{\jmath} & \hat{\jmath}_1 \cdot \hat{\jmath} & \hat{k}_1 \cdot \hat{\jmath} \\ \hat{\imath}_1 \cdot \hat{k} & \hat{\jmath}_1 \cdot \hat{k} & \hat{k}_1 \cdot \hat{k} \end{bmatrix} = \begin{bmatrix} a_{11} & a_{12} & a_{13} \\ a_{21} & a_{22} & a_{23} \\ a_{31} & a_{32} & a_{33} \end{bmatrix}.$$

Then, equation (2.1) becomes

$$\begin{bmatrix} x \\ y \\ z \end{bmatrix} = \begin{bmatrix} X \\ Y \\ Z \end{bmatrix} + \begin{bmatrix} a_{11} & a_{12} & a_{13} \\ a_{21} & a_{22} & a_{23} \\ a_{31} & a_{32} & a_{33} \end{bmatrix} \begin{bmatrix} x_1 \\ y_1 \\ z_1 \end{bmatrix}.$$

That is,

$$^0r_{QP} = {^0R_{QQ_1}} + {^0A_1}\,{^1r_{Q_1P}}. \tag{2.2}$$

The elements of the matrix $^0A_1$ are the direction cosines between $\hat{\imath}$, $\hat{\jmath}$, $\hat{k}$ and $\hat{\imath}_1$, $\hat{\jmath}_1$, $\hat{k}_1$. Thus, given the coordinates of a point in frame 1, the matrix $^0A_1$ can be used to determine the coordinates of that point in frame 0.

Figure 2.3 can be used to clarify the meaning of the elements of the direction cosine matrix $^0A_1$. The figure shows the unit vector $\hat{\imath}_1$ in relation to the unit vectors $\hat{\imath}$, $\hat{\jmath}$, and $\hat{k}$. (Note that there is no loss in generality by translating the unit vector $\hat{\imath}_1$ to Q.) In Fig. 2.3, α is the angle between $\hat{\imath}$ and $\hat{\imath}_1$, β is the angle between $\hat{\jmath}$ and $\hat{\imath}_1$, and γ is the angle between $\hat{k}$ and $\hat{\imath}_1$. Therefore,

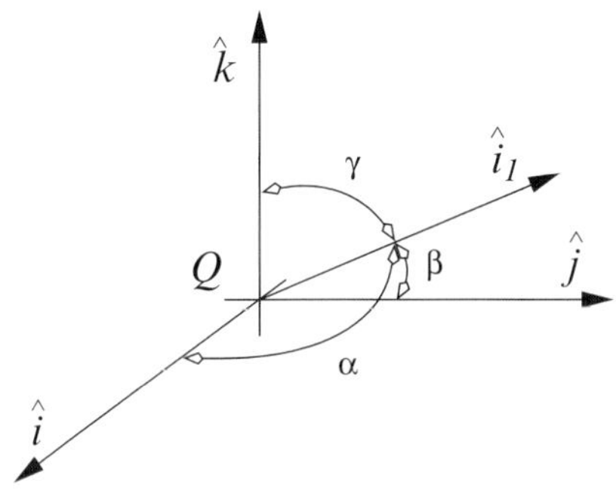

Fig. 2.3 Direction cosine

$$\hat{\imath}_1 = \cos\alpha\,\hat{\imath} + \cos\beta\,\hat{\jmath} + \cos\gamma\,\hat{k}.$$

Then the dot product of $\hat{\imath}_1$ with respect to $\hat{\imath}$, $\hat{\jmath}$, and $\hat{k}$, gives

$$\begin{aligned} a_{11} &= \hat{\imath}\cdot\hat{\imath}_1 = \cos\alpha, \\ a_{21} &= \hat{\jmath}\cdot\hat{\imath}_1 = \cos\beta, \\ a_{31} &= \hat{k}\cdot\hat{\imath}_1 = \cos\gamma, \end{aligned}$$

which are the elements of the first column of $^0A_1$. The other elements of $^0A_1$ are defined in a similar manner.

Example 2.2.

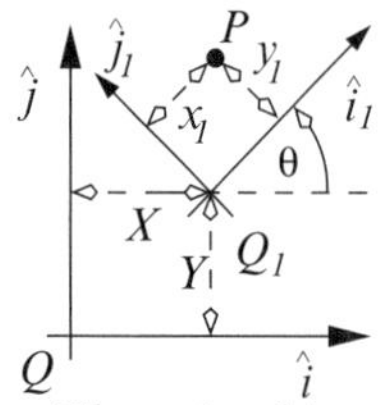

This figure shows a fixed reference frame with origin Q and mutually orthogonal unit vectors $\hat{\imath}$, $\hat{\jmath}$, and $\hat{k}$ in the x, y and z directions, respectively. The translating frame has origin Q_1, and mutually orthogonal unit vectors $\hat{\imath}_1$, $\hat{\jmath}_1$, and $\hat{k}_1$ in the x_1, y_1 and z_1 directions, respectively. The coordinate systems are aligned such that for all time, $\hat{k} = \hat{k}_1$, and the angle between $\hat{\imath}$ and $\hat{\imath}_1$ is θ.
Thus, in this case the elements of the direction cosine matrix are

$$
\begin{aligned}
&a_{11} = \hat{\imath}_1 \cdot \hat{\imath} = \cos\theta,\ a_{21} = \hat{\imath}_1 \cdot \hat{\jmath} = \cos\left(\frac{\pi}{2} - \theta,\ \right) = \sin\theta,\\
&a_{31} = \hat{\imath}_1 \cdot \hat{k} = \cos\frac{\pi}{2} = 0,\ a_{12} = \hat{\jmath}_1 \cdot \hat{\imath} = \cos\left(\frac{\pi}{2} + \theta\right) = -\sin\theta,\\
&a_{22} = \hat{\jmath}_1 \cdot \hat{\jmath} = \cos\theta,\ a_{32} = \hat{\jmath}_1 \cdot \hat{k} = \cos\frac{\pi}{2} = 0,\ a_{13} = \hat{k}_1 \cdot \hat{\imath} = \cos\frac{\pi}{2} = 0,\\
&a_{23} = \hat{k}_1 \cdot \hat{\jmath} = \cos\frac{\pi}{2} = 0,\ a_{33} = \hat{k}_1 \cdot \hat{k} = 1
\end{aligned}
$$

Therefore, using (2.2), the position of P in the fixed frame x-y-z is

$$
\begin{aligned}
{}^0r_{QP} &= {}^0R_{QQ_1} + {}^0A_1{}^1r_{Q_1P}\\
\begin{bmatrix} x \\ y \\ z \end{bmatrix} &= \begin{bmatrix} X \\ Y \\ Z \end{bmatrix} + \begin{bmatrix} \cos\theta & -\sin\theta & 0 \\ \sin\theta & \cos\theta & 0 \\ 0 & 0 & 1 \end{bmatrix} \begin{bmatrix} x_1 \\ y_1 \\ z_1 \end{bmatrix}.
\end{aligned}
$$

The terms in these equations have the following meaning;

- ${}^0r_{QP} = [x,\ y,\ z]^T$: are the components of the vector from Q to P in the x-y-z coordinate system (frame 0).
- ${}^0R_{QQ_1} = [X,\ Y,\ Z]^T$: are the components of the vector from Q to Q_1 in the x-y-z coordinate system.
- ${}^0A_1$: is the direction cosine matrix relating the x_1-y_1-z_1 coordinate system to the x-y-z coordinate system.
- ${}^1r_{Q_1P} = [x_1,\ y_1,\ z_1]^T$: are the components of the vector from Q_1 to P in the x_1-y_1-z_1 coordinate system (frame 1).

Given a point in the x_1-y_1-z_1 system, equation (2.2) can be used to find it's coordinates in the x-y-z system. An inverse transformation is also possible, that is, given a point in the x-y-z system there is an equation that will determine it's coordinates in the x_1-y_1-z_1 system. To derive this inverse transformation rearrange equation (2.2) to get

$$
\begin{aligned}
{}^0A_1{}^1r_{Q_1P} &= {}^0r_{QP} - {}^0R_{QQ_1},\\
{}^1r_{Q_1P} &= \left({}^0A_1\right)^{-1}\left({}^0r_{QP} - {}^0R_{QQ_1}\right).
\end{aligned}
$$

Here $\left({}^0A_1\right)^{-1}$ is the inverse of the direction cosine matrix ${}^0A_1$.

For the linear transformations described above, the matrix $\left({}^{0}A_1\right)^{-1}$ can be easily determined if ${}^{0}A_1$ is known. From the vector equation (2.1) it can be seen that

$$
\begin{aligned}
{}^{1}\bar{r}_{Q_1P} &= {}^{0}\bar{r}_{QP} - {}^{0}\bar{R}_{QQ_1} \\
x_1\,\hat{\imath}_1 + y_1\,\hat{\jmath}_1 + z_1\,\hat{k}_1 &= (x - X)\,\hat{\imath} + (y - Y)\,\hat{\jmath} + (z - Z)\,\hat{k}.
\end{aligned}
$$

Using the property of the dot product shows that the components of the vector ${}^{1}\bar{r}_{Q_1P}$ satisfy

$$
\begin{aligned}
x_1 &= {}^{1}\bar{r}_{Q_1P} \cdot \hat{\imath}_1 = (x - X)\,\hat{\imath} \cdot \hat{\imath}_1 + (y - Y)\,\hat{\jmath} \cdot \hat{\imath}_1 + (z - Z)\,\hat{k} \cdot \hat{\imath}_1, \\
y_1 &= {}^{1}\bar{r}_{Q_1P} \cdot \hat{\jmath}_1 = (x - X)\,\hat{\imath} \cdot \hat{\jmath}_1 + (y - Y)\,\hat{\jmath} \cdot \hat{\jmath}_1 + (z - Z)\,\hat{k} \cdot \hat{\jmath}_1, \\
z_1 &= {}^{1}\bar{r}_{Q_1P} \cdot \hat{k}_1 = (x - X)\,\hat{\imath} \cdot \hat{k}_1 + (y - Y)\,\hat{\jmath} \cdot \hat{k}_1 + (z - Z)\,\hat{k} \cdot \hat{k}_1.
\end{aligned}
$$

In matrix notation these equations are

$$
\begin{bmatrix} x_1 \\ y_1 \\ z_1 \end{bmatrix} = \begin{bmatrix} \hat{\imath}_1 \cdot \hat{\imath} & \hat{\imath}_1 \cdot \hat{\jmath} & \hat{\imath}_1 \cdot \hat{k} \\ \hat{\jmath}_1 \cdot \hat{\imath} & \hat{\jmath}_1 \cdot \hat{\jmath} & \hat{\jmath}_1 \cdot \hat{k} \\ \hat{k}_1 \cdot \hat{\imath} & \hat{k}_1 \cdot \hat{\jmath} & \hat{k}_1 \cdot \hat{k} \end{bmatrix} \begin{bmatrix} x - X \\ y - Y \\ z - Z \end{bmatrix} = \begin{bmatrix} a_{11} & a_{21} & a_{31} \\ a_{12} & a_{22} & a_{32} \\ a_{13} & a_{23} & a_{33} \end{bmatrix} \begin{bmatrix} x - X \\ y - Y \\ z - Z \end{bmatrix}.
$$

That is,

$$
{}^{1}\bar{r}_{Q_1P} = ({}^{0}A_1)^T({}^{0}r_{QP} - {}^{0}R_{QQ_1}). \tag{a}
$$

Hence, $({}^{0}A_1)^{-1} = ({}^{0}A_1)^T$. If the inverse of a matrix is equal to its transpose then the matrix is said to be *orthogonal*. Therefore, the direction cosine matrix, ${}^{0}A_1$, defined above is orthogonal. It is easy to verify that the direction cosine matrix, ${}^{0}A_1$, in Example 2.2 is indeed orthogonal.

Finally, from equation (a), note that $({}^{0}r_{QP} - {}^{0}R_{QQ_1})$ represents the components of a vector that is in frame 0, while ${}^{1}r_{Q_1P}$ represents the components of a vector that is in frame 1. Hence, $({}^{0}A_1)^T$ is a transformation matrix from frame 0 to frame 1, and is written as

$$
({}^{0}A_1)^T = ({}^{0}A_1)^{-1} = {}^{1}A_0.
$$

Moreover, equation (a) can be written as

$$
{}^{1}\bar{r}_{Q_1P} = {}^{1}A_0({}^{0}r_{QP} - {}^{0}R_{QQ_1}).
$$

Example 2.3.

The points Q_1, S_1, T_1 and U_1 have the following coordinates with respect to the right-handed rectangular system x-y-z (frame 0):

$$
{}^{0}Q_1 = \begin{bmatrix} 1 \\ 4.3 \\ -1.7 \end{bmatrix}, \quad {}^{0}S_1 = \begin{bmatrix} 1.6010 \\ 5.8755 \\ -1.4844 \end{bmatrix},
$$

$$^0T_1 = \begin{bmatrix} -1.0208 \\ 4.7185 \\ 0.8751 \end{bmatrix}, \quad ^0U_1 = \begin{bmatrix} 4.6062 \\ 2.4969 \\ 1.4231 \end{bmatrix}.$$

The point Q_1 is the origin of the origin of the right-handed rectangular system x_1-y_1-z_1 (frame 1), the point S_1 is a point on the x_1-axis, the point T_1 is a point on the y_1-axis, and the point U_1 is a point on the z_1-axis.

(a) Show that the vectors $^0\bar{r}_{Q_1S_1}$, $^0\bar{r}_{Q_1T_1}$ and $^0\bar{r}_{Q_1U_1}$ are orthogonal, and form a right-handed rectangular coordinate system.
(b) Find the direction cosine matrix relating the x_1-y_1-z_1 system to the x-y-z system.
(c) The point P_1 has coordinate $^1P_1 = [1, \; 1, \; 1]^T$ with respect to the x_1-y_1-z_1 system. Find the coordinates of P_1 with respect to the x-y-z system.
(d) The point P_2 has coordinate $^0P_2 = [1, \; 1, \; 1]^T$ with respect to the x-y-z system. Find the coordinates of P_2 with respect to the x_1-y_1-z_1 system.

Solution:

(a) Let $\hat{\imath}$, $\hat{\jmath}$ and $\hat{k}$ be the unit vectors along the x-axis, y-axis and z-axis, respectively. Then the vectors $^0\bar{r}_{Q_1S_1}$, $^0\bar{r}_{Q_1T_1}$ and $^0\bar{r}_{Q_1U_1}$ can be written as

$$\begin{aligned} ^0\bar{r}_{Q_1S_1} &= (1.601 - 1)\,\hat{\imath} + (5.8755 - 4.3)\,\hat{\jmath} + (-1.4844 + 1.7)\,\hat{k} \\ &= 0.601\,\hat{\imath} + 1.5755\,\hat{\jmath} + 0.2156\,\hat{k}, \\ ^0\bar{r}_{Q_1T_1} &= (-1.0208 - 1)\,\hat{\imath} + (4.7185 - 4.3)\,\hat{\jmath} + (0.8751 + 1.7)\,\hat{k} \\ &= -2.0208\,\hat{\imath} + 0.4185\,\hat{\jmath} + 2.5751\,\hat{k}, \\ ^0\bar{r}_{Q_1U_1} &= (4.6062 - 1)\,\hat{\imath} + (2.4969 - 4.3)\,\hat{\jmath} + (1.4231 + 1.7)\,\hat{k} \\ &= 3.6062\,\hat{\imath} - 1.8031\,\hat{\jmath} + 3.1231\,\hat{k}. \end{aligned}$$

Let $^0\hat{r}_{Q_1S_1}$, $^0\hat{r}_{Q_1T_1}$ and $^0\hat{r}_{Q_1U_1}$ be the unit vectors (normalized vectors) associated with $^0\bar{r}_{Q_1S_1}$, $^0\bar{r}_{Q_1T_1}$ and $^0\bar{r}_{Q_1U_1}$ respectively. That is,

$$\begin{aligned} ^0\hat{r}_{Q_1S_1} &= {}^0\bar{r}_{Q_1S_1}/|{}^0\bar{r}_{Q_1S_1}| = 0.35354\,\hat{\imath} + 0.92678\,\hat{\jmath} + 0.12683\,\hat{k}, \\ ^0\hat{r}_{Q_1T_1} &= {}^0\bar{r}_{Q_1T_1}/|{}^0\bar{r}_{Q_1T_1}| = -0.61237\,\hat{\imath} + 0.12682\,\hat{\jmath} + 0.78034\,\hat{k}, \\ ^0\hat{r}_{Q_1U_1} &= {}^0\bar{r}_{Q_1U_1}/|{}^0\bar{r}_{Q_1U_1}| = 0.70710\,\hat{\imath} - 0.35355\,\hat{\jmath} + 0.61238\,\hat{k}, \end{aligned}$$

where $|^0\bar{r}_{Q_1S_1}|$ denotes magnitude of the vector $^0\bar{r}_{Q_1S_1}$, etc.
If the vectors $^0\hat{r}_{Q_1S_1}$, $^0\hat{r}_{Q_1T_1}$ and $^0\hat{r}_{Q_1U_1}$ are mutually orthogonal then $^0\hat{r}_{Q_1S_1} \cdot {}^0\hat{r}_{Q_1T_1} = {}^0\hat{r}_{Q_1S_1} \cdot {}^0\hat{r}_{Q_1U_1} = {}^0\hat{r}_{Q_1T_1} \cdot {}^0\hat{r}_{Q_1U_1} = 0$. Indeed this can be easily verified using the results above.
Finally, if the unit vectors $^0\hat{r}_{Q_1S_1}$, $^0\hat{r}_{Q_1T_1}$ and $^0\hat{r}_{Q_1U_1}$ form a right-handed rectangular system then $(^0\hat{r}_{Q_1S_1} \times {}^0\hat{r}_{Q_1T_1}) \cdot {}^0\hat{r}_{Q_1T_1} = 1$. Using the results above it can be shown that this identity holds.

(b) Since ${}^0\hat{r}_{Q_1S_1}$, ${}^0\hat{r}_{Q_1T_1}$ and ${}^0\hat{r}_{Q_1U_1}$ are unit vectors in the x_1, y_1 and z_1 directions we have $\hat{\imath}_1 = {}^0\hat{r}_{Q_1S_1}$, $\hat{\jmath}_1 = {}^0\hat{r}_{Q_1T_1}$, and $\hat{k}_1 = {}^0\hat{r}_{Q_1U_1}$. The direction cosine matrix relating the x_1-y_1-z_1 system to the x-y-z system can be written as

$$ {}^0A_1 = \begin{bmatrix} a_{11} & a_{12} & a_{13} \\ a_{21} & a_{22} & a_{23} \\ a_{31} & a_{32} & a_{33} \end{bmatrix}, $$

where

$$ \begin{aligned} \hat{\imath}_1 &= a_{11}\,\hat{\imath} + a_{21}\,\hat{\jmath} + a_{31}\,\hat{k}, \\ \hat{\jmath}_1 &= a_{12}\,\hat{\imath} + a_{22}\,\hat{\jmath} + a_{32}\,\hat{k}, \\ \hat{k}_1 &= a_{13}\,\hat{\imath} + a_{23}\,\hat{\jmath} + a_{33}\,\hat{k}. \end{aligned} $$

Using the results from (a) it can be seen that

$$ {}^0A_1 = \begin{bmatrix} 0.35354 & -0.61237 & 0.70710 \\ 0.92678 & 0.12682 & -0.35355 \\ 0.12683 & 0.78034 & 0.61238 \end{bmatrix}. $$

(c) The coordinate of the point P_1 with respect to the x-y-z system is given by

$$ \begin{aligned} {}^0P_1 &= {}^0Q_1 + {}^0A_1\,{}^1P_1 \\ &= \begin{bmatrix} 1 \\ 4.3 \\ -1.7 \end{bmatrix} + \begin{bmatrix} 0.35354 & -0.61237 & 0.70710 \\ 0.92678 & 0.12682 & -0.35355 \\ 0.12683 & 0.78034 & 0.61238 \end{bmatrix} \begin{bmatrix} 1 \\ 1 \\ 1 \end{bmatrix} \\ &= \begin{bmatrix} 1.44827 \\ 5.0 \\ -0.18046 \end{bmatrix}. \end{aligned} $$

(d) The coordinate of the point P_2 with respect to the x_1-y_1-z_1 system is given by

$$ \begin{aligned} {}^1P_2 &= {}^1A_0({}^0P_1 - {}^0Q_1) \\ &= ({}^0A_1)^T({}^0P_1 - {}^0Q_1) \\ &= \begin{bmatrix} 0.35354 & 0.92678 & 0.12683 \\ -0.61237 & 0.12682 & 0.78034 \\ 0.70710 & -0.35355 & 0.61238 \end{bmatrix} \left(\begin{bmatrix} 1 \\ 1 \\ 1 \end{bmatrix} - \begin{bmatrix} 1 \\ 4.3 \\ -1.7 \end{bmatrix} \right) \\ &= \begin{bmatrix} -2.7160 \\ 1.6884 \\ 2.8201 \end{bmatrix}. \end{aligned} $$

Velocity analysis

The velocity of the point P in the translating frame can be computed from (2.1) as

$$^0\bar{v}_{QP} = \frac{d}{dt}\,(^0\bar{r}_{QP}) = \frac{d}{dt}\,(^0\bar{R}_{QQ_1}) + \frac{d}{dt}\,(^1\bar{r}_{Q_1P}).$$

Here, the term

$$\frac{d}{dt}\,(^0\bar{r}_{QP}) = \dot{x}\,\hat{\imath} + \dot{y}\,\hat{\jmath} + \dot{z}\,\hat{k}$$

is the velocity of P relative to Q as seen from frame 0, the term

$$\frac{d}{dt}\,(^0\bar{R}_{QQ_1}) = \dot{X}\,\hat{\imath} + \dot{Y}\,\hat{\jmath} + \dot{Z}\,\hat{k}$$

is the velocity of Q_1 relative to Q as seen from 0, and the term

$$\frac{d}{dt}\,(^1\bar{r}_{Q_1P}) = \dot{x}_1\,\hat{\imath}_1 + \dot{y}_1\,\hat{\jmath}_1 + \dot{z}_1\,\hat{k}_1$$

is the velocity of P relative to Q_1 as seen from frame 1. The term $d(^1\bar{r}_{Q_1P})/dt$ uses the fact that $d\hat{\imath}_1/dt = d\hat{\jmath}_1/dt = d\hat{k}_1/dt = 0$, i.e., the unit vectors $\hat{\imath}_1$, $\hat{\jmath}_1$, and $\hat{k}_1$, do not change direction.

Using matrix notation the velocity equation takes the form,

$$^0\dot{r}_{QP} = {}^0\dot{R}_{QQ_1} + {}^0A_1\,{}^1\dot{r}_{Q_1P}, \tag{2.3}$$

where $^0\dot{r}_{QP} = [\dot{x},\ \dot{y},\ \dot{z}]^T$, $^0\dot{R}_{QQ_1} = [\dot{X},\ \dot{Y},\ \dot{Z}]^T$, $^1\dot{r}_{Q_1P} = [\dot{x}_1,\ \dot{y}_1,\ \dot{z}_1]^T$, and $^0A_1$ is the direction cosine matrix relating the translating frame, x_1-y_1-z_1, to the fixed frame, x-y-z.

Example 2.4.

In Example 2.2 suppose the displacement of Q_1 relative to Q is given by

$$\begin{bmatrix} X \\ Y \\ Z \end{bmatrix} = \begin{bmatrix} a_1 t^2 + X_0 \\ a_2 t + Y_0 \\ Z_0 \end{bmatrix},$$

where a_1, a_2, X_0, Y_0, Z_0, are constants, and t is the time. In addition, suppose the displacement of P relative to Q_1 is defined by

$$\begin{bmatrix} x_1 \\ y_1 \\ z_1 \end{bmatrix} = \begin{bmatrix} \rho\cos\lambda t \\ \rho\sin\lambda t \\ t \end{bmatrix},$$

where ρ and λ are constants.

Then, the velocity of the point P is

$$
\begin{aligned}
{}^0\dot{r}_{QP} &= {}^0\dot{R}_{QQ_1} + {}^0A_1\,{}^1\dot{r}_{Q_1P} \\
\begin{bmatrix} \dot{x} \\ \dot{y} \\ \dot{z} \end{bmatrix} &= \begin{bmatrix} \dot{X} \\ \dot{Y} \\ \dot{Z} \end{bmatrix} + \begin{bmatrix} \cos\theta & -\sin\theta & 0 \\ \sin\theta & \cos\theta & 0 \\ 0 & 0 & 1 \end{bmatrix} \begin{bmatrix} \dot{x}_1 \\ \dot{y}_1 \\ \dot{z}_1 \end{bmatrix} \\
&= \begin{bmatrix} 2a_1t \\ a_2 \\ 0 \end{bmatrix} + \begin{bmatrix} \cos\theta & -\sin\theta & 0 \\ \sin\theta & \cos\theta & 0 \\ 0 & 0 & 1 \end{bmatrix} \begin{bmatrix} -\rho\lambda\sin\lambda t \\ \rho\lambda\cos\lambda t \\ 1 \end{bmatrix} \\
&= \begin{bmatrix} 2a_1t - \rho\lambda(\cos\theta\sin\lambda t + \sin\theta\cos\lambda t) \\ a_2 - \rho\lambda(\sin\theta\sin\lambda t - \cos\theta\cos\lambda t) \\ 1 \end{bmatrix} .
\end{aligned}
$$

2.1.3 A point fixed in a rotating frame

This section considers the kinematics of a point P, that is fixed in a rotating frame (see Fig. 2.4). As in the previous section we will let the rectangular coordinate system x-y-z, with origin Q, be a fixed reference frame, while the rectangular coordinate system x_1-y_1-z_1, with origin Q_1, will represent the rotating frame. Throughout this section Q and Q_1 will remain coincident however, the x_1-y_1-z_1 coordinate system is allowed to rotate with respect to the x-y-z coordinate system. In addition, the point P, is in a fixed position relative to the x_1-y_1-z_1 system. At any instant we let ${}^0A_1$ denote the direction cosine matrix that relates the x_1-y_1-z_1 coordinate system to the x-y-z coordinate system.

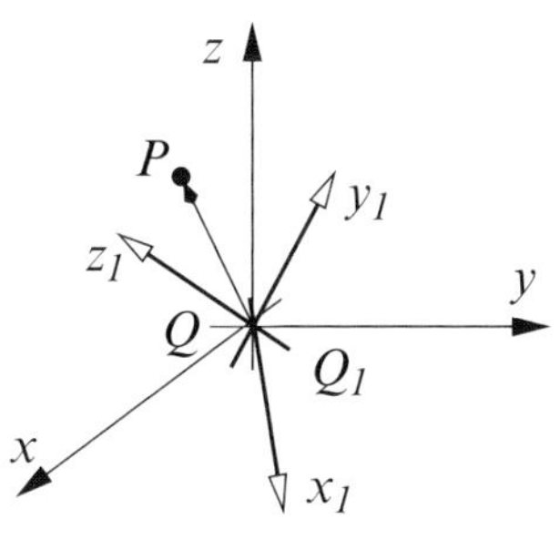

Fig. 2.4 A fixed point in a rotating frame

Displacement analysis

Let the unit vectors $\hat{\imath}$, $\hat{\jmath}$, and $\hat{k}$ be aligned with the x, y and z directions, respectively. Let the unit vectors $\hat{\imath}_1$, $\hat{\jmath}_1$ and $\hat{k}_1$ be aligned with the x_1, y_1 and z_1 directions, respectively. Then, the position of the point P is given by the vector

$$ {}^0\bar{r}_{QP} = {}^1\bar{r}_{Q_1P}, \tag{2.4} $$

where ${}^0\bar{r}_{QP} = x\,\hat{\imath} + y\,\hat{\jmath} + z\,\hat{k}$ is the vector from Q to P, and ${}^1\bar{r}_{Q_1P} = x_1\,\hat{\imath}_1 + y_1\,\hat{\jmath}_1 + z_1\,\hat{k}_1$ is the vector from Q_1 to P. Note that Q and Q_1 are always coincident thus, ${}^0\bar{r}_{QP}$ and ${}^1\bar{r}_{Q_1P}$ are the same vector but with different coordinate system representations.

Velocity analysis

The coordinates of the point P in the x_1-y_1-z_1 frame are ${}^1r_{Q_1P}=[x_1,\ y_1,\ z_1]^T$. Then, P has coordinates ${}^0r_{QP} = {}^0A_1{}^1r_{Q_1P}$, in the fixed frame, where ${}^0r_{QP} = [x,\ y,\ z]^T$. In this case the velocity of P, as seen from the fixed frame, is

$$ {}^0v_{QP} = \frac{d}{dt}({}^0r_{QP}) = \frac{d}{dt}({}^0A_1{}^1r_{Q_1P}) = {}^0\dot{A}_1{}^1r_{Q_1P}. \tag{a} $$

where ${}^0v_{QP} = [\dot{x},\ \dot{y},\ \dot{z}]^T$. Here, we have used the fact that $\frac{d}{dt}({}^1r_{Q_1P}) = 0$, because P is fixed in the x_1-y_1-z_1 frame. Thus, the velocity, ${}^0v_{QP}$, is due solely to the rotation of the x_1-y_1-z_1 frame.

We next develop an expression for the matrix ${}^0\dot{A}_1$. Since the direction cosine matrix, ${}^0A_1$, is orthogonal, we have $({}^0A_1)^{-1} = ({}^0A_1)^T$, and $({}^0A_1)^T({}^0A_1) = I$, where I denotes the 3×3 identity matrix. Therefore,

$$ \frac{d}{dt}(({}^0A_1)^T({}^0A_1)) = ({}^0\dot{A}_1)^T({}^0A_1) + ({}^0A_1)^T({}^0\dot{A}_1) = \frac{d}{dt}I = 0. \tag{b} $$

Let ${}^1\tilde{\omega} = ({}^0A_1)^T({}^0\dot{A}_1)$, then equation (b) implies that

$$ {}^1\tilde{\omega} = -({}^1\tilde{\omega})^T. $$

That is, ${}^1\tilde{\omega}$ is a skew-symmetric matrix, and it has the form

$$ {}^1\tilde{\omega} = \begin{bmatrix} 0 & -\omega_3 & \omega_2 \\ \omega_3 & 0 & -\omega_1 \\ -\omega_2 & \omega_1 & 0 \end{bmatrix}. \tag{2.5} $$

Using the orthogonality property of ${}^0A_1$ it can be seen that

$$ {}^0A_1({}^0A_1)^T({}^0\dot{A}_1) = {}^0A_1{}^1\tilde{\omega} = {}^0\dot{A}_1. $$

Hence, the velocity of P can be written as

$$^0v_{QP} = {}^0A_1 {}^1\tilde{\omega}({}^1r_{Q_1P}). \tag{2.6}$$

Equation (2.6) can be written as a vector equation as follows. Let the vector $^0\bar{v}_{QP} = \dot{x}\,\hat{i} + \dot{y}\,\hat{j} + \dot{z}\,\hat{k}$, represent the velocity of P as seen from the fixed frame. Let the vector $^1\bar{r}_{Q_1P} = x_1\,\hat{i}_1 + y_1\,\hat{j}_1 + z_1\,\hat{k}_1$, represent the position of P relative to the rotating frame. Finally, define the vector $^1\bar{\omega} = \omega_1\,\hat{i}_1 + \omega_2\,\hat{j}_1 + \omega_3\,\hat{k}_1$, as the *angular velocity* of the rotating frame. Then, from equation (2.6) it can be seen that

$$^0\bar{v}_{QP} = {}^1\bar{\omega} \times {}^1\bar{r}_{Q_1P}. \tag{2.7}$$

The vector on the left-hand side of equation (2.7) is in terms of the unit vectors $\hat{i}$, $\hat{j}$, and $\hat{k}$, while the vector on the right-hand side of the equation is in terms of the unit vectors $\hat{i}_1$, $\hat{j}_1$, and $\hat{k}_1$. Thus, an appropriate coordinate transformation is required to equate the terms on either side of equation (2.7).

Suppose that the position of the point P is given by $^0\bar{r}_{QP} = \hat{i}_1 = {}^1\bar{r}_{Q_1P}$ then, equation (2.7) implies that

$$\frac{d}{dt}({}^0\bar{r}_{QP}) = \frac{d\hat{i}_1}{dt} = {}^1\bar{\omega} \times \hat{i}_1. \tag{2.8}$$

Similarly, if the position of P is given by $^0\bar{r}_{QP} = \hat{j}_1 = {}^1\bar{r}_{Q_1P}$, or $^0\bar{r}_{QP} = \hat{k}_1 = {}^1\bar{r}$, then

$$\frac{d\hat{j}_1}{dt} = {}^1\bar{\omega} \times \hat{j}_1, \tag{2.9}$$

$$\frac{d\hat{k}_1}{dt} = {}^1\bar{\omega} \times \hat{k}_1, \tag{2.10}$$

respectively. Equations (2.8), (2.9) and (2.10) describe the velocities of the unit vectors $\hat{i}_1$, $\hat{j}_1$, and $\hat{k}_1$, due to the rotation of the x_1-y_1-z_1 frame.

Example 2.5.

Point C denotes the center of the disk shown in Fig. (a), and point P is on the edge of the disk. The disk rotates about the shaft QC with angular velocity ω_C. The shaft QC is perpendicular to CP, and it has a fixed angle α relative to the horizontal plane. The shaft QC also rotates about the line QZ with angular velocity ω_Z. The radius of the disk CP is ρ, and the length of the shaft QC is l. An analysis of the position and velocity of the point P proceeds as follows.

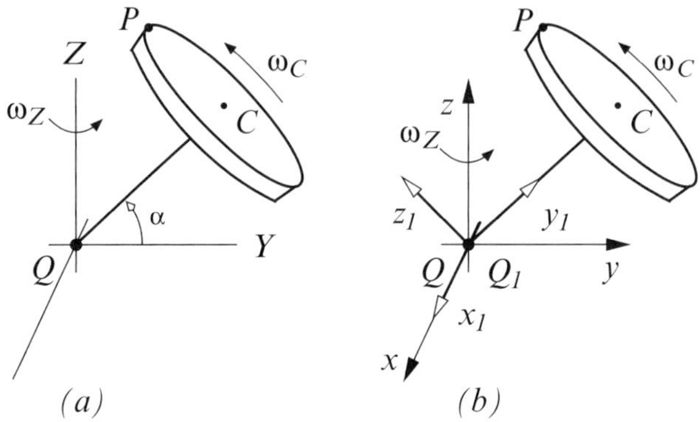

First, we establish a fixed reference frame defined by the rectangular coordinates x, y, and z, with its origin at Q. The z axis is along the line QZ, and the y axis is along the line QY. A right handed rectangular coordinate system is formed by selecting the x axis perpendicular to the y-z plane, as shown in Fig. (b). The unit vectors along the x, y, and z axes are $\hat{\imath}$, $\hat{\jmath}$, and $\hat{k}$, respectively.

Next, a moving (rotating) frame is defined by the rectangular coordinates system x_1-y_1-z_1, with origin Q_1 that is coincident with Q. This frame is attached to the disk and undergoes the same rotations as the disk. The unit vectors along the x_1, y_1, and z_1 axes are $\hat{\imath}_1$, $\hat{\jmath}_1$, and $\hat{k}_1$, respectively.

A vector approach

Displacement analysis

At the instant shown, in the figures above, the relationship between the unit vectors of the fixed frame and the unit vectors of the rotating frame is

$$\begin{bmatrix} \hat{\imath} \\ \hat{\jmath} \\ \hat{k} \end{bmatrix} = \begin{bmatrix} 1 & 0 & 0 \\ 0 & \cos\alpha & -\sin\alpha \\ 0 & \sin\alpha & \cos\alpha \end{bmatrix} \begin{bmatrix} \hat{\imath}_1 \\ \hat{\jmath}_1 \\ \hat{k}_1 \end{bmatrix}. \tag{a}$$

Let the vector ${}^0\bar{r}$ denote the position of P relative to Q in the x-y-z coordinate system (i.e., frame 0). Let the vector ${}^1\bar{r}$ denote the position of P relative to Q_1 in the x_1-y_1-z_1 frame. Then,

$$\begin{aligned} {}^0\bar{r}_{QP} &= {}^1\bar{r}_{Q_1P} \\ &= l\,\hat{\jmath}_1 + \rho\,\hat{k}_1 \\ &= l(\cos\alpha\,\hat{\jmath} + \sin\alpha\,\hat{k}) + \rho(-\sin\alpha\,\hat{\jmath} + \cos\alpha\,\hat{k}) \\ &= (l\cos\alpha \;-\rho\sin\alpha)\,\hat{\jmath} + (l\sin\alpha + \rho\cos\alpha)\,\hat{k}, \end{aligned}$$

where ${}^0\bar{r}_{QP} = x\,\hat{\imath} + y\,\hat{\jmath} + z\,\hat{k}$. Therefore, the coordinates of P in the x-y-z coordinate system are

$$\begin{aligned} x &= 0 \\ y &= l\cos\alpha \ - \rho\sin\alpha \\ z &= l\sin\alpha + \rho\cos\alpha. \end{aligned}$$

Velocity analysis

Since the x_1-y_1-z_1 coordinate system is attached to the disk, the point P can be treated as a fixed point in a rotating frame. Thus, the velocity of P is given by

$${}^0\bar{v}_{QP} = {}^1\bar{\omega} \times {}^1\bar{r}_{Q_1P},$$

where ${}^0\bar{v} = \dot{x}\hat{\imath} + \dot{y}\hat{\jmath} + \dot{z}\hat{k}$, is the velocity of P relative to the fixed frame, and the vector ${}^1\bar{\omega}$ denotes the angular velocity of the rotating frame. From the problem description it can be seen that ${}^1\bar{\omega}$ is given by

$${}^1\bar{\omega} = \omega_C\hat{\jmath}_1 + \omega_Z\hat{k}.$$

The term $\omega_C\hat{\jmath}_1$ is due to the rotation of the disk about the line QC, and the term $\omega_Z\hat{k}$ is due to the rotation of the line QC about the line QZ.

Using equation (a) then gives

$${}^1\bar{\omega} = \omega_C\hat{\jmath}_1 + \omega_Z(\sin\alpha\,\hat{\jmath}_1 + \cos\alpha\,\hat{k}_1) = (\omega_C + \omega_Z\sin\alpha)\,\hat{\jmath}_1 + \omega_Z\cos\alpha\,\hat{k}_1. \quad (b)$$

Hence,

$$\begin{aligned} {}^0\bar{v}_{QP} &= ((\omega_C + \omega_Z\sin\alpha)\,\hat{\jmath}_1 + \omega_Z\cos\alpha\,\hat{k}_1) \times (l\,\hat{\jmath}_1 + \rho\,\hat{k}_1) \\ &= (\rho(\omega_C + \omega_Z\sin\alpha) - l\omega_Z\cos\alpha)\,\hat{\imath}_1 \\ &= (\rho(\omega_C + \omega_Z\sin\alpha) - l\omega_Z\cos\alpha)\,\hat{\imath}. \end{aligned}$$

Therefore, the components of the velocity vector are

$$\begin{aligned} \dot{x} &= \rho(\omega_C + \omega_Z\sin\alpha) - l(\omega_Z\cos\alpha) \\ \dot{y} &= 0 \\ \dot{z} &= 0. \end{aligned}$$

A matrix approach

The results given above can also be be obtained using a matrix algebra approach. Specifically, let the orientation between the x-y-z frame and the x_1-y_1-z_1 frame be given by the direction cosine matrix ${}^0A_1$. The coordinates of P in the fixed frame are

$${}^0r_{QP} = {}^0A_1\,{}^1r_{Q_1P}, \quad (c)$$

where ${}^0r_{QP} = [x,\ y,\ z]^T$, ${}^1r_{Q_1P} = [0,\ l,\ \rho]^T$, and

$$^0A_1 = \begin{bmatrix} 1 & 0 & 0 \\ 0 & \cos\alpha & -\sin\alpha \\ 0 & \sin\alpha & \cos\alpha \end{bmatrix}.$$

Thus, equation (c) yields results identical to those obtained above.

The velocity of P is given by

$$^0v_{QP} = {}^0\dot{A}_1\,{}^1r_{Q_1P} = {}^0A_1\,{}^1\tilde{\omega}({}^1r_{Q_1P}), \tag{d}$$

where $^0v_{QP} = [\dot{x},\ \dot{y},\ \dot{z}]^T$ and $^1\tilde{\omega}$ is defined by equation (2.5). The matrix $^1\tilde{\omega}$ contains the coefficients of the angular velocity vector, $^1\bar{\omega} = \omega_1\hat{\imath}_1 + \omega_2\hat{\jmath}_1 + \omega_3\hat{k}_1$. From equation (b) it can be seen that

$$\begin{aligned} \omega_1 &= 0, \\ \omega_2 &= \omega_C + \omega_Z \sin\alpha, \\ \omega_3 &= \omega_Z \cos\alpha. \end{aligned}$$

Using these terms in (d) yields the same velocity computed via the vector approach.

2.1.4 A point moving in a moving frame

This section considers the kinematics of a point that is moving in a frame that can translate and rotate relative to a fixed frame.

Displacement analysis

Using Fig. 2.2, let the rectangular coordinate system x-y-z, with origin Q, represent the fixed reference frame. Let the rectangular coordinate system x_1-y_1-z_1, with origin Q_1, represent the moving frame that can translate and rotate with respect to the x-y-z coordinate system. The position of Q_1 with respect to Q is given by $^0\bar{R}_{QQ_1} = X\,\hat{\imath} + Y\,\hat{\jmath} + Z\,\hat{k}$, where $\hat{\imath}$, $\hat{\jmath}$, and $\hat{k}$ are unit vectors in the x, y, and z directions, respectively. The position of the point P relative to Q_1 is given by $^1\bar{r}_{Q_1P} = x_1\,\hat{\imath}_1 + y_1\,\hat{\jmath}_1 + z_1\,\hat{k}_1$, where $\hat{\imath}_1$, $\hat{\jmath}_1$, and $\hat{k}_1$ are unit vectors in the x_1, y_1, and z_1 directions, respectively. Then using vector calculus it can be seen that the position of P with relative to Q is given by

$$^0\bar{r}_{QP} = {}^0\bar{R}_{QQ_1} + {}^1\bar{r}_{Q_1P} \tag{2.11}$$

where $^1\bar{r}_{Q_1P} = x_1\,\hat{\imath}_1 + y_1\,\hat{\jmath}_1 + z_1\,\hat{k}_1$.

Equation (2.11) can be written in matrix form as

$$ {}^0r_{QP} = {}^0R_{QQ_1} + {}^0A_1\,{}^1r_{Q_1P}, \tag{2.12} $$

where ${}^0r_{QP} = [x,\ y,\ z]^T$, ${}^0R_{QQ_1} = [X,\ Y,\ Z]^T$, ${}^1r_{Q_1P} = [x_1,\ y_1,\ z_1]^T$, and ${}^0A_1$ is the direction cosine matrix that relates the x_1-y_1-z_1 coordinate system to the x-y-z coordinate system.

Velocity analysis

The velocity of P is given by

$$
\begin{aligned}
{}^0\bar{v}_{QP} &= \frac{d}{dt}({}^0\bar{r}_{QP}) \\
&= \frac{d}{dt}({}^0\bar{R}_{QQ_1}) + \frac{d}{dt}({}^1\bar{r}_{Q_1P}) \\
&= \dot{X}\,\hat{\imath} + \dot{Y}\,\hat{\jmath} + \dot{Z}\,\hat{k} + \dot{x}_1\,\hat{\imath}_1 + \dot{y}_1\,\hat{\jmath}_1 + \dot{z}_1\,\hat{k}_1 \\
&\qquad + x_1\frac{d\hat{\imath}_1}{dt} + y_1\frac{d\hat{\jmath}_1}{dt} + z_1\frac{d\hat{k}_1}{dt} \\
&= \dot{X}\,\hat{\imath} + \dot{Y}\,\hat{\jmath} + \dot{Z}\,\hat{k} + \dot{x}_1\,\hat{\imath}_1 + \dot{y}_1\,\hat{\jmath}_1 + \dot{z}_1\,\hat{k}_1 \\
&\qquad + x_1({}^1\bar{\omega} \times \hat{\imath}_1) + y_1({}^1\bar{\omega} \times \hat{\jmath}_1) + z_1({}^1\bar{\omega} \times \hat{k}_1) \\
&= {}^0\dot{\bar{R}}_{QQ_1} + {}^1\dot{\bar{r}}_{Q_1P} + {}^1\bar{\omega} \times {}^1\bar{r}_{Q_1P}.
\end{aligned} \tag{2.13}
$$

Here, the term ${}^0\dot{\bar{R}}_{QQ_1} = \dot{X}\,\hat{\imath} + \dot{Y}\,\hat{\jmath} + \dot{Z}\,\hat{k}$ is the velocity of Q_1 relative to Q. The term ${}^1\dot{\bar{r}}_{Q_1P} = \dot{x}_1\,\hat{\imath}_1 + \dot{y}_1\,\hat{\jmath}_1 + \dot{z}_1\,\hat{k}_1$ is the velocity of P relative to Q_1. The term ${}^1\bar{\omega} \times {}^1\bar{r}_{Q_1P}$ is the velocity of P due to the rotation of x_1-y_1-z_1. The angular velocity of x_1-y_1-z_1 is given by the vector ${}^1\bar{\omega} = \omega_1\,\hat{\imath}_1 + \omega_2\,\hat{\jmath}_1 + \omega_3\,\hat{k}_1$.

A matrix representation of the velocity of P can be obtained by differentiating (2.12) to get

$$
\begin{aligned}
{}^0v_{QP} &= \frac{d}{dt}{}^0r_{QP} \\
&= \frac{d}{dt}{}^0R_{QQ_1} + \frac{d}{dt}({}^0A_1\,{}^1r_{Q_1P}) \\
&= {}^0\dot{R}_{QQ_1} + {}^0\dot{A}_1\,{}^1r_{Q_1P} + {}^0A_1\,{}^1\dot{r}_{Q_1P} \\
&= {}^0\dot{R}_{QQ_1} + {}^0A_1\,{}^1\tilde{\omega}({}^1r_{Q_1P}) + {}^0A_1\,{}^1\dot{r}_{Q_1P},
\end{aligned} \tag{2.14}
$$

where ${}^0v_{QP} = [\dot{x},\ \dot{y},\ \dot{z}]^T$, ${}^0\dot{R}_{QQ_1} = [\dot{X},\ \dot{Y},\ \dot{Z}]^T$, ${}^1\dot{r}_{Q_1P} = [\dot{x}_1,\ \dot{y}_1,\ \dot{z}_1]^T$, ${}^0A_1$ is the direction cosine matrix relating x_1-y_1-z_1 to x-y-z, and ${}^1\tilde{\omega}$ is defined in equation (2.5).

Example 2.6.

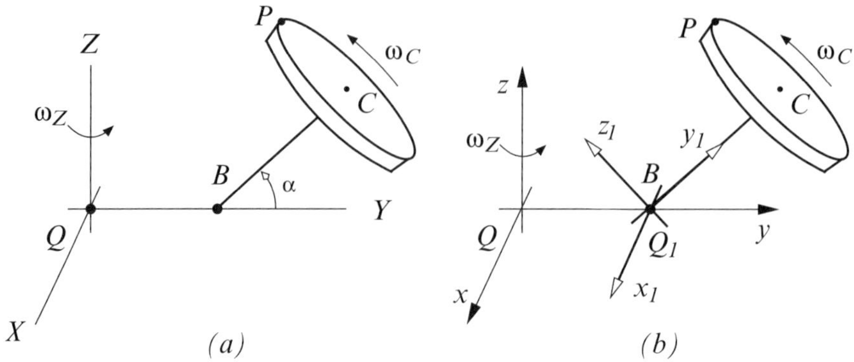

Figure (a) shows a thin disk, with center C, that is attached to the rigid link BC. The point P is on the edge of the disk, and the line CP is perpendicular to the line BC. The disk rotates about the line BC with angular velocity ω_C. The point B moves in a circular path in the X-Y plane with angular velocity ω_Z. The length of QB is l_1, the length of BC is l_2, and the radius of the disk is ρ. The line BC is at a fixed angle α relative to the line QY. A kinematic analysis of the point P, at the instant shown, proceeds as follows.

First, establish a fixed reference frame. The rectangular coordinate system x-y-z, with origin at Q, will be used as the fixed reference frame in this analysis (see Fig. (b)). The x-axis is directed along the line QX, the y-axis is directed along the line QY, and the z-axis is directed alone the line QZ. The unit vectors along the x, y, and z axes are $\hat{\imath}$, $\hat{\jmath}$, and $\hat{k}$, respectively.

Next, a moving frame, x_1-y_1-z_1, is attached to point B such that the origin of the moving frame, Q_1, is coincident with B, the y_1-axis is directed along the line BC, the x_1-axis is parallel to the line QX, and the z_1-axis is selected to form a right-handed rectangular coordinate system. The unit vectors along the x_1, y_1, and z_1 axes are $\hat{\imath}_1$, $\hat{\jmath}_1$, and $\hat{k}_1$, respectively.

Displacement analysis

Using the fixed and moving frames described above, the point P can be treated as a point that us moving in a moving frame. Let ${}^0R_{QQ_1} = [X,\ Y,\ Z]^T$ be the coordinates of Q_1 relative to Q, in the x-y-z frame. Let ${}^1r_{Q_1P} = [x_1,\ y_1,\ z_1]^T$ be the coordinates of P relative to Q_1, in the x_1-y_1-z_1 frame. Let ${}^0A_1$ be the direction cosine matrix relating the x_1-y_1-z_1 frame to the x-y-z frame. Then the coordinate of the point P relative to Q in the x-y-z frame is given by

$$ {}^0r_{QP} = {}^0R_{QQ_1} + {}^0A_1\,{}^1r_{Q_1P}, \qquad (a) $$

where ${}^0r_{QP} = [x,\ y,\ z]^T$. For the configuration shown in Fig. (b), ${}^0R_{QQ_1} = [0,\ l_1,\ 0]^T$, ${}^1r_{Q_1P} = [0,\ l_2,\ \rho]^T$, and

$$
{}^0A_1 = \begin{bmatrix} 1 & 0 & 0 \\ 0 & \cos\alpha & -\sin\alpha \\ 0 & \sin\alpha & \cos\alpha \end{bmatrix}.
$$

Using these terms in (a) gives

$$
\begin{bmatrix} x \\ y \\ z \end{bmatrix} = \begin{bmatrix} 0 \\ l_1 \\ 0 \end{bmatrix} + \begin{bmatrix} 1 & 0 & 0 \\ 0 & \cos\alpha & -\sin\alpha \\ 0 & \sin\alpha & \cos\alpha \end{bmatrix} \begin{bmatrix} 0 \\ l_2 \\ \rho \end{bmatrix}
= \begin{bmatrix} 0 \\ l_1 + l_2\cos\alpha - \rho\sin\alpha \\ l_2\sin\alpha + \rho\cos\alpha \end{bmatrix}.
$$

Velocity analysis

The velocity of P relative to the fixed frame is

$$
{}^0v_{QP} = {}^0\dot{R}_{QQ_1} + {}^0A_1\,{}^1\dot{r}_{Q_1P} + {}^0A_1\,{}^1\tilde{\omega}({}^1r_{Q_1P}), \tag{b}
$$

where ${}^0v_{QP} = [\dot{x},\ \dot{y},\ \dot{z}]^T$, ${}^0\dot{R}_{QQ_1} = [\dot{X},\ \dot{Y},\ \dot{Z}]^T$, is the velocity of Q_1 relative to Q in the x-y-z frame, and ${}^1\dot{r}_{Q_1P} = [\dot{x}_1,\ \dot{y}_1,\ \dot{z}_1]^T$ is the velocity of P relative to Q_1 in the x_1-y_1-z_1 frame. The matrix ${}^1\tilde{\omega}$ is given by equation (2.5), and it contains the coefficients of ${}^1\bar{\omega} = \omega_1\hat{\imath}_1 + \omega_2\hat{\jmath}_1 + \omega_3\hat{k}_1$, which represents the angular velocity of the moving frame.

For the configuration shown in Fig. (b) we get

$$
{}^0\dot{R}_{QQ_1} = \begin{bmatrix} -l_1\omega_Z \\ 0 \\ 0 \end{bmatrix}, \quad {}^1\dot{r}_{Q_1P} = \begin{bmatrix} \rho\omega_C \\ 0 \\ 0 \end{bmatrix}, \text{ and } \begin{bmatrix} \omega_1 \\ \omega_2 \\ \omega_3 \end{bmatrix} = \begin{bmatrix} 0 \\ \omega_Z\sin\alpha \\ \omega_Z\cos\alpha \end{bmatrix}.
$$

Here, ${}^0\dot{R}_{QQ_1}$ is obtained by noting that Q_1 moves in a circular path with angular velocity ω_Z in the x-y plane. The term ${}^1\dot{r}_{Q_1P}$ is obtained by noting that P moves in a circular path with angular velocity ω_C in the x_1-z_1 plane. Finally, the angular velocity of the moving frame is ${}^1\bar{\omega} = \omega_Z\,\hat{k} = \omega_Z\sin\alpha\,\hat{\jmath}_1 + \omega_Z\cos\alpha\,\hat{k}_1$.

Applying these terms to equation (b) gives

$$
\begin{bmatrix} \dot{x} \\ \dot{y} \\ \dot{z} \end{bmatrix} = \begin{bmatrix} -l_1\omega_Z \\ 0 \\ 0 \end{bmatrix} + \begin{bmatrix} 1 & 0 & 0 \\ 0 & \cos\alpha & -\sin\alpha \\ 0 & \sin\alpha & \cos\alpha \end{bmatrix} \begin{bmatrix} \rho\omega_C \\ 0 \\ 0 \end{bmatrix}
+ \begin{bmatrix} 1 & 0 & 0 \\ 0 & \cos\alpha & -\sin\alpha \\ 0 & \sin\alpha & \cos\alpha \end{bmatrix} \begin{bmatrix} 0 & -\omega_Z\cos\alpha & \omega_Z\sin\alpha \\ \omega_Z\cos\alpha & 0 & 0 \\ -\omega_Z\sin\alpha & 0 & 0 \end{bmatrix} \begin{bmatrix} 0 \\ l_2 \\ \rho \end{bmatrix}
$$

$$= \begin{bmatrix} \rho(\omega_C + \omega_Z \sin\alpha) - l_1\omega_Z - l_2\omega_Z\cos\alpha \\ 0 \\ 0 \end{bmatrix}.$$

2.1.5 Euler angles

It is perhaps apparent from the previous sections that the nine elements of the direction cosine matrix, ${}^0A_1$, are not all independent. In fact since the columns of ${}^0A_1$ are orthogonal we have the following three constraints;

$$\begin{aligned} a_1^T a_2 &= a_{11}a_{12} + a_{21}a_{22} + a_{31}a_{32} = 0, \\ a_1^T a_3 &= a_{11}a_{13} + a_{21}a_{23} + a_{31}a_{33} = 0, \\ a_2^T a_3 &= a_{12}a_{13} + a_{22}a_{23} + a_{32}a_{33} = 0. \end{aligned}$$

Here, a_i is the i-th column of ${}^0A_1$.

Moreover, since the columns of ${}^0A_1$ represents unit vectors we have the three additional constraints;

$$\begin{aligned} |a_1| &= \sqrt{a_{11}^2 + a_{21}^2 + a_{31}^2} = 1, \\ |a_2| &= \sqrt{a_{12}^2 + a_{22}^2 + a_{32}^2} = 1, \\ |a_3| &= \sqrt{a_{13}^2 + a_{23}^2 + a_{33}^2} = 1. \end{aligned}$$

These six constraints imply that only three variables are required to determine the orientation of one rectangular system with respect to another.

In this text we use three *Euler-angles* as the independent variables that determine the direction cosine matrix. The description of the Euler angles utilizes rectangular coordinate systems called an intermediate frames. These intermediate frames are used to show how the x-y-z system can be brought into alignment with the x_1-y_1-z_1 system. The Euler angles are three successive angular displacements that causes an intermediate frame, that is initially aligned with frame 0, to be aligned with frame 1. These rotations take place, in a particular order, about the x-axis, y-axis, or z-axis of the intermediate frame.

To denote the axis of rotation, the angle of rotation, and the sequence of rotations we use the following notation.

- X_α indicates that the x-axis of the intermediate frame rotates an amount α. Similarly, Y_β indicates that the y-axis of the intermediate frame rotates an amount β, and Z_γ indicates that the z-axis of the intermediate frame rotates an amount γ.

- Z_α-X_β-Z_γ denotes the following sequence of rotations are performed on the intermediate frame; (i) rotate the z-axis an amount α, then (ii) rotate the x-axis an amount β, finally (iii) rotate the z-axis an amount γ.

There are twelve possible sequences of such rotations that can be used to obtain the direction cosine matrix. For example, X_α-Y_β-Z_γ, X_α-Y_β-X_γ, X_α-Z_β-Y_γ, etc. Here, we will consider the two sequences that are used frequently. Namely, the sequence X_α-Y_β-Z_γ and the sequence Z_α-X_β-Z_γ.

X_α-Y_β-Z_γ Euler angles

In Fig. 2.5a the origin of the fixed frame is labeled Q. Let the unit vectors $\hat{i}$, $\hat{j}$, and $\hat{k}$, be directed along the x-axis, y-axis, and z-axis respectively. To establish the Euler angles let x_a-y_a-z_a be a rectangular coordinate system with origin Q_a that coincides with Q. Moreover, the x_a, y_a, and z_a axes are aligned with x, y, and z directions respectively (see Fig. 2.5a). The intermediate frame x_a-y_a-z_a can only rotate about the point Q.

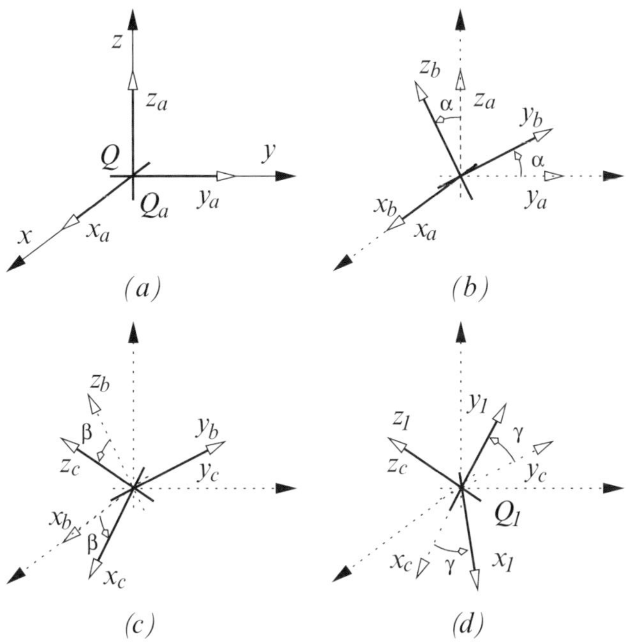

Fig. 2.5 Finite rotations

Now consider the consequence of three successive finite rotations of x_a-y_a-z_a. First, rotate the axis x_a an angle α in the counterclockwise direction to obtain the new orientation x_b-y_b-z_b of the moving frame (see Fig. 2.5b).

Thus, after rotation by the angle α, the x_a axis becomes the x_b axis, the y_a axis becomes the y_b axis, and the z_a axis becomes the z_b axis.

A point ${}^b r = [x_b,\ y_b,\ z_b]^T$ in the x_b-y_b-z_b frame has coordinate ${}^0 r = [x,\ y,\ z]^T$ in the x-y-z frame where

$$ {}^0 r = {}^0 A_b {}^b r, \quad {}^0 A_b = \begin{bmatrix} 1 & 0 & 0 \\ 0 & c_\alpha & -s_\alpha \\ 0 & s_\alpha & c_\alpha \end{bmatrix}, $$

$c_\alpha = \cos\alpha$, and $s_\alpha = \sin\alpha$. The matrix ${}^0 A_b$ is the direction cosine matrix that relates the x_b-y_b-z_b frame to the x-y-z frame.

Next, rotate the axis y_b an angle β in the counterclockwise direction to obtain the new orientation x_c-y_c-z_c of the moving frame (see Fig. 2.5c). Thus, after rotation by the angle β, the x_b axis becomes the x_c axis, the y_b axis becomes the y_c axis, and the z_b axis becomes the z_c axis.

A point ${}^c r = [x_c,\ y_c,\ z_c]^T$ in the x_c-y_c-z_c frame has coordinate $r_b = [x_b,\ y_b,\ z_b]^T$ in the x_b-y_b-z_b frame where

$$ {}^b r = {}^b A_c {}^c r, \quad {}^b A_c = \begin{bmatrix} c_\beta & 0 & s_\beta \\ 0 & 1 & 0 \\ -s_\beta & 0 & c_\beta \end{bmatrix}, $$

$c_\beta = \cos\beta$, and $s_\beta = \sin\beta$. The matrix ${}^b A_c$ is the direction cosine matrix that relates the x_c-y_c-z_c frame to the x_b-y_b-z_b frame.

Finally, rotate the axis z_c an angle γ in the counterclockwise direction to obtain the new orientation x_1-y_1-z_1 of the moving frame (see Fig. 2.5d). Thus, after rotation by the angle γ, the x_c axis becomes the x_1 axis, the y_c axis becomes the y_1 axis, and the z_c axis becomes the z_1 axis.

A point ${}^1 r = [x_1,\ y_1,\ z_1]^T$ in the x_1-y_1-z_1 frame has coordinate ${}^c r = [x_c,\ y_c,\ z_c]^T$ in the x_c-y_c-z_c frame where

$$ {}^c r = {}^c A_1 {}^1 r, \quad {}^c A_1 = \begin{bmatrix} c_\gamma & -s_\gamma & 0 \\ s_\gamma & c_\gamma & 0 \\ 0 & 0 & 1 \end{bmatrix}, $$

$c_\gamma = \cos\gamma$, and $s_\gamma = \sin\gamma$. The matrix ${}^c A_1$ is the direction cosine matrix that relates the x_1-y_1-z_1 frame to the x_c-y_c-z_c frame.

Thus, after three successive rotations, (α about x_a, β about y_b and, γ about z_c), it can be seen that the point ${}^1 r = [x_1,\ y_1,\ z_1]^T$ in the x_1-y_1-z_1 frame has coordinate ${}^0 r = [x,\ y,\ z]^T$ in the x-y-z frame where

$$ {}^0 r = {}^0 A_b {}^b r = {}^0 A_b {}^b A_c {}^c r = {}^0 A_b {}^b A_c {}^c A_1 {}^1 r = {}^0 A_1 {}^1 r. \tag{2.15} $$

The coordinate transformation matrix ${}^0 A_1$ is

$$
{}^0A_1 = \begin{bmatrix} 1 & 0 & 0 \\ 0 & c_\alpha & -s_\alpha \\ 0 & s_\alpha & c_\alpha \end{bmatrix} \begin{bmatrix} c_\beta & 0 & s_\beta \\ 0 & 1 & 0 \\ -s_\beta & 0 & c_\beta \end{bmatrix} \begin{bmatrix} c_\gamma & -s_\gamma & 0 \\ s_\gamma & c_\gamma & 0 \\ 0 & 0 & 1 \end{bmatrix}
$$
$$
= \begin{bmatrix} c_\beta c_\gamma & -c_\beta s_\gamma & s_\beta \\ s_\alpha s_\beta c_\gamma + c_\alpha s_\gamma & -s_\alpha s_\beta s_\gamma + c_\alpha c_\gamma & -s_\alpha c_\beta \\ -c_\alpha s_\beta c_\gamma + s_\alpha s_\gamma & c_\alpha s_\beta s_\gamma + s_\alpha c_\gamma & c_\alpha c_\beta \end{bmatrix}. \tag{2.16}
$$

The elements of ${}^0A_1$ are the direction cosines of the x_1, y_1, and z_1, axes with respect to the x, y, and z axes, in terms of the X_α-Y_β-Z_γ Euler angles.

Given a direction cosine matrix ${}^0A_1$ the corresponding Euler angles can be determined using

$$
\alpha = \tan^{-1}\left(\frac{-a_{23}}{a_{33}}\right), \ \beta = \sin^{-1} a_{13}, \ \gamma = \tan^{-1}\left(\frac{-a_{12}}{a_{11}}\right),
$$

where a_{ij} are the ij-th elements of ${}^0A_1$.

Example 2.7.

Compute the X_α-Y_β-Z_γ Euler angles associated with the direction cosine matrix ${}^0A_1$ computed in Example 2.3.

Solution:

The direction cosine matrix is

$$
{}^0A_1 = \begin{bmatrix} 0.35354 & -0.61237 & 0.70710 \\ 0.92678 & 0.12682 & -0.35355 \\ 0.12683 & 0.78034 & 0.61238 \end{bmatrix}.
$$

Therefore,

$$
\alpha = \tan^{-1}\left(\frac{-a_{23}}{a_{33}}\right) = \tan^{-1}\left(\frac{0.35355}{0.61238}\right) = 0.5236 \text{ radians},
$$
$$
\beta = \sin^{-1} a_{13} = \sin^{-1} 0.70710 = 0.7854 \text{ radians},
$$
$$
\gamma = \tan^{-1}\left(\frac{-a_{12}}{a_{11}}\right) = \tan^{-1}\left(\frac{0.61237}{0.35354}\right) = 1.0472 \text{ radians}.
$$

Unlike finite translations of the moving frame, the finite rotations of the moving frame can not be treated as vector quantities. In particular, the order in which the finite rotations occur has significance. To see this consider the same three rotations, described above, but in reverse order. First, a rotation γ about the z_a axis to produce the x_b-y_b-z_b orientation. Then, a rotation β

about the y_b axis to produce the x_c-y_c-z_c orientation. Finally, a rotation α about the x_c axis to produce the x_1-y_1-z_1 orientation. If ${}^1r = [x_1,\ y_1,\ z_1]^T$ is a point in the x_1-y_1-z_1 frame then its coordinate in the x-y-z frame is

$$ {}^0r = {}^0A_b{}^bA_c{}^cA_1{}^1r = {}^0A_1{}^1r, $$

where

$$ {}^0A_1 = \begin{bmatrix} c_\beta c_\gamma & s_\alpha s_\beta c_\gamma - c_\alpha s_\gamma & c_\alpha s_\beta c_\gamma + s_\alpha s_\gamma \\ c_\beta s_\gamma & s_\alpha s_\beta s_\gamma + c_\alpha c_\gamma & c_\alpha s_\beta s_\gamma - s_\alpha c_\gamma \\ -s_\beta & s_\alpha c_\beta & c_\alpha c_\beta \end{bmatrix}. \tag{2.17} $$

Here, ${}^0A_1$ is the direction cosine matrix obtained from the Z_γ-Y_β-X_α Euler angle sequence. It is clear that, in general, the transformation ${}^0A_1$ in equation (2.16) is not equal the transformation in equation (2.17). Hence, successive finite rotations can not be represented as simple vector summations.

Example 2.8.

This figure shows the effect of reversing the order in which finite rotations take place.

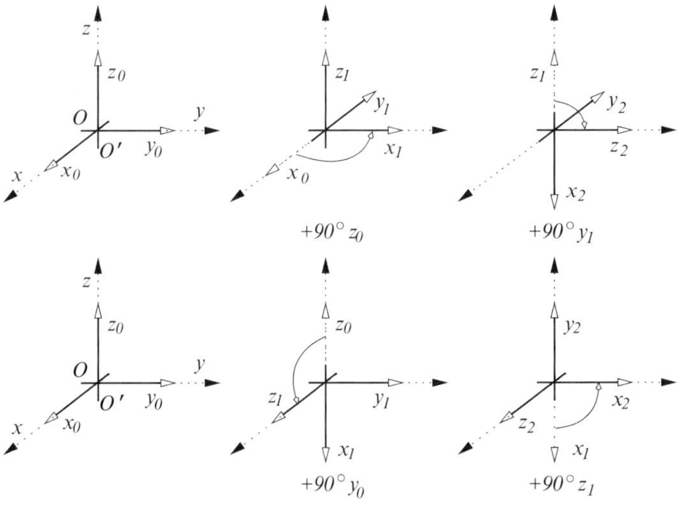

In the first sequence the frame x_0-y_0-z_0 rotates 90° counterclockwise about the z_0 axis. This is followed by a rotation 90° counterclockwise about the y_1 axis.

In the second sequence the frame x_0-y_0-z_0 rotates 90° counterclockwise about the y_0 axis. This is followed by a rotation 90° counterclockwise about the z_1 axis.

As can be seen the final orientation of the frame is different if the order of the rotations is changed.

Angular velocity

A kinematic analysis requires the angular velocity of the moving frame. If the orientation of the moving frame is described using Euler angles then the angular velocities should also be expressed in terms of the Euler angles. Our objective here is to determine the angular velocities of the x_1-axis, y_1-axis, and z_1-axis in the case where the orientation of the moving frame is described using the X_α-Y_β-Z_γ Euler angles.

By construction (see Fig. 2.5) the angular velocity of the moving frame can be written as

$$^1\bar{\omega} = \dot{\alpha}\,\hat{\imath}_a + \dot{\beta}\,\hat{\jmath}_b + \dot{\gamma}\,\hat{k}_c, \tag{a}$$

where $\hat{\imath}_a$ is the unit vector along the x_a-axis, $\hat{\jmath}_b$ is the unit vector along the y_b-axis, and $\hat{k}_c$ is the unit vector along the z_c-axis,

Now, the direction cosine matrix relating the x_1-y_1-z_1 system to the x_a-y_a-z_a system is $^0A_1$. (Note that x_a-y_a-z_a is aligned with x-y-z.) Thus,

$$\hat{\imath}_a = \hat{\imath} = c_\beta c_\gamma\,\hat{\imath}_1 - c_\beta s_\gamma\,\hat{\jmath}_1 + s_\beta\,\hat{k}_1. \tag{b}$$

The direction cosine matrix relating the x_1-y_1-z_1 system to the x_b-y_b-z_b system is

$$^bA_1 = {^bA_c}{^cA_1} = \begin{bmatrix} c_\beta c_\gamma & -c_\beta s_\gamma & s_\beta \\ s_\gamma & c_\gamma & 0 \\ -s_\beta c_\gamma & s_\beta s_\gamma & c_\beta \end{bmatrix}.$$

Therefore,

$$\hat{\jmath}_b = s_\gamma\,\hat{\imath}_1 + c_\gamma\,\hat{\jmath}_1. \tag{c}$$

The direction cosine matrix relating the x_1-y_1-z_1 system to the x_c-y_c-z_c system is $^cA_1$, which shows that

$$\hat{k}_c = \hat{k}_1. \tag{d}$$

Using (b), (c) and (d) in (a) gives

$$\begin{aligned} ^1\bar{\omega} &= \dot{\alpha}(c_\beta c_\gamma\,\hat{\imath}_1 - c_\beta s_\gamma\,\hat{\jmath}_1 + s_\beta\,\hat{k}_1) \\ &\quad + \dot{\beta}(s_\gamma\,\hat{\imath}_1 + c_\gamma\,\hat{\jmath}_1) + \dot{\gamma}\,\hat{k}_1, \\ &= (\dot{\alpha}c_\beta c_\gamma + \dot{\beta}s_\gamma)\,\hat{\imath}_1 + (-\dot{\alpha}c_\beta s_\gamma + \dot{\beta}c_\gamma)\,\hat{\jmath}_1 \\ &\quad + (\dot{\alpha}s_\beta + \dot{\gamma})\,\hat{k}_1 \\ &= {^1\omega_1}\,\hat{\imath}_1 + {^1\omega_2}\,\hat{\jmath}_1 + {^1\omega_3}\,\hat{k}_1. \end{aligned}$$

Here, $^1\omega_1$ is the angular velocity of the x_1-axis, $^1\omega_2$ is the angular velocity of the y_1-axis, and $^1\omega_3$ is the angular velocity of the z_1-axis. This result can be put in matrix form to get

$$\begin{bmatrix} {}^1\omega_1 \\ {}^1\omega_2 \\ {}^1\omega_3 \end{bmatrix} = \begin{bmatrix} c_\beta c_\gamma & s_\gamma & 0 \\ -c_\beta s_\gamma & c_\gamma & 0 \\ s_\beta & 0 & 1 \end{bmatrix} \begin{bmatrix} \dot{\alpha} \\ \dot{\beta} \\ \dot{\gamma} \end{bmatrix}. \tag{2.18}$$

Given the angular velocities ${}^1\omega_1$, ${}^1\omega_2$, and ${}^1\omega_3$, equation (2.18) can be inverted to find the rate of change of the Euler angles. Specifically,

$$\begin{bmatrix} \dot{\alpha} \\ \dot{\beta} \\ \dot{\gamma} \end{bmatrix} = \frac{1}{c_\beta} \begin{bmatrix} c_\gamma & -s_\gamma & 0 \\ c_\beta s_\gamma & c_\beta c_\gamma & 0 \\ -s_\beta c_\gamma & s_\beta s_\gamma & c_\beta \end{bmatrix} \begin{bmatrix} {}^1\omega_1 \\ {}^1\omega_2 \\ {}^1\omega_3 \end{bmatrix}. \tag{2.19}$$

Note that the coefficient matrix in equation (2.18) is singular when $\beta = \pm\pi/2$. Thus, at orientations of the moving frame where $\beta = \pm\pi/2$ we will be unable to compute the angular velocities.

All Euler angle sequence of rotations have singular points in the configuration space. To avoid computational difficulties we will select a sequence of Euler angles that are not singular near the nominal configuration of the system.

2.2 Mechanisms

Mechanisms are mechanical systems that consists of rigid bodies called links that are connected at points called joints. Figure 2.6 presents schematics of some well known mechanisms. The links of a mechanism are classified according to the number of joints that are on the link. As shown in Fig. 2.7 binary links have two joints, ternary links have three joints, quaternary links have four joints, etc.

If each link in the mechanism is connected to at least two other links then the mechanism forms a *closed kinematic chain*. Otherwise, the mechanism has an *open kinematic chain*. In Fig. 2.6 the R-R robot has an open-kinematic chain, all the other mechanisms are closed kinematic chains.

The joints in the mechanism are also called kinematic pairs. These kinematic pairs permit relative motion between the links in the mechanism. Moreover, the joints are classified according to the type of motion allowed between the links. The kinematic pairs considered in this text are as follows.

- **Spherical joint**

  A schematic of a spherical joint is shown below. This kinematic pair allows link 1 and link 2 to rotate relative to each other about three axes.

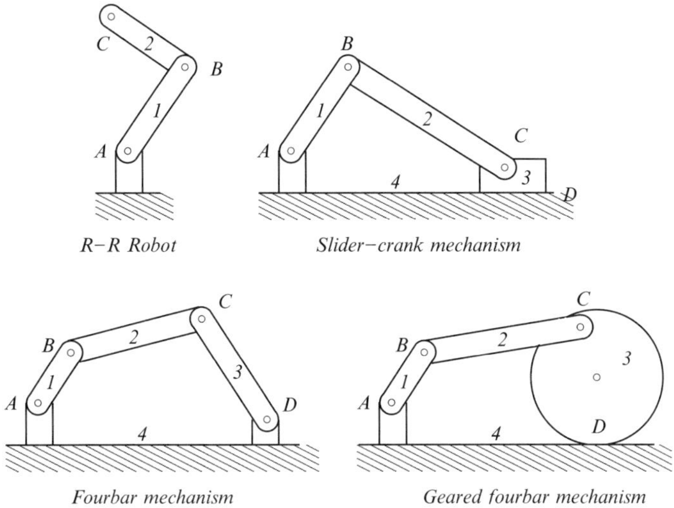

Fig. 2.6 Mechanisms

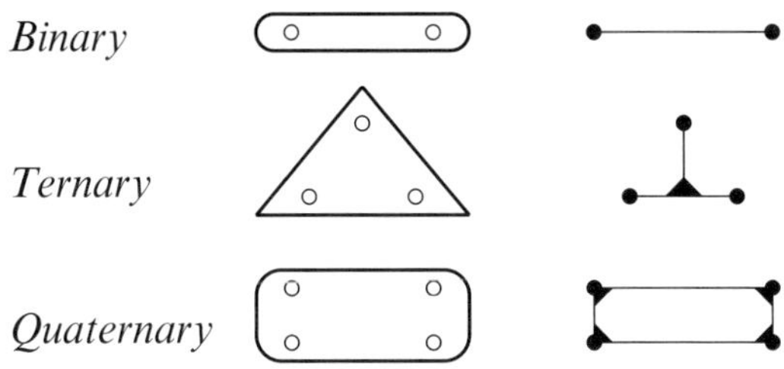

Fig. 2.7 Types of links

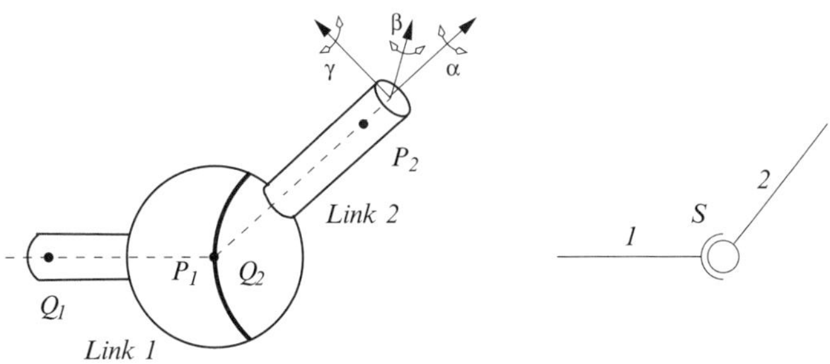

At each instant the point P_1 on link 1 and the point Q_2 on link 2 are coincident. The symbol S is used to denote a spherical joint.

- **Universal joint**

A schematic of a universal joint is shown below. This kinematic pair allows link 1 and link 2 to rotate relative to each other about two orthogonal axes.

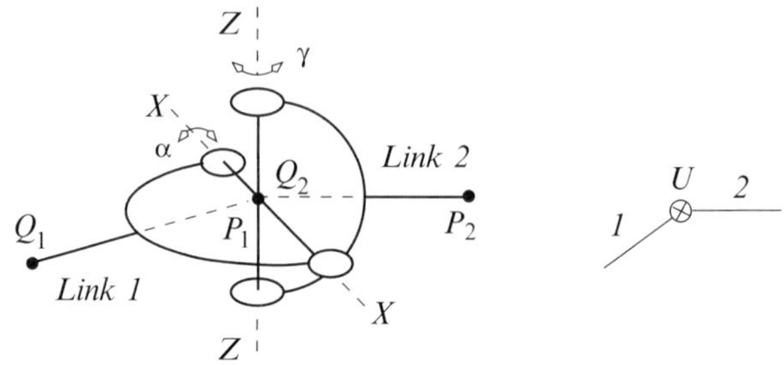

At each instant the point P_1 on link 1 and the point Q_2 on link 2 are coincident. In addition the links can rotate relative to each other about the orthogonal lines XX and ZZ. The symbol U is used to denote a universal joint.

- **Cylindrical joint**

The cylindrical joint allows rotation and translation of the links relative to each other. A sketch of a cylindrical joint is shown below. Let L_1 be the line that passes through the point P_1 on link 1 and the point Q_2 on link 2. Then at a cylindrical joint, link 2 can simultaneously rotate about the line L_1 and translate along the line L_1.

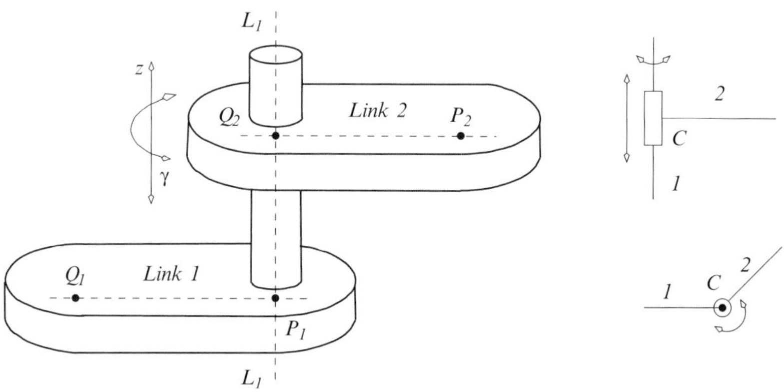

The symbol C is used to denote a cylindrical joint.

- **Revolute joint**

The revolute joint allows rotation of the links relative to each other. A sketch of a revolute joint is shown below. Let L_1 be the line that passes through the point P_1 on link 1 and the point Q_2 on link 2. Then at a revolute joint, link 2 can only rotate about the line L_1.

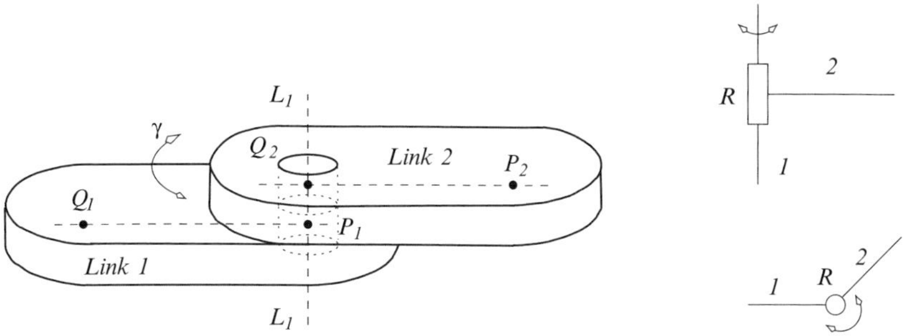

The symbol R is used to denote a revolute joint.

- **Prismatic joint**

The prismatic joint allows translation (sliding) of the links relative to each other. A sketch of a prismatic joint is shown below. Let L_1 be the line that passes through the point P_1 on link 1 and the point Q_2 on link 2. Then, at a prismatic joint, link 2 can only translate along the line L_1.

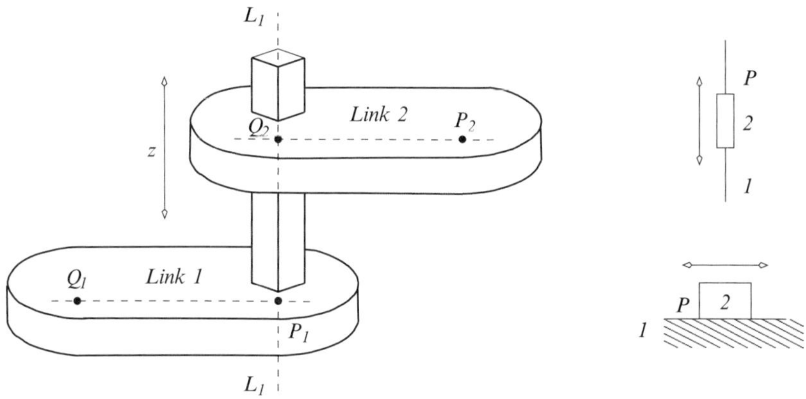

The symbol P is used to denote a prismatic joint.

- **Rolling joint**

The rolling pair shown below allows link 2 to roll on link 1 without slipping.

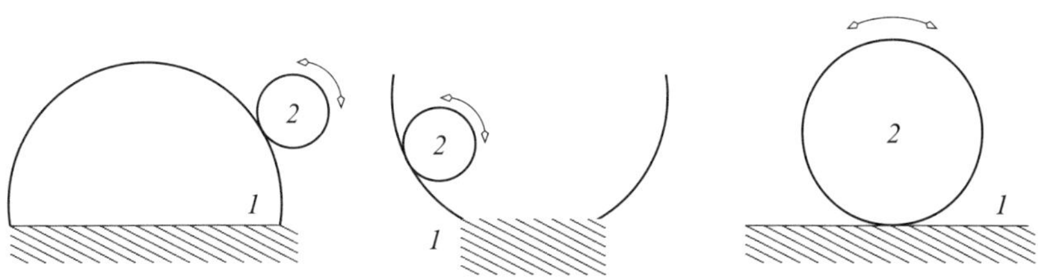

- **Cam joint**

Typical examples of cam pairs are shown below. At the point of contact links 2 and 3 can roll and slide relative to each other.

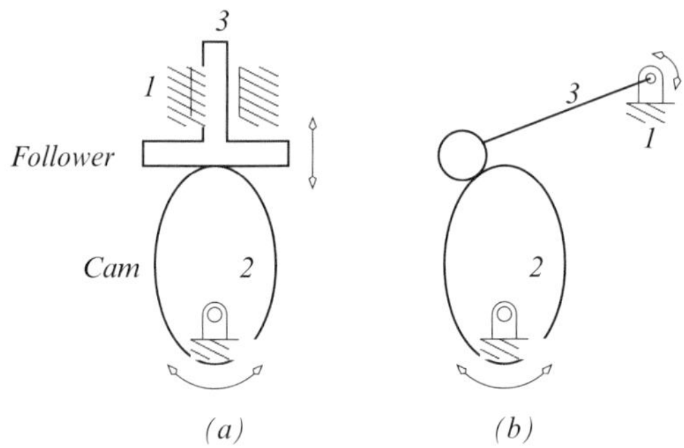

2.2.1 Mobility analysis

Fundamental to the dynamic analysis of mechanical devices is the determination of the mobility of the system, i.e., can the device move? This question is answered by finding the number of *degrees of freedom* for the system. The degrees of freedom are the minimum number of coordinates required to determine the position and orientation of the mechanism.

To develop a methodology for determining the number of degrees of freedom in a particular mechanism consider the mobility of the following systems.

- **Spatial rigid-body motion**

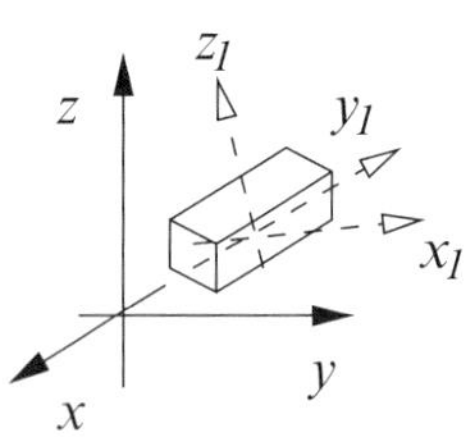

Consider a rigid-body that is free to move in space. In this case we require six coordinates to determine the position and orientation of the body. Specifically, if the rectangular coordinate system, x_1-y_1-z_1, is attached to the body then, three coordinates are required to determine the location of the origin of the x_1-y_1-z_1 system with respect to the x-y-z system.

In addition three coordinates are required to establish the direction cosine matrix relating x_1-y_1-z_1 to x-y-z. Thus, this system has 6 degrees of freedom.

- **Planar rigid-body motion**

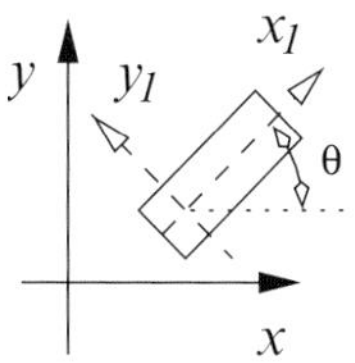

Consider a rigid-body that is constrained to move in the x-y plane. In this case we require three coordinates to determine the position and orientation of the body. If the rectangular coordinate system, x_1-y_1, is attached to the body then, two coordinates are required to determine the location of the origin of the x_1-y_1 system with respect to the x-y system.
The orientation of the x_1-y_1 system relative to the x-y system is determine by a single angle. Thus, a rigid body that is constrained to move in a plane has 3 degrees of freedom.

- **A revolute joint**

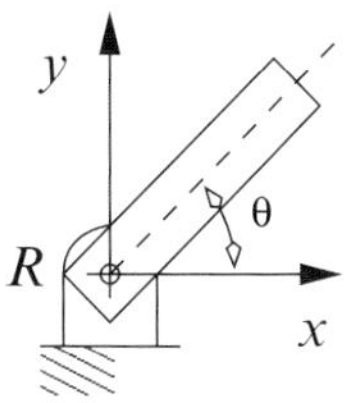

Consider the motion of a link about a revolute joint. In this system one link is fixed (i.e., the ground) and the other is free to rotate. If the angle θ is known then we can determine the position of ever point on the link. Thus, this system has 1 degree of freedom. When compared to spatial motion of a rigid-body it can be seen that the revolute joint *removes* 5 degrees of freedom.
When compared to plane motion of a rigid-body it can be seen that the revolute joint removes 2 degrees of freedom.

- **A prismatic joint**

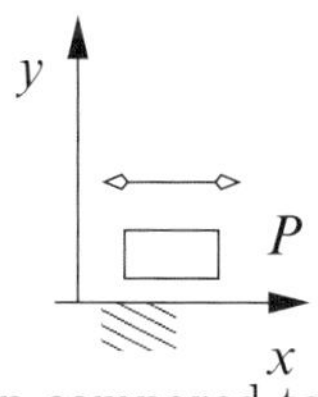

Consider the motion of a link about a prismatic joint. In this system one link is fixed (i.e., the ground) and the other is free to slide (translate). Hence, the displacement x is all that is required to determine the position and orientation of the link. Thus, this system has 1 degree of freedom.
When compared to spatial motion of a rigid-body it can be seen that the prismatic joint removes 5 degrees of freedom. When compared to plane motion of a rigid-body it can be seen that the prismatic joint removes 2 degrees of freedom.

- **Pure rolling**

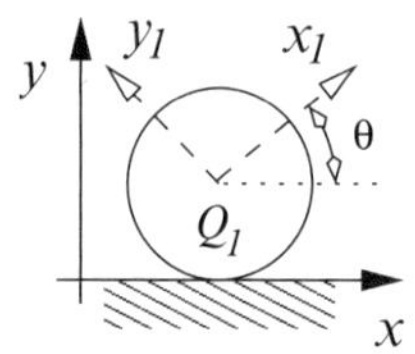

For this system the disk is constrained to roll without slipping. In which case the translation of the center of the disk is directly related to the angle of rotation. Thus, in pure rolling the system has 1 degree of freedom.

If the disk can simultaneously roll and slip then, two coordinates are required to determine the position and orientation of the system.

Using these examples we can develop an expression to determine the mobility of a mechanical system. Let l be the number of links in the system, and j be the number of joints. If all the links in the system are unconstrained the system will have $F = \lambda l$ degrees of freedom, where $\lambda = 6$ for spatial motion, and $\lambda = 3$ for planar motion. However, the i-th joint in the mechanism removes $\lambda - f_i$ degrees of freedom, where f_i is the number of degrees of freedom allowed at the joint. Moreover, since the ground link is fixed, λ degrees must also be removed.

Thus, a general formula for determining the degrees of freedom in a mechanism is given by

$$F = \lambda(l - j - 1) + \sum_{i=1}^{j} f_i, \tag{2.20}$$

where F is the number of degrees of freedom, $\lambda = 6$ for spatial mechanisms, $\lambda = 3$ for planar mechanisms, l is the number of links, j is the number of joints, and f_i is the number of degrees of freedom allowed by the i-th joint. The formula (2.20) is known as Gruebler's equation. When using Gruebler's equation note the following:

1. The ground is treated as a link.
2. All joints are assumed to connect only two links. If more than two links are connected at a joint, then the joint is counted as $n_l - 1$ joints, where n_l is the number of links connected to the joint.

Example 2.9.

This example applies equation 2.20 to the mechanisms shown in Fig. 2.6.

- **R-R robot**

This planar mechanism has $l = 3$ links and $j = 2$ joints. The revolute joints (at A and B) each allow 1 degree of freedom. Thus, equation (2.20) gives

$$F = \lambda(l - j - 1) + \sum_{i=1}^{j} f_i$$

$$
\begin{aligned}
&= 3(3-2-1)+1+1 \\
&= 2.
\end{aligned}
$$

Therefore, we require two coordinates to specify, uniquely, the position and orientation of this mechanism. Note that robot mechanisms are often classified by the types of joints in the system. Hence the term 'R-R' is used to indicate that the robot has two revolute joints in sequence.

- **Slider crank**

The planar slider crank mechanism has $l = 4$ links, and $j = 4$ joints. The joints at A, B and C are revolute joints, and each allow 1 degree of freedom. The prismatic joint at D allows 1 degree of freedom. The number of degrees of freedom for the mechanism is

$$
\begin{aligned}
F &= \lambda(l-j-1)+\sum_{i=1}^{j} f_i \\
&= 3(4-4-1)+1+1+1+1 \\
&= 1.
\end{aligned}
$$

- **Fourbar mechanism**

The planar fourbar mechanism has $l = 4$ links and $j = 4$ joints. All of the joints are revolute joints. Thus the degrees of freedom for the mechanism is given by

$$
\begin{aligned}
F &= \lambda(l-j-1)+\sum_{i=1}^{j} f_i \\
&= 3(4-4-1)+1+1+1+1 \\
&= 1.
\end{aligned}
$$

- **Geared fourbar mechanism**

The planar mechanism has $l = 4$ links and $j = 4$ joints. The joints at A, B and C are revolute joints, and each allow 1 degree of freedom. The pure rolling joint at D allows 1 degree of freedom. Thus the degrees of freedom for the mechanism is given by

$$
\begin{aligned}
F &= \lambda(l-j-1)+\sum_{i=1}^{j} f_i \\
&= 3(4-4-1)+1+1+1+1 \\
&= 1.
\end{aligned}
$$

For certain mechanisms equation (2.20) may indicate that there are fewer degrees of freedom than actually present. Typically, such mechanism require a specific geometry in order to have mobility. This is illustrated in the next example.

Example 2.10.

The figure below shows two circular rollers in contact. This system has $l = 3$ links and $j = 3$ joints (two revolute joints and a pure rolling joint).

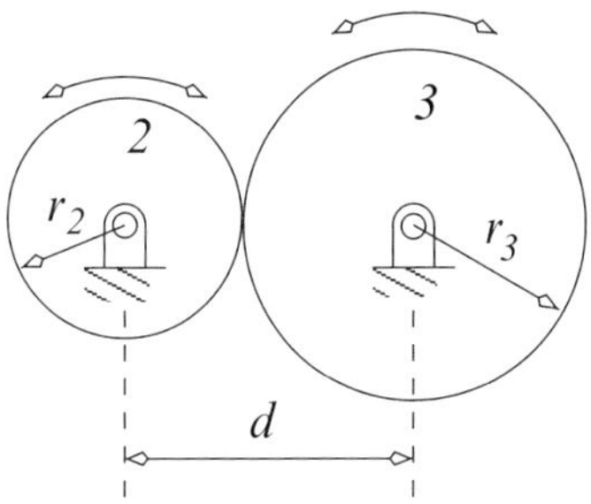

The application of equation (2.20) gives the degrees of freedom as

$$\begin{aligned} F &= \lambda(l - j - 1) + \sum_{i=1}^{j} f_i \\ &= 3(3 - 3 - 1) + 1 + 1 + 1 \\ &= 0. \end{aligned}$$

However, if the rollers are both precisely circular, and $r_2 + r_3 = d$ then, the system will have 1 degree of freedom. If the rollers are not precisely manufactured then the system will jam. (Gears/rollers that are noncircular can also be made to undergo pure rolling by satisfying the criteria that the sum of the radii of both gears must be equal to the distance between the centers of rotation.)

Equation (2.20) can sometimes reveal passive degrees of freedom that exist in certain mechanisms. These are coordinates whose variation does not produce any change in the other system coordinates. This is illustrated in the next example.

Example 2.11.

The device shown here is a spatial fourbar mechanism. The systems has revolute joints at A and D, and spherical joints at B and C.

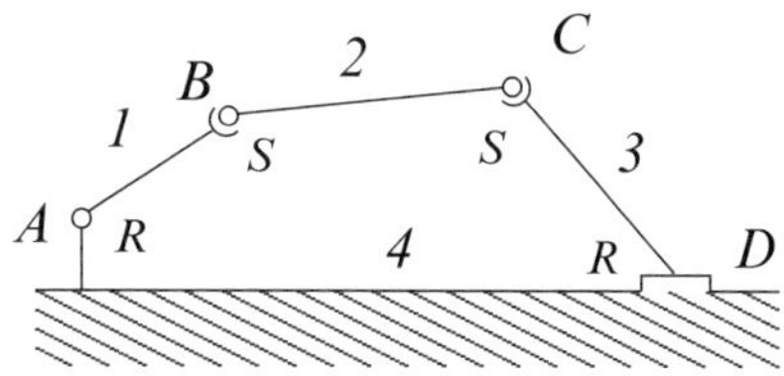

The application of equation (2.20) gives,

$$
\begin{aligned}
F &= \lambda(l - j - 1) + \sum_{i=1}^{j} f_i \\
&= 6(4 - 4 - 1) + 1 + 3 + 3 + 1 \\
&= 2.
\end{aligned}
$$

One of these degrees of freedom is the rotation of the link 2 about the line BC. Such a rotation will not cause any other motion thus, this degree of freedom is considered to be passive.

2.2.2 Kinematic analysis

With the results from the previous sections we are now well equipped to perform the kinematic analysis of mechanisms. Here we take a direct approach that is summarized by the following steps;

1. Perform a structural analysis of the mechanism to determine its mobility, i.e., the number of degrees of freedom, F.
2. Attach a coordinate system to each moving link (rigid body) in the mechanism. As a result there will be $\lambda(l - 1)$ coordinates associated with the system ($\lambda = 6$ for spatial mechanisms, and $\lambda = 3$ for planar mechanisms).
3. Use the kinematic condition at the joints to determine $\lambda(l - 1) - F$ constraint equations. These constraint equations ensure the proper behavior of the joint.

This procedure is illustrated via several examples.

Example 2.12.

A schematic of an R-R robot is shown below. Here we are interested in finding the equations required to determine the position and velocity of the point C. The distance from A to B is l_1, and the distance from B to C is l_2.

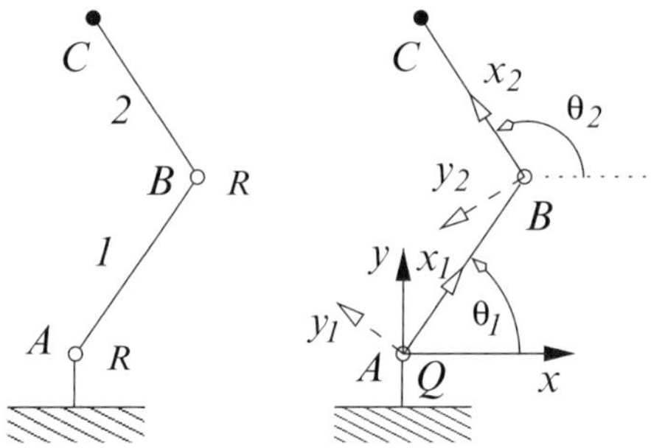

The rectangular coordinate system x-y with origin Q represents the fixed system. The rectangular coordinate system x_1-y_1 is attached to the link 1, with origin at A, and the rectangular coordinate system x_2-y_2 is attached to the link 2, with origin at B, as shown in the figure. The orientation of the x_1-y_1 system relative to the x-y system is determined by the angle θ_1. The angle θ_2 measures the angular displacement if the x_2-y_2 system relative to the fixed frame.

As noted above the system has 2 degrees of freedom however, the two moving frames require 6 coordinates i.e.,

$^0r_{QA} = \begin{bmatrix} x_A \\ y_A \end{bmatrix}$: the position of the origin of the x_1-y_1 frame,

θ_1 : the orientation of the x_1-y_1 frame,

$^0r_{QB} = \begin{bmatrix} x_B \\ y_B \end{bmatrix}$: the position of the origin of the x_2-y_2 frame,

θ_2 : the orientation of the x_2-y_2 frame.

Let θ_1 and θ_2 be the independent coordinates for the system, i.e., the degrees of freedom. Then 4 constraint equations are required to determine x_A, y_A, x_B, and y_B in terms of θ_1 and θ_2. These equations are determined using the properties of the revolute joints. Specifically, the point A and Q are coincident at all times hence, we have the constraints

$$\phi_1 = x_A = 0,$$
$$\phi_2 = y_A = 0.$$

Also, the coordinate of the point B is $^0r_{QB} = {}^0r_{QA} + {}^0A_1{}^1r_{AB}$ which gives the two constraints

$$\phi_3 = x_B - l_1 \cos\theta_1 = 0,$$
$$\phi_4 = y_B - l_1 \sin\theta_1 = 0.$$

Here, ${}^0A_1$ is the direction cosine matrix relating the x_1-y_1 coordinate system to the x-y coordinate system, and ${}^1r_{AB}$ is the coordinate of B relative to A in the x_1-y_1 coordinate system.

The coordinate of the point C is given by

$$\begin{aligned}
{}^0r_{QC} &= {}^0r_{QA} + {}^0A_1{}^1r_{AB} + {}^0A_2{}^2r_{BC} \\
\begin{bmatrix} x_C \\ y_C \end{bmatrix} &= \begin{bmatrix} x_A \\ y_A \end{bmatrix} + \begin{bmatrix} \cos\theta_1 & -\sin\theta_1 \\ \sin\theta_1 & \cos\theta_1 \end{bmatrix} \begin{bmatrix} l_1 \\ 0 \end{bmatrix} + \begin{bmatrix} \cos\theta_2 & -\sin\theta_2 \\ \sin\theta_2 & \cos\theta_2 \end{bmatrix} \begin{bmatrix} l_2 \\ 0 \end{bmatrix} \\
&= \begin{bmatrix} l_1\cos\theta_1 + l_2\cos\theta_2 \\ l_1\sin\theta_1 + l_2\sin\theta_2 \end{bmatrix},
\end{aligned}$$

where ${}^0A_2$ is the direction cosine matrix relating the x_2-y_2 coordinate system to the x-y coordinate system, and ${}^2r_{BC}$ is the coordinate of C relative to B in the x_2-y_2 coordinate system.

Finally, the velocity of the point C can be computed via a direct differentiation of the position of C, i.e.,

$$\begin{aligned}
{}^0v_{QC} &= \frac{d}{dt}({}^0r_{QC}) = {}^0\dot{r}_{QA} + {}^0A_1{}^1\tilde{\omega}({}^1r_{AB}) + {}^0A_2{}^2\tilde{\omega}({}^2r_{BC}) \\
\begin{bmatrix} \dot{x}_C \\ \dot{y}_C \end{bmatrix} &= \begin{bmatrix} \cos\theta_1 & -\sin\theta_1 \\ \sin\theta_1 & \cos\theta_1 \end{bmatrix} \begin{bmatrix} 0 & -\dot{\theta}_1 \\ \dot{\theta}_1 & 0 \end{bmatrix} \begin{bmatrix} l_1 \\ 0 \end{bmatrix} + \begin{bmatrix} \cos\theta_2 & -\sin\theta_2 \\ \sin\theta_2 & \cos\theta_2 \end{bmatrix} \begin{bmatrix} 0 & -\dot{\theta}_2 \\ \dot{\theta}_2 & 0 \end{bmatrix} \begin{bmatrix} l_2 \\ 0 \end{bmatrix} \\
&= \begin{bmatrix} -\dot{\theta}_1 l_1\sin\theta_1 - \dot{\theta}_2 l_2\sin\theta_2 \\ \dot{\theta}_1 l_1\cos\theta_1 + \dot{\theta}_2 l_2\cos\theta_2 \end{bmatrix}.
\end{aligned}$$

The plots below show the position and velocity of the point C for the R-R robot with the proportions, $l_1 = 2$, and $l_2 = 1$. The angles $\theta_1 = 2\pi t$, and $\theta_2 = 20\pi t$, where t is the time. The plots show the trajectory of the system in the interval $0 \le t \le 1$. The first plot shows the path traced by the point C, the second and third plots show, respectively, the x-axis and y-axis components of the velocity of C.

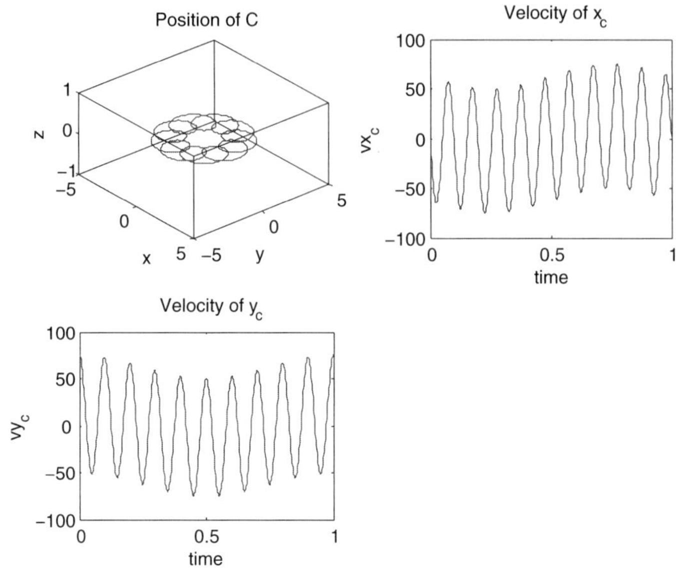

Example 2.13.

This example performs a kinematic analysis of a spatial R-R robot. A sketch of the mechanism is shown in the diagram labeled (a). The motion generated by this mechanism is spatial because the revolute joint at A allows rotation about the vertical axis, while the revolute joint at B allows rotation about an axis in the horizontal plane.

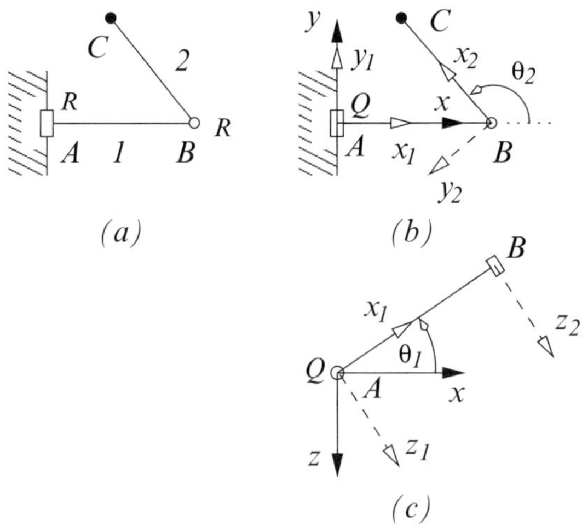

The diagrams (*b*) and (*c*) show the coordinate system assignment for the mechanism. The rectangular coordinate system x-y-z with origin Q represents the fixed frame. The rectangular coordinate system x_1-y_1-z_1 is attached to the link AB with origin at A. The rectangular coordinate system x_2-y_2-z_2 is attached to the link BC with origin at B. The link AB has length l_1, and the link BC has length l_2.

The diagram (*b*) shows the mechanism in the x-y plane, and the diagram (*c*) shows the link AB displaced in the x-z plane. The revolute joint at A allows angular displacement θ_1 of the y_1-axis only while, the revolute joint at B allows angular displacement θ_2 of the z_2-axis only. Note that the angle θ_2 is measured from the x_1-axis.

A mobility analysis shows that the mechanism has 2 degrees of freedom. However, the two moving frames require 12 variables to specify their position and orientation. Of these 12 variables we will select θ_1 and θ_2 as the two independent variables. The kinematic properties of the joints must be used to determine 10 constraints that will account for the excess variables. As will be seen many of these constraints are trivial.

First note that at all times the point Q, (the origin of the fixed frame), is coincident with A, the origin of the x_1-y_1-z_1 frame. This gives the three constraints

$$\begin{aligned}\phi_1 &= x_A = 0,\\ \phi_2 &= y_A = 0,\\ \phi_3 &= z_A = 0.\end{aligned}$$

Here, the orientation of the x_1-y_1-z_1 frame relative to the fixed frame will be described using the X_{α_1}-Y_{θ_1}-Z_{γ_1} Euler angles. Due to the revolute joint constraint it can be seen that the following two constraints must be satisfied;

$$\begin{aligned}\phi_4 &= \alpha_1 = 0,\\ \phi_5 &= \gamma_1 = 0.\end{aligned}$$

The coordinate of the point B (the origin of the moving frame x_2-y_2-z_2) is given by

$$\begin{aligned}{}^0r_{QB} &= {}^0r_{QA} + {}^0A_1\,{}^1r_{AB}\\ \begin{bmatrix} x_B\\ y_B\\ z_B\end{bmatrix} &= \begin{bmatrix} x_A\\ y_A\\ z_A\end{bmatrix} + \begin{bmatrix}\cos\theta_1 & 0 & \sin\theta_1\\ 0 & 1 & 0\\ -\sin\theta_1 & 0 & \cos\theta_1\end{bmatrix}\begin{bmatrix} l_1\\ 0\\ 0\end{bmatrix}.\end{aligned}$$

Here, ${}^0A_1$ is the direction cosine matrix relating x_1-y_1-z_1 to the fixed frame, and ${}^1r_{AB}$ is the coordinate of B relative to A in the x_1-y_1-z_1 coordinate system. These equations lead to the three constraints

$$\phi_6 = x_B - l_1\cos\theta_1 = 0,$$

$$\phi_7 = y_B = 0,$$
$$\phi_8 = z_B + l_1 \sin\theta_1 = 0.$$

The orientation of the x_2-y_2-z_2 frame relative to the x_1-y_1-z_1 frame will be described using the X_{α_2}-Y_{β_2}-Z_{θ_2} Euler angles. Thus, to satisfy the revolute joint constraint at B we require

$$\phi_9 = \alpha_2 = 0,$$
$$\phi_{10} = \beta_2 = 0.$$

Given the independent variables θ_1 and θ_2 the 10 constraints $\phi_1, \phi_2, \cdots, \phi_{10}$, can be used to find the position and orientation of the moving frames.

The coordinate of the point C is given by

$$
\begin{aligned}
{}^0r_{QC} &= {}^0r_{QA} + {}^0A_1{}^1r_{AB} + {}^0A_1{}^1A_2{}^2r_{BC}, \\
\begin{bmatrix} x_C \\ y_C \\ z_C \end{bmatrix} &= \begin{bmatrix} \cos\theta_1 & 0 & \sin\theta_1 \\ 0 & 1 & 0 \\ -\sin\theta_1 & 0 & \cos\theta_1 \end{bmatrix} \begin{bmatrix} l_1 \\ 0 \\ 0 \end{bmatrix} \\
&\quad + \begin{bmatrix} \cos\theta_1 & 0 & \sin\theta_1 \\ 0 & 1 & 0 \\ -\sin\theta_1 & 0 & \cos\theta_1 \end{bmatrix} \begin{bmatrix} \cos\theta_2 & -\sin\theta_2 & 0 \\ \sin\theta_2 & \cos\theta_2 & 0 \\ 0 & 0 & 1 \end{bmatrix} \begin{bmatrix} l_2 \\ 0 \\ 0 \end{bmatrix}, \\
&= \begin{bmatrix} (l_1 + l_2\cos\theta_2)\cos\theta_1 \\ l_2 \sin\theta_2 \\ -(l_1 + l_2\cos\theta_2)\sin\theta_1 \end{bmatrix}.
\end{aligned}
$$

Here, ${}^1A_2$ is the direction cosine matrix relating x_2-y_2-z_2 to x_1-y_1-z_1, and ${}^2r_{BC}$ is the coordinate of C relative to B in the x_2-y_2-z_2 coordinate system.

Differentiation of the last equation with respect to time gives the velocity of the point C as

$$
\begin{bmatrix} \dot{x}_C \\ \dot{y}_C \\ \dot{z}_C \end{bmatrix} = \begin{bmatrix} -(l_1 + l_2\cos\theta_2)\dot{\theta}_1 \sin\theta_1 - l_2\dot{\theta}_2 \cos\theta_1 \sin\theta_2 \\ l_2\dot{\theta}_2 \cos\theta_2 \\ -(l_1 + l_2\cos\theta_2)\dot{\theta}_1 \cos\theta_1 + l_2\dot{\theta}_2 \sin\theta_1 \sin\theta_2 \end{bmatrix}.
$$

The plots below show the position and velocity of the point C for the R-R robot with the proportions, $l_1 = 2$, and $l_2 = 1$. The angles $\theta_1 = 2\pi t$, and $\theta_2 = 20\pi t$ where t is the time. The plots show the trajectory of the system in the interval $0 \le t \le 1$. The first plot shows the path traced by the point C, and the remaining plots show the components of the velocity of C.

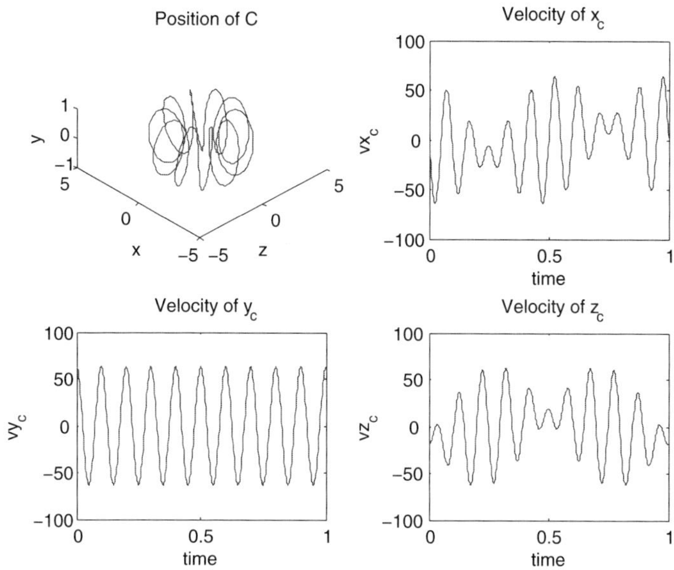

Example 2.14.

The diagram below shows a disk, with radius r, that rolls on horizontal surface without slipping. If the disk rotates counterclockwise with a constant angular velocity ω, find the equations that describe the position and velocity of a point C that is on the edge of the disk.

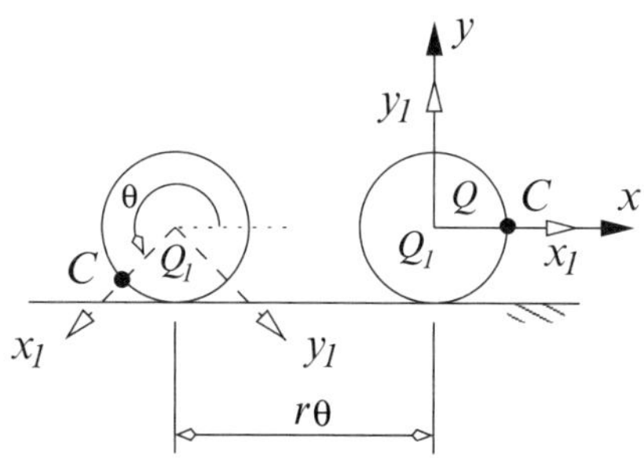

Let the rectangular coordinate system x-y represent the fixed frame, and attach the rectangular coordinate system x_1-y_1 to the center of the disk. Initially, the x-y frame and the x_1-y_1 frames are aligned with common origins Q and Q_1 respectively.

A mobility analysis indicates the this system has 1 degree of freedom. Let, θ, the angular orientation of the disk, be the independent coordinate. Then the constraints that specify the location of Q_1 relative to the fixed frame are

$$\phi_1 = x_{Q_1} + r\theta = 0,$$
$$\phi_2 = y_{Q_1} = 0.$$

The coordinate of the point C relative to the Q in the fixed frame is

$$
\begin{aligned}
{}^0r_{QC} &= {}^0r_{QQ_1} + {}^0A_1\,{}^1r_{Q_1C} \\
\begin{bmatrix} x_C \\ y_C \end{bmatrix} &= \begin{bmatrix} -r\theta \\ 0 \end{bmatrix} + \begin{bmatrix} \cos\theta & -\sin\theta \\ \sin\theta & \cos\theta \end{bmatrix} \begin{bmatrix} r \\ 0 \end{bmatrix} \\
&= \begin{bmatrix} -r\theta + r\cos\theta \\ r\sin\theta \end{bmatrix}.
\end{aligned}
$$

Here, ${}^0r_{QQ_1} = [x_{Q_1},\ \ y_{Q_1}]^T$ is the coordinate of Q_1 relative to Q in the fixed frame, ${}^0A_1$ is the direction cosine matrix relating the x_1-y_1 coordinate system to the fixed frame, and ${}^1r_{Q_1C} = [r,\ \ 0]^T$ is the coordinate of C relative to Q_1 in the x_1-y_1 coordinate system.

Differentiating the last equation with respect to time gives the velocity of the point C, i.e.,

$$\begin{bmatrix} \dot{x}_C \\ \dot{y}_C \end{bmatrix} = \begin{bmatrix} -r\omega(1+\sin\theta) \\ r\omega\cos\theta \end{bmatrix},$$

where $\omega = \dot{\theta}$.

The trajectory of a point C on a disk with radius $r = 1$ and constant angular velocity $\omega = 2\pi$ is shown below. As can be seen the point C traces the path of a cycloid.

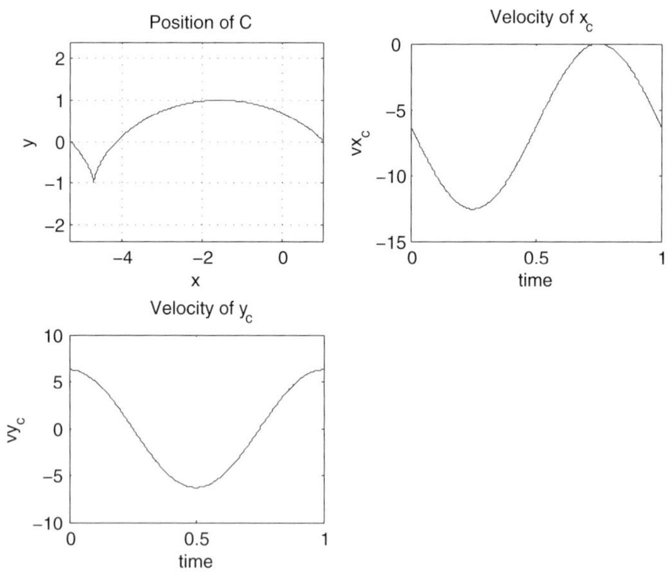

Example 2.15.

The diagram below shows a disk, with radius r, that rolls without slipping on a fixed circular cylinder with radius R. If the disk rotates around the cylinder in a counterclockwise direction with constant angular velocity $\dot{\phi}$, find the equations that determine the position and velocity of the point C on the edge of the disk.

The fixed frame is represented by the rectangular coordinate system x-y, with origin Q at the center of the cylinder. Attached to the disk is the rectangular coordinate system x_1-y_1, with origin Q_1 at the center of the disk.

The disk is shown in an initial position 1. After a rotation of angle θ the disk moves to position 2. As a result of this rotation the line QQ_1, in position 2, makes an angle ϕ with the x-axis.

A mobility analysis of the mechanism shows that the system has 1 degree of freedom. Here, we select ϕ as the independent coordinate. Therefore, constraint equations must be found to determine (x_{Q_1}, y_{Q_1}) the location of the center of the disk, and the angle θ.

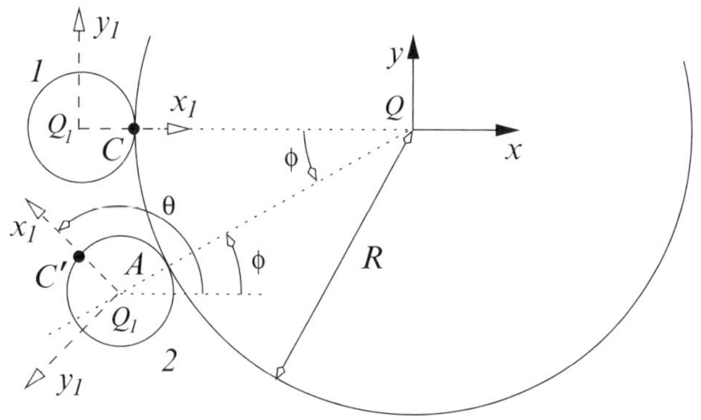

The coordinate of the point Q_1 relative to Q in the fixed frame is given by

$$^0r_{QQ_1} = \begin{bmatrix} x_{Q_1} \\ y_{Q_1} \end{bmatrix} = \begin{bmatrix} -(R+r)\cos\phi \\ -(R+r)\sin\phi \end{bmatrix}.$$

In position 2, the point A is the point of contact between the disk and the cylinder. Since the disk rolls without slipping the arc length AC (on the cylinder) is equal to the arc length AC' (on the disk), i.e.,

$$R\phi = r(\theta - \phi).$$

Hence, $\theta = ((R+r)/r)\phi$.

The coordinate of the point C relative to Q in the fixed frame is given by

$$^0r_{QC} = {}^0r_{QQ_1} + {}^0A_1\,{}^1r_{Q_1C}$$

$$\begin{bmatrix} x_C \\ y_C \end{bmatrix} = \begin{bmatrix} -(R+r)\cos\phi \\ -(R+r)\sin\phi \end{bmatrix} + \begin{bmatrix} \cos\theta & -\sin\theta \\ \sin\theta & \cos\theta \end{bmatrix} \begin{bmatrix} r \\ 0 \end{bmatrix}$$
$$= \begin{bmatrix} -(R+r)\cos\phi + r\cos((R/r+1)\phi) \\ -(R+r)\sin\phi + r\sin((R/r+1)\phi) \end{bmatrix}.$$

Here, ${}^0A_1$ is the direction cosine matrix relating x_1-y_1 to x-y, and ${}^1r_{Q_1C}$ is the coordinate of C relative to Q_1 in the x_1-y_1 coordinate system.

The velocity of the point C can be obtained by differentiating the last equation with respect to time to get

$$\begin{bmatrix} \dot{x}_C \\ \dot{y}_C \end{bmatrix} = \begin{bmatrix} (R+r)\dot{\phi}\,[\sin\phi + \sin((R/r+1)\phi)] \\ -(R+r)\dot{\phi}\,[-\cos\phi + \cos((R/r+1)\phi)] \end{bmatrix}.$$

The trajectory of a point C on a system with radii $R = 3$, $r = 1$ and constant angular velocity $\dot{\phi} = 2\pi$ is shown below. In the first plot, the dotted line represents the cylinder, and the solid line shows the path of the point C that is on the disk. The second and third plots show the components of the velocity of the point C.

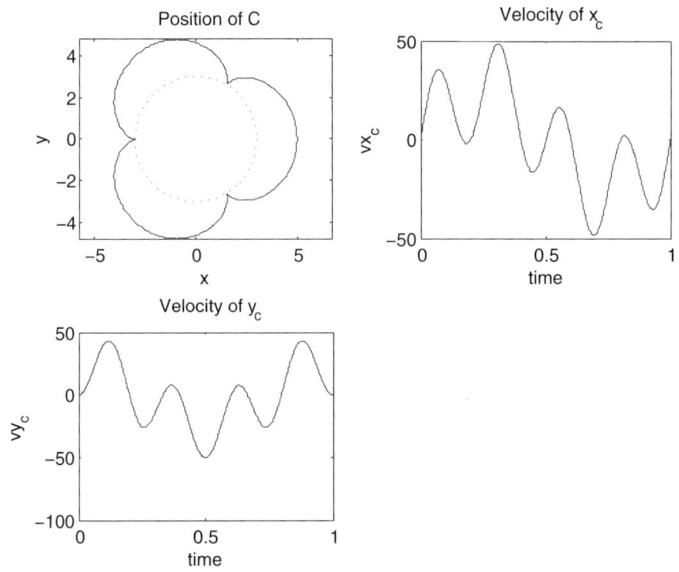

Example 2.16.

The diagram below shows a planar fourbar mechanism. The proportions of the mechanism is given in the table below. If the crank AB has a constant angular velocity $\dot{\theta}_1$, derive the equations necessary to determine the position and velocity of the point E.

QA	l_A	3
AB	l_1	1.5
BC	l_2	5
BE	l_E	2.5
QD	l_D	4.5
DC	l_3	3

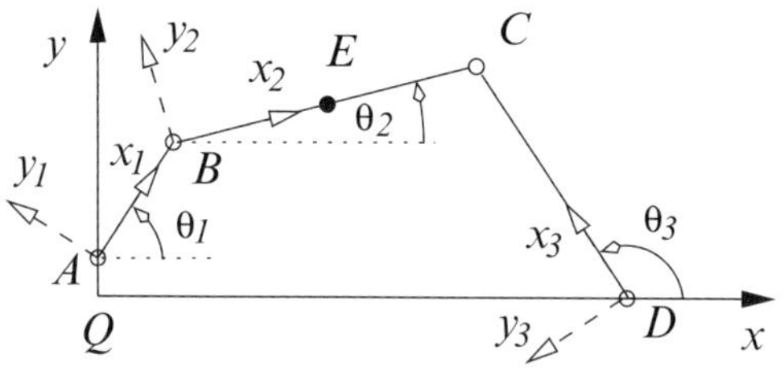

Fig. 2.8 Planar fourbar mechanism

The rectangular coordinate system x-y with origin at Q represents the fixed frame for the system. The rectangular coordinate system x_1-y_1 is attached to the link AB with origin at A. The rectangular coordinate system x_2-y_2 is attached to the link BC with origin at B. The rectangular coordinate system x_3-y_3 is attached to the link DC with origin at D.

A mobility analysis shows that the mechanism has 1 degree of freedom. However, the 3 moving coordinates require 9 variables to specify their position and orientation. Therefore, 8 constraint equations are required to account for the excess coordinates. Here, we will use θ_1 as the independent coordinate.

The coordinates of the points A, B and D, with respect to the fixed frame, yield the following 6 constraints;

$$\begin{aligned}
\phi_1 &= x_A = 0,\\
\phi_2 &= y_A - l_A = 0,\\
\phi_3 &= x_B - l_1 \cos\theta_1 = 0,\\
\phi_4 &= y_B - l_A - l_1 \sin\theta_1 = 0,\\
\phi_5 &= x_D - l_D = 0,\\
\phi_6 &= y_D = 0.
\end{aligned}$$

The remaining constraints are determined by noting that the mechanism forms a closed kinematic chain hence,

$$^0r_{QA} + {}^0A_1{}^1r_{AB} + {}^0A_2{}^2r_{BC} = {}^0r_{QD} + {}^0A_3{}^3r_{DC}. \tag{a}$$

The right hand side of this equation is the vector summation ${}^0\bar{r}_{QA} + {}^1\bar{r}_{AB} + {}^2\bar{r}_{BC}$, and the left hand side of the equations is the vector summation ${}^0\bar{r}_{QD} + {}^3\bar{r}_{DC}$. The term ${}^0A_i$ is the direction cosine matrix that relates the i-th coordinate system to the fixed frame, ${}^1r_{AB}$ is the coordinate of B relative to A in the x_1-y_1 coordinate system, ${}^2r_{BC}$ is the coordinate of C relative to B in the x_2-y_2 coordinate system, and ${}^3r_{DC}$ is the coordinate of C relative to D in the x_3-y_3 coordinate system.

Using the definitions for the coordinate systems equation (a) can be written as

$$\begin{aligned}&\begin{bmatrix}0\\l_A\end{bmatrix}+\begin{bmatrix}\cos\theta_1 & -\sin\theta_1\\ \sin\theta_1 & \cos\theta_1\end{bmatrix}\begin{bmatrix}l_1\\0\end{bmatrix}+\begin{bmatrix}\cos\theta_2 & -\sin\theta_2\\ \sin\theta_2 & \cos\theta_2\end{bmatrix}\begin{bmatrix}l_2\\0\end{bmatrix}\\ &\quad=\begin{bmatrix}l_D\\0\end{bmatrix}+\begin{bmatrix}\cos\theta_3 & -\sin\theta_3\\ \sin\theta_3 & \cos\theta_3\end{bmatrix}\begin{bmatrix}l_3\\0\end{bmatrix}.\end{aligned}$$

Which yields the two constraint equations

$$\begin{aligned}\phi_7 &= l_1\cos\theta_1 + l_2\cos\theta_2 - l_3\cos\theta_3 - l_D = 0,\\ \phi_8 &= l_1\sin\theta_1 + l_2\sin\theta_2 - l_3\sin\theta_3 + l_A = 0.\end{aligned}$$

Given θ_1, the 8 constraint equations, $\phi_1, \phi_2, \cdots, \phi_8$, can used to compute the coordinates x_A, y_A, x_B, y_B, x_D, y_D, θ_2, and θ_3. Note however, that ϕ_7 and ϕ_8 are nonlinear equations that must be solved to determine θ_2 and θ_3.

The coordinate of the point E with respect to the fixed frame is given by

$$\begin{aligned}{}^0r_{QE} &= {}^0r_{QA} + {}^0A_1\,{}^1r_{AB} + {}^0A_2\,{}^2r_{BE}\\ \begin{bmatrix}x_E\\y_E\end{bmatrix} &= \begin{bmatrix}0\\l_A\end{bmatrix}+\begin{bmatrix}\cos\theta_1 & -\sin\theta_1\\ \sin\theta_1 & \cos\theta_1\end{bmatrix}\begin{bmatrix}l_1\\0\end{bmatrix}+\begin{bmatrix}\cos\theta_2 & -\sin\theta_2\\ \sin\theta_2 & \cos\theta_2\end{bmatrix}\begin{bmatrix}l_E\\0\end{bmatrix}\\ &= \begin{bmatrix}l_1\cos\theta_1 + l_E\cos\theta_2\\ l_A + l_1\sin\theta_1 + l_E\sin\theta_2\end{bmatrix}.\end{aligned}$$

Differentiation of the last two equations with respect to time gives the velocity of the point E as

$$\begin{bmatrix}\dot{x}_E\\ \dot{y}_E\end{bmatrix} = \begin{bmatrix}-l_1\dot{\theta}_1\sin\theta_1 - l_E\dot{\theta}_2\sin\theta_2\\ l_1\dot{\theta}_1\cos\theta_1 + l_E\dot{\theta}_2\cos\theta_2\end{bmatrix}.$$

Note that the computation of the velocity of E requires the angular velocity $\dot{\theta}_2$ which must be computed from ϕ_7 and ϕ_8.

The differentiation of ϕ_7 and ϕ_8 with respect to the time gives

$$\begin{aligned}\frac{d\phi_7}{dt} &= -l_1\dot{\theta}_1\sin\theta_1 - l_2\dot{\theta}_2\sin\theta_2 + l_3\dot{\theta}_3\sin\theta_3 = 0,\\ \frac{d\phi_8}{dt} &= l_1\dot{\theta}_1\cos\theta_1 + l_2\dot{\theta}_2\cos\theta_2 - l_3\dot{\theta}_3\cos\theta_3 = 0.\end{aligned}$$

These equations can be rewritten as

$$\begin{bmatrix} -l_2 \sin\theta_2 & l_3 \sin\theta_3 \\ l_2 \cos\theta_2 & -\, l_3 \cos\theta_3 \end{bmatrix} \begin{bmatrix} \dot{\theta}_2 \\ \dot{\theta}_3 \end{bmatrix} = \begin{bmatrix} l_1 \dot{\theta}_1 \sin\theta_1 \\ -l_1 \dot{\theta}_1 \cos\theta_1 \end{bmatrix}. \qquad (b)$$

Given a trajectory $\theta_1(t)$, $\theta_2(t)$, $\theta_3(t)$ and $\dot{\theta}_1(t)$, the linear equation (b) can be solved to obtain the angular velocities $\dot{\theta}_2(t)$ and $\dot{\theta}_3(t)$. The coefficient matrix on the left hand side of (b) is the Jacobian of the constraints (ϕ_7 and ϕ_8), and the Jacobian may be singular at certain configurations of the mechanism.

The figures below show the trajectory of the mechanism using the proportions stated above. The crank angle is given by $\theta_1 = 2\pi t$ where, t is the time. The first and second plots show the the coupler angle, θ_2, and the follower angle, θ_3, as a function of the crank angle θ_1. The third plot shows the path traced by the point E on the coupler curve. The fourth plot shows the configuration of the mechanism in 10 different positions. Note that the follower link DC rocks backward and forward as the crank AB makes complete revolutions. Such fourbar mechanisms are called crank and rocker mechanisms.

The remaining plots show the velocities of some of the system coordinates as a function of the crank angle θ_1.

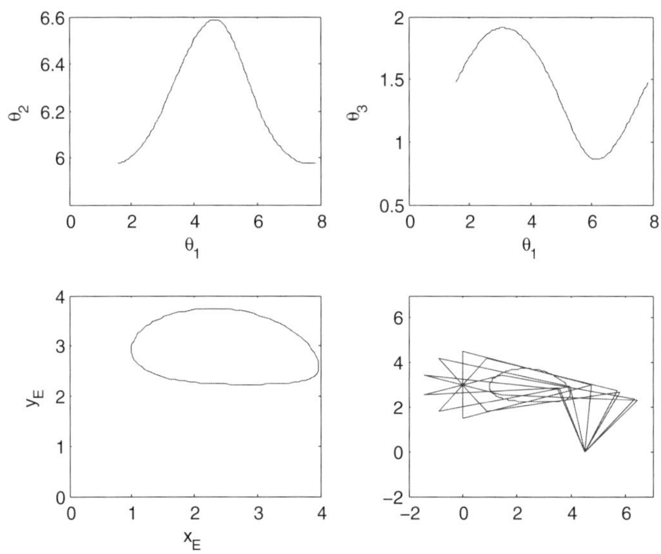

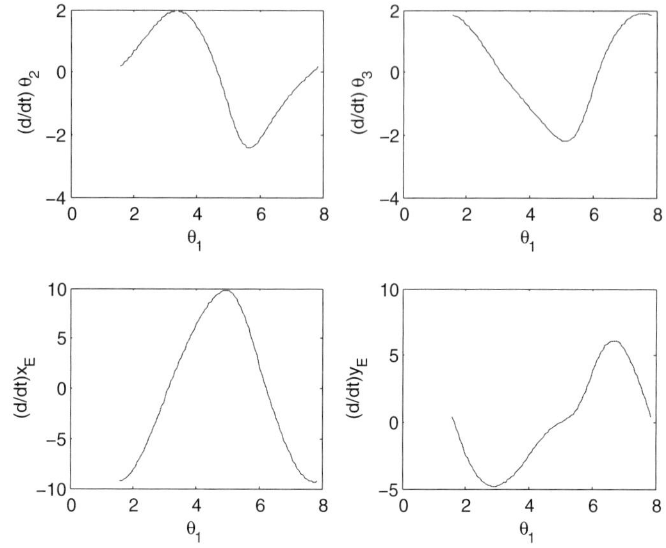

Example 2.17.

The figure below shows a planar slider crank mechanism. The joints at A, B and C are revolute, and the joint at D is prismatic. This example develops the equations necessary to determine the position and velocity of the point C.

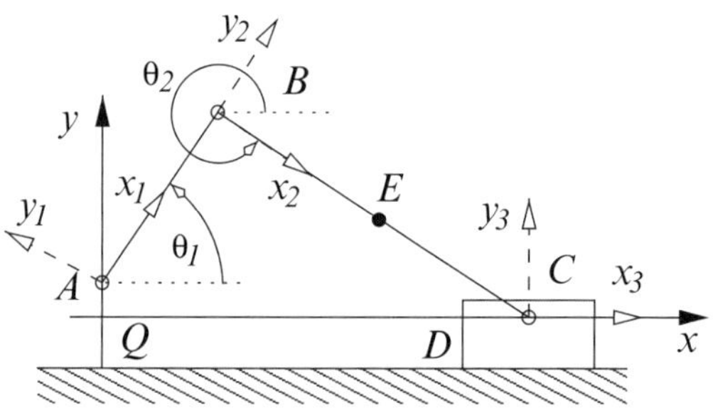

The proportions of the mechanism are given in the table below.

QA	l_A	0.25
AB	l_1	1.5
BE	l_E	2.5
BC	l_2	5

The rectangular coordinate system x-y with origin at Q is the fixed reference frame. Also shown in the figure are three moving rectangular coordinate systems; (i) x_1-y_1 attached to the link AB with origin at A, (ii) x_2-y_2 attached to the link BC with origin at B, and (iii) x_3-y_3 attached to the slider CD at point C.

A mobility analysis shows that the mechanism has 1 degree of freedom. However, the 3 moving coordinate systems require 9 variables to establish their position and orientation. Thus, 8 constraint equations must be developed to account for the excess variables. Using the figure above, we select θ_1, the orientation of the x_1-y_1 coordinate system, as the independent variable.

The required constraints are determined by writing the position equations for the origin of each of the moving frames, making use of the fact that the mechanism forms a closed kinematic chain. The coordinate of the point A with respect to the fixed frame, ${}^0r_{QA}$, yields the two constraint equations,

$$\begin{aligned}\phi_1 &= x_A = 0,\\ \phi_2 &= y_A - l_A = 0.\end{aligned}$$

The coordinate of the point B with respect to the fixed frame, ${}^0r_{QB} = {}^0r_{QA} + {}^0A_1{}^1r_{AB}$, yields the two constraint equations

$$\begin{aligned}\phi_3 &= x_B - l_1\cos\theta_1 = 0,\\ \phi_4 &= y_A - l_A - l_1\sin\theta_1 = 0.\end{aligned}$$

Here, ${}^0A_1$ is the direction cosine matrix relating the x_1-y_1 coordinate system to fixed frame, and ${}^1r_{AB}$ is the coordinate of B in the x_1-y_1 coordinate system.

To obtain the constrains related to the x_3-y_3 coordinate system we must consider the kinematic behavior of the prismatic joint. By construction the slider can only translate along the x-axis, and since the x_3-y_3 coordinate system is fixed to the slider at point C we have the constraints,

$$\begin{aligned}\phi_5 &= y_C = 0,\\ \phi_6 &= \theta_3 = 0.\end{aligned}$$

The last equation indicates that there is no rotation of the x_3-y_3 coordinate system.

Finally, we make use of the fact that the mechanism forms a closed kinematic chain to get

$$ {}^0r_{QA} + {}^0A_1{}^1r_{AB} + {}^0A_2{}^2r_{BC} = {}^0r_{QC} \tag{a}$$

$$\begin{bmatrix}0\\ l_A\end{bmatrix} + \begin{bmatrix}\cos\theta_1 & -\sin\theta_1\\ \sin\theta_1 & \cos\theta_1\end{bmatrix}\begin{bmatrix}l_1\\ 0\end{bmatrix} + \begin{bmatrix}\cos\theta_2 & -\sin\theta_2\\ \sin\theta_2 & \cos\theta_2\end{bmatrix}\begin{bmatrix}l_2\\ 0\end{bmatrix} = \begin{bmatrix}x_C\\ y_C\end{bmatrix},$$

where ${}^0A_2$ is the direction cosine matrix relating the x_2-y_2 coordinate system to fixed frame, and ${}^2r_{BC}$ is the coordinate of C in the x_2-y_2 coordinate system. This equation yields the constraints,

$$\phi_7 = l_1 \cos\theta_1 + l_2 \cos\theta_2 - x_C = 0,$$
$$\phi_8 = l_A + l_1 \sin\theta_1 + l_2 \sin\theta_2 = 0.$$

Given a value for the crank angle, θ_2, the 8 equations $\phi_1, \phi_2, \cdots, \phi_8$ can be used to solve for the coordinates x_A, y_A, x_B, y_B, θ_2, x_C, y_C, and θ_3. Clearly many of the equations are trivial however, ϕ_7 and ϕ_8 are coupled nonlinear equations that must be solved to determine θ_2 and x_C.

To find the velocity of the point C we differentiate equation (a) with respect to time to get

$$\begin{bmatrix} \dot{x}_C \\ \dot{y}_C \end{bmatrix} = \begin{bmatrix} -l_1\dot{\theta}_1 \sin\theta_1 - l_2\dot{\theta}_2 \sin\theta_2 \\ l_1\dot{\theta}_1 \cos\theta_1 + l_2\dot{\theta}_2 \cos\theta_2 \end{bmatrix}.$$

However, we know from ϕ_5 that $\dot{y}_C = 0$. These last two equations yield the linear system

$$\begin{bmatrix} -l_2 \sin\theta_2 & 1 \\ l_2 \cos\theta_2 & 0 \end{bmatrix} \begin{bmatrix} \dot{\theta}_2 \\ \dot{x}_C \end{bmatrix} = \begin{bmatrix} l_1\dot{\theta}_1 \sin\theta_1 \\ -l_1\dot{\theta}_1 \cos\theta_1 \end{bmatrix},$$

which must be solved to determine $\dot{\theta}_2$ and $\dot{x}_C$.

Finally, it can be seen that the position and velocity of the point E are given by

$$\begin{bmatrix} x_E \\ y_E \end{bmatrix} = \begin{bmatrix} l_1 \cos\theta_1 + l_E \cos\theta_2 \\ l_A + l_1 \sin\theta_1 + l_E \sin\theta_2 \end{bmatrix},$$

and

$$\begin{bmatrix} \dot{x}_E \\ \dot{y}_E \end{bmatrix} = \begin{bmatrix} -l_1\dot{\theta}_1 \sin\theta_1 - l_E\dot{\theta}_2 \sin\theta_2 \\ l_1\dot{\theta}_1 \cos\theta_1 + l_E\dot{\theta}_2 \cos\theta_2 \end{bmatrix},$$

respectively.

The trajectory of the slider crank mechanism, described above, is shown in the figures below. These results are obtained using a crank angular velocity $\dot{\theta}_1 = 2\pi$. The first plot shows the coupler angle θ_2 versus the crank angle θ_1. The second plot shows the slider position x_C versus θ_1. The third plot shows the path traced by the point E. The fourth plot shows the configuration of the mechanism in ten different positions. This plot also shows the path traced by the point E. The remaining plots shows the velocities of the system variables as a function of θ_1.

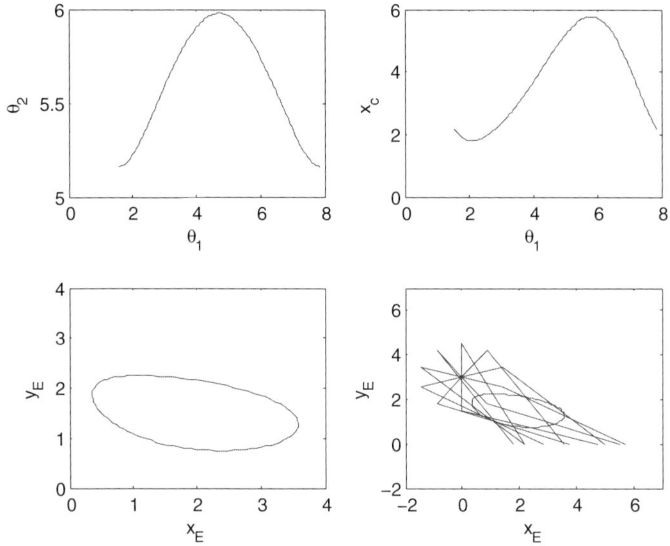

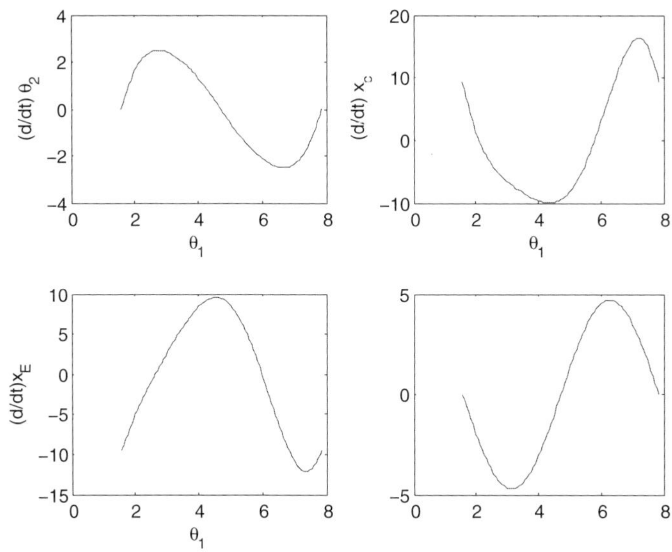

Example 2.18.

The diagrams below show the front and top views of a spatial slider crank mechanism. The device has moving binary links at AB, CB and DC. The joint at A is revolute and the direction of rotation is along an axis parallel to the line YY. In addition the line AB makes a fixed angle, ψ, with respect to the line ZZ (in the front view). The joint at B is a spherical joint, the joint at C is a universal joint, and the joint at D is a prismatic joint. The translation in the prismatic joint takes place parallel to the line YY. The link AB has length l_1, and the link CB has length l_2.

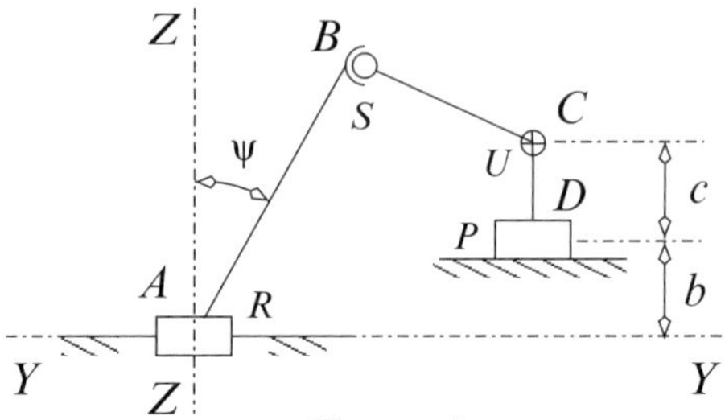

Front view

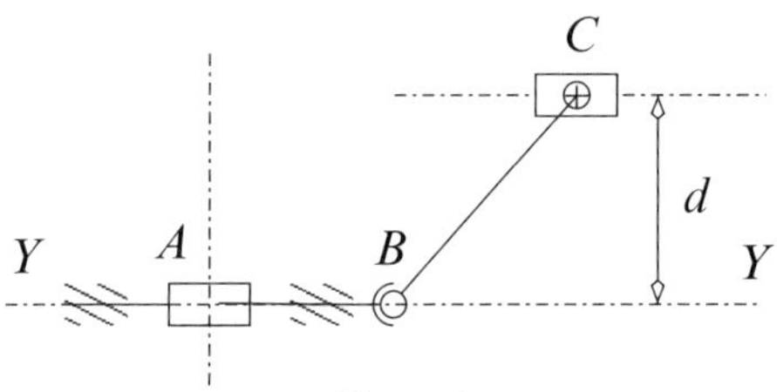

Top view

A mobility analysis of the mechanism shows that there are $l = 4$ links, $j = 4$ joints and

$$F = \lambda(l - j - 1) + \sum_{i=1}^{j} f_i = 6(4 - 4 - 1) + (1 + 3 + 2 + 1) = 1$$

degree of freedom.

The rectangular coordinate system x-y-z, shown in the figure below, represents the fixed frame for the system. The rectangular coordinate system x_1-y_1-z_1 is attached to the link AB with origin at A, and y-axis parallel to the line YY. The rectangular coordinate system x_2-y_2-z_2 is attached to the link DC with origin at D, and y-axis parallel to the line YY (in the direction of translation for the prismatic joint). The rectangular coordinate system x_3-y_3-z_3 is attached to the link CB with origin at C, and y-axis along the line CB.

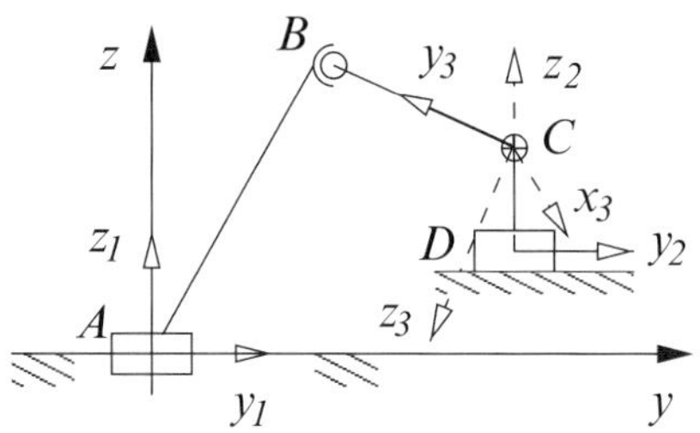

The three moving frames require a total of 18 variables to describe the position and orientation of the system however, there is only 1 degree of freedom. Hence, 17 constraint equations are required to determine the excess variables in terms of the independent coordinate. Here, we will take the angular displacement of the link AB as the independent variable.

To determine the needed constraints we will establish the position and orientation of each coordinate frame ensuring that the kinematic conditions allowed at each joint is satisfied. Consider the x_1-y_1-z_1 coordinate system. The position of A gives the three constraints

$$\phi_1 = x_A = 0, \quad \phi_2 = y_A - l_A = 0, \quad \phi_3 = z_A = 0.$$

The orientation of x_1-y_1-z_1 is defined by the X_{α_1}-Y_{β_1}-Z_{γ_1} Euler angles. Thus, to satisfy the kinematic conditions, for the revolute joint at A, we must have the two constraints

$$\phi_4 = \alpha_1 = 0, \quad \phi_5 = \gamma_1 = 0.$$

Consider the x_2-y_2-z_2 coordinate system. To satisfy the kinematic conditions for the prismatic joint at D we must have the two constraints

$$\phi_6 = x_D + d = 0, \ \phi_7 = z_D - b = 0.$$

The orientation of the x_2-y_2-z_2 coordinate system is defined using the X_{α_2}-Y_{β_2}-Z_{γ_2} Euler angles. Since the x_2-y_2-z_2 coordinate system remains parallel to the x-y-z coordinate system we have the three constraints

$$\phi_8 = \alpha_2 = 0, \quad \phi_9 = \beta_2 = 0, \quad \phi_{10} = \gamma_2 = 0.$$

Consider the x_3-y_3-z_3 coordinate system. The coordinates of the origin of the system (point C) satisfy the three constraints

$$\phi_{11} = x_C - x_D = 0, \quad \phi_{12} = y_C - y_D = 0, \quad \phi_{13} = z_C - z_D - c = 0.$$

The orientation of the x_3-y_3-z_3 coordinate system is defined using the X_{α_3}-Y_{β_3}-Z_{γ_3} Euler angles. Since the joint at C is a universal joint it allows two rotations about orthogonal axes. Here, we assume that the two rotations are α_3 and γ_3. Thus, the third angle must satisfy the constraint

$$\phi_{14} = \beta_3 = 0.$$

The remaining three constraints are determined by noting that the mechanism forms a closed kinematic chain, i.e.,

$$^0A_1{}^1r_{AB} = {}^0r_{QD} + {}^0A_2{}^2r_{DC} + {}^0A_2{}^2A_3{}^3r_{CB}, \tag{a}$$

where

$$^0A_1 = \begin{bmatrix} c_{\beta_1} & 0 & s_{\beta_1} \\ 0 & 1 & 0 \\ -s_{\beta_1} & 0 & c_{\beta_1} \end{bmatrix}, \; {}^1r_{AB} = \begin{bmatrix} 0 \\ l_1 s_\psi \\ l_1 c_\psi \end{bmatrix}, {}^0r_{QD} = \begin{bmatrix} -d \\ y_D \\ b \end{bmatrix},$$

$$^0A_2 = \begin{bmatrix} 1 & 0 & 0 \\ 0 & 1 & 0 \\ 0 & 0 & 1 \end{bmatrix}, \; {}^2r_{DC} = \begin{bmatrix} 0 \\ 0 \\ c \end{bmatrix}, \; {}^2A_3 = \begin{bmatrix} c_{\gamma 3} & -s_{\gamma 3} & 0 \\ c_{\alpha 3} s_{\gamma 3} & c_{\alpha 3} c_{\gamma 3} & -s_{\alpha 3} \\ s_{\alpha 3} s_{\gamma 3} & s_{\alpha 3} c_{\gamma 3} & c_{\alpha 3} \end{bmatrix},$$

$$^3r_{CB} = \begin{bmatrix} 0 \\ l_2 \\ 0 \end{bmatrix}.$$

Here, $^0A_1$ is the direction cosine matrix relating the x_1-y_1-z_1 coordinate system to the fixed frame, $^1r_{AB}$ is the coordinate of B relative to A in the x_1-y_1-z_1 coordinate system, $^0r_{QD}$ is the coordinate of D relative to Q in the fixed frame, $^0A_2$ is the direction cosine matrix relating the x_2-y_2-z_2 coordinate system to the fixed frame, $^2r_{DC}$ is the coordinate of C relative to D in the x_2-y_2-z_2 coordinate system, $^2A_3$ is the direction cosine matrix relating the x_3-y_3-z_3 coordinate system to the x_2-y_2-z_2 coordinate system, and $^3r_{CB}$ is the coordinate of B relative to C in the x_3-y_3-z_3 coordinate system.

Equation (a) yields the three constraints

$$\begin{aligned} \phi_{15} &= l_1 c_\psi s_{\beta_1} + d + l_2 s_{\gamma_3} = 0, \\ \phi_{16} &= l_1 s_\psi - y_D - l_2 c_{\alpha_3} c_{\gamma_3} = 0, \\ \phi_{17} &= l_1 c_\psi c_{\beta_1} - b - c - l_2 s_{\alpha_3} c_{\gamma_3} = 0. \end{aligned}$$

Given a crank angle β_1 the 17 constraints $\phi_1, \phi_2, \cdots, \phi_{17}$ can be used to determine the system coordinates. Note that all but the last three constraints are trivial. The constraints ϕ_{15}, ϕ_{16}, and ϕ_{17} form a system of nonlinear equations that determine y_D, α_3 and γ_3.

To determine the velocity of slider, $\dot{y}_D$, we differentiate the last three constraints with respect to time to get

$$\begin{bmatrix} l_2 c_{\gamma 3} & 0 & 0 \\ l_2 c_{\alpha 3} s_{\gamma 3} & l_2 s_{\alpha 3} c_{\gamma 3} & -1 \\ l_2 s_{\alpha 3} s_{\gamma 3} & -l_2 c_{\alpha 3} c_{\gamma 3} & 0 \end{bmatrix} \begin{bmatrix} \dot{\gamma}_3 \\ \dot{\alpha}_3 \\ \dot{y}_D \end{bmatrix} = \begin{bmatrix} -\dot{\beta}_1 l_1 c_\psi c_{\beta_1} \\ 0 \\ \dot{\beta}_1 l_1 c_\psi s_{\beta_1} \end{bmatrix}. \tag{b}$$

Thus, given β_1, α_3, γ_3, and $\dot{\beta}_1$, the linear equation (b) can be used to determine the slider velocity $\dot{y}_D$, and the rate of change of angular position $\dot{\alpha}_3$ and $\dot{\gamma}_3$. Note that $\dot{\alpha}_3$ and $\dot{\gamma}_3$ are not the angular velocities of the x_3-axis and z_3-axis, respectively. In fact using equation (2.18) it can be seen that the angular velocities of the x_3-y_3-z_3 coordinate system satisfies

$$\begin{bmatrix} {}^3\omega_1 \\ {}^3\omega_2 \\ {}^3\omega_3 \end{bmatrix} = \begin{bmatrix} c_{\beta_3} c_{\gamma_3} & s_{\gamma_3} & 0 \\ -c_{\beta_3} s_{\gamma_3} & c_{\gamma_3} & 0 \\ s_{\beta_3} & 0 & 1 \end{bmatrix} \begin{bmatrix} \dot{\alpha}_3 \\ \dot{\beta}_3 \\ \dot{\gamma}_3 \end{bmatrix}.$$

Here, ${}^3\omega_1$ is the angular velocity of the x_3-axis, ${}^3\omega_2$ is the angular velocity of the y_3-axis, and ${}^3\omega_3$ is the angular velocity of the z_3-axis. Due to the kinematic constraints at joint C (a universal joint), the rotation $\beta_3 = 0$, and $\dot{\beta}_3 = 0$. Thus, the angular velocities of the x_3-y_3-z_3 coordinate system are ${}^3\omega_1 = c_{\gamma_3}\dot{\alpha}_3$, ${}^3\omega_2 = -s_{\gamma_3}\dot{\alpha}_3$, and ${}^3\omega_3 = \dot{\gamma}_3$.

The behavior of the spatial slider crank mechanism analyzed above is illustrated in the figures below. In these plots the constraint equations are solved for a mechanism with the following proportions; $\psi=\pi/6.0$, $l_1=1.0$, $l_2=4.0$, $d=2.0$, $b=1.0$, and $c=0.25$. The angular velocity of the crank, AB, is $\dot{\beta}_1=2\pi$ radians/second. The nonlinear equations ϕ_{15}, ϕ_{16}, and ϕ_{17} are solved using Newton's method.

The first plot shows the Euler angle α_3 as a function of the crank angle β_1. The second plot shows the Euler angle γ_3 as a function of the crank angle. The third plot shows the the slider position as a function of the crank angle. The fourth plot shows the configuration in 10 different positions.

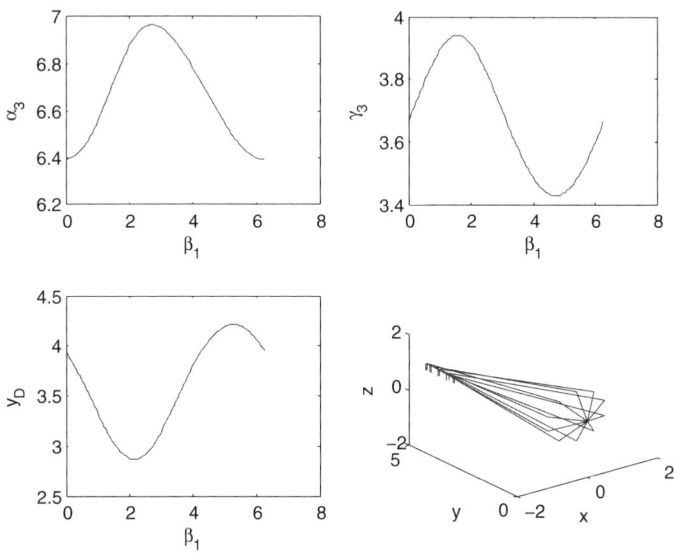

The plots below show the angular velocities $\dot{\alpha}_3$, $\dot{\gamma}_3$, and the slider velocity $\dot{y}_D$ as a function of the crank angle.

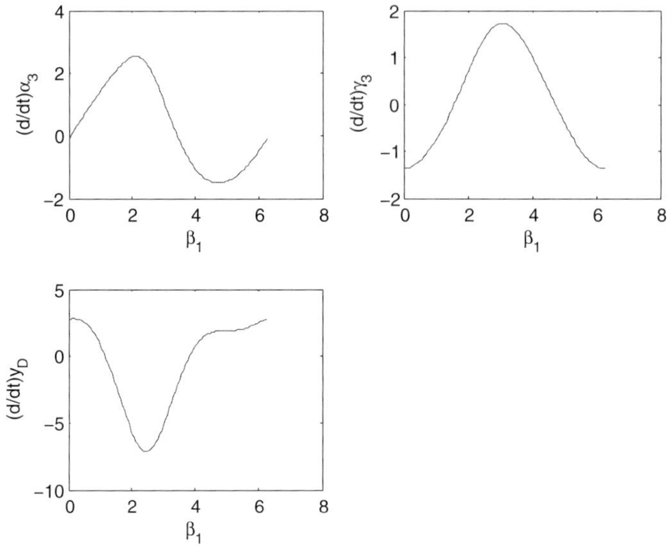

The plots below show the angular velocities ${}^3\omega_1$, ${}^3\omega_2$, and ${}^3\omega_3$ as a function of the crank angle.

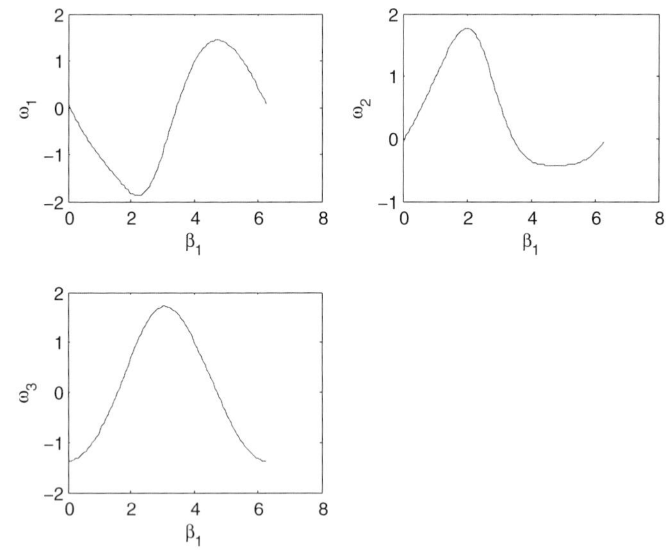

2.3 Network Systems

The models for electrical, fluid and thermal systems can be represented as networks. The structure (topology) of these networks define the relationship between the fundamental variables associated with the model. Here, we will use some terminology and results from graph theory to describe the kinematic constraints that exists in network systems.

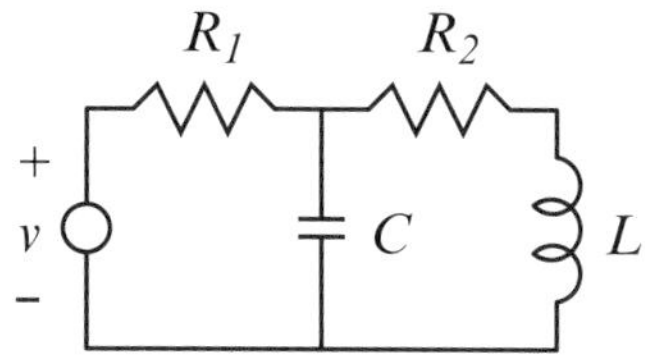

Fig. 2.9 A network system

The electrical circuit shown in Fig. 2.9 will be used to illustrate the procedure used to determine the kinematic behavior of network systems. The basic steps of the method are as follows.

1. Node assignment:
   The first step in the modeling technique is to assign a node to the points of interconnection between the circuit components, as shown below.

   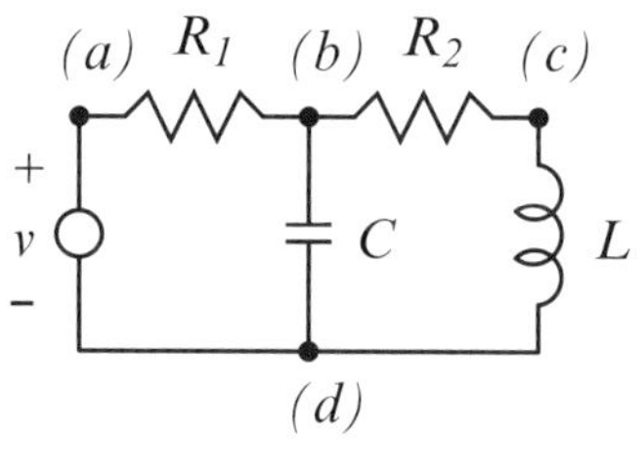

   Node assignment

   Thus, a node should be assigned to the terminals of each element in the network. In this case we have nodes a, b, c and d.
2. Flow assignment:
   A flow variable and a positive reference direction is assigned to each element of the system. For flow sources the direction of positive flow is assigned a priori. In the case of an effort source the positive direction of flow is taken from the negative terminal to the positive terminal of the source. For all other system elements (ideal resistors, capacitors and inductors) the reference direction of positive flow can be assigned arbitrarily.

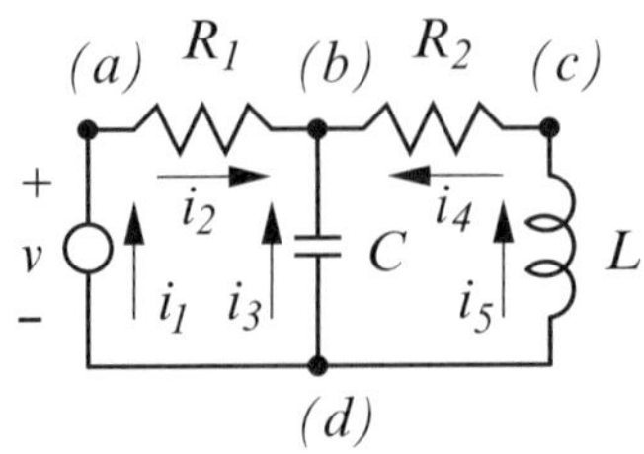

Flow assignment

In the figure shown above we have assigned the flow i_1 through the voltage source, i_2 through the resistor R_1, i_3 through the capacitor, i_4 through the resistor R_2, and i_5 through the inductor. The actual flow directions, for these passive elements, are determined by solving the dynamic equations of motion.

Annotating a network model as described above can lead to a large number of flow variables. However, due to the interconnections in the system not all these flow variables are independent. In fact if we assume continuity of flow at the nodes in the network then we can develop a set of equations that relate the flow variables in the model. If continuity of flow is satisfied then the sum of all the flows into a node must be zero. In the analysis of electrical circuits this continuity principle is known as *Kirchhoff's current law.*

As in mechanical systems it is important to know how many of the flow variables are independent, i.e., how many degrees of freedom exists in the network. To determine the number of independent flow variables in a network we can borrow some results from graph theory.

Networks contain entities called *nodes*, *branches*, *trees* and *chords*. These terms will be defined with the aid of the electrical circuit shown in Fig. 2.9. This circuit contains the following system elements; an effort (voltage) source, $v(t)$, resistors R_1 and R_2, a capacitor, C, and an inductor, L.

- Nodes:

  The nodes in the network have already been defined above. These are the points at the terminals of the system elements. Hence, a, b, c and d represent nodes in the network. Let N denote the number of nodes in the network. (Therefore, $N = 4$ for the circuit in Fig. 2.9).

- Branches:

  The system elements themselves are called branches in the network. Thus, $v(t)$, R_1, R_2, C and L are the branches of the network. Let B denote the number of branches in the network. (Therefore, $B = 5$ for the circuit in Fig. 2.9).

- Trees:

  Given a network, if we remove the minimum number of branches such that no closed loops remain, then the result is a tree. Thus, for the network shown in Fig. 2.9 removing the branches R_1 and R_2 will give the tree shown below.

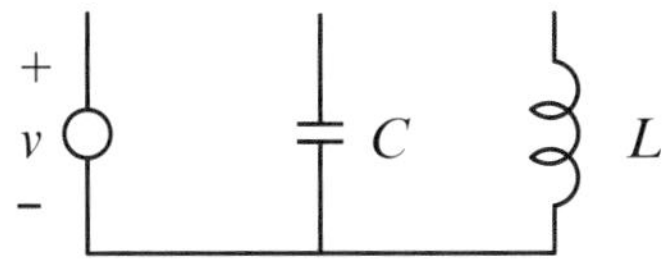

A tree

  Let T denote the number of branches in a tree, then it should be clear that for any network with N nodes,

$$T = N - 1.$$

- Chords:

  The chords are the branches that must be added to the tree so that the network is complete. Let C denote the number of chords in the network. Hence,

$$C = B - T = B - (N - 1) = B - N + 1.$$

The number of chords actually identify the number of independent closed-loops in the network. Since each closed-loop only requires one flow variable, C is equal to the number of independent flow variables in the network.

Applying this equation to the circuit in Fig. 2.9 shows that the system has $C = 5 - 4 + 1 = 2$ degrees of freedom. However, there are 5 flow variables assigned to the model. This indicates that there are three constraint equations that relate the 'excess' variables to the independent variables. Let i_1 and i_4 be the independent flow variables then, assuming continuity of flow at nodes a, b and c gives the three constraints

$$\begin{aligned}\psi_1 &= i_1 - i_2 = 0,\\ \psi_2 &= i_2 + i_3 - i_4 = 0,\\ \psi_3 &= i_5 - i_4 = 0.\end{aligned}$$

The flow constraints ψ_1, ψ_2 and ψ_3 can be used to eliminate the variables i_2, i_3 and i_5 from the model description.

References

1. H. Goldstein, *Classical Mechanics*, 2nd ed., Addison-Wesley, 1980.
2. D. T. Greenwood, *Principles of Dynamics*, 2nd ed., Prentice-Hall, 1988.
3. E. J. Haug, *Intermediate Dynamics*. Prentice-Hall, 1992.
4. J. L. Junkins and J. D. Turner, *Optimal Spacecraft Rotational Maneuvers*. Elsevier, 1986.
5. S. W. McCuskey, *Introduction to Advanced Dynamics*, Addison-Wesley, 1959.
6. L. Meirovitch, *Methods of Analytical Dynamics*, McGraw-Hill, 1970.
7. B. Paul, *Kinematics and Dynamics of Planar Machinery*, Prentice-Hall, 1979.
8. G. Sandor and A. Erdman, *Advanced Mechanism Design: Analysis and Synthesis, Volume 2*, Prentice-Hall, 1984.
9. A. Shabana, *Computational Dynamics*, John Wiley & Sons, 2001.
10. M. R. Spiegel, *Theory and Problems of Theoretical Mechanics*, Schaum's Outline Series, McGraw-Hill, 1967.
11. W. T. Thomson, *Introduction to Space Dynamics*, Dover Publications, 1986.
12. A. H. von Flotow and D. Rosenthal, *Multi-Body Dynamics; An Algorithmic Approach Based Upon Kane's Equations*. Lecture notes. MIT: 1990.
13. D. A. Wells, *Theory and Problems of Lagrangian Dynamics*, Schaum's Outline Series, McGraw-Hill, 1967.
14. J. H. Williams, Jr., *Fundamentals of Applied Dynamics*, John Wiley & Sons, 1996.
15. C. E. Wilson and J. P. Sadler *Kinematic and Dynamics of Machinery*, Harper Colins, 1991.

Problems

1. The cylindrical coordinates of a point are given by

$$\begin{bmatrix} \rho \\ \theta \\ \zeta \end{bmatrix} = \begin{bmatrix} 1.253 \\ \frac{4\pi}{9} \cos 5t \\ 1 - \cos 2t \end{bmatrix},$$

where t denotes the time. (a) Find the position of the point using rectangular coordinates. (b) Plot the position and velocity of the point in rectangular coordinates for $0 \le t \le 2\pi$. (c) Let $\lambda = \sqrt{x^2 + y^2}$, where x and y are the position of the point along the x-axis and y-axis respectively. Plot λ for $0 \le t \le 2\pi$.

2. The spherical coordinates of a point are given by

$$\begin{bmatrix} \rho \\ \theta \\ \phi \end{bmatrix} = \begin{bmatrix} 1 \\ 3e^{-t} \\ \frac{3\pi}{2}e^{-2t} \end{bmatrix},$$

where t denotes the time. (a) Find the position of the point using rectangular coordinates. (b) Plot the position and velocity of the point in rectangular coordinates for $0 \le t \le 2\pi$. (c) At what value of t does the point come to rest.

3. Consider a point that is free to move in a fixed frame that is assigned cylindrical coordinates show that the acceleration of the point is given by

$$\bar{a} = \frac{d}{dt}\bar{v} = (\ddot{\rho} - \rho\dot{\theta}^2)\,\hat{e}_r + (\rho\ddot{\theta} + 2\dot{\rho}\dot{\theta})\,\hat{e}_\theta + \ddot{\zeta}\,\hat{e}_z,$$

where ρ, θ, and ζ are described in Fig. 2.1b.

4. Consider a point that is free to move in a fixed frame that is assigned spherical coordinates show that the acceleration of the point is given by

$$\begin{aligned} \bar{a} = \frac{d}{dt}\bar{v} &= (\ddot{\rho} - \rho\dot{\phi}^2 - \rho\dot{\theta}^2\sin^2\phi)\,\hat{e}_\rho \\ &+ \quad (2\dot{\rho}\dot{\phi} + \rho\ddot{\phi} - \rho\dot{\theta}^2\sin\phi\cos\phi)\,\hat{e}_\phi \\ &+ \quad (2\rho\dot{\phi}\dot{\theta}\cos\phi + 2\dot{\rho}\dot{\theta}\sin\phi + \rho\ddot{\theta}\sin\phi)\,\hat{e}_\theta, \end{aligned}$$

where ρ, θ, and ϕ are described in Fig. 2.1c.

5. Shown below are two right-handed rectangular coordinate systems x-y-z (frame 0) and x_1-y_1-z_1 (frame 1). The origin of the x-y-z coordinate system is Q, and the origin of the x_1-y_1-z_1 coordinate system is Q_1. The point Q_1 has coordinate ${}^0R_{QQ_1} = [3 \quad 4.5 \quad 3.5]^T$ with respect to the x-y-z coordinate system, and the angle $\theta = 28.51°$.

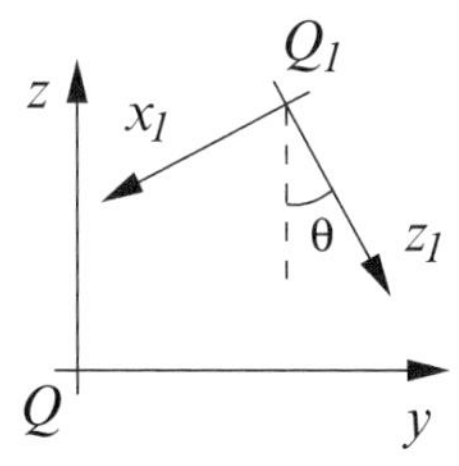

(a) Find the direction cosine matrix, ${}^0A_1$, that relates the x_1-y_1-z_1 coordinate system to the x-y-z coordinate system. (b) A point P has coordinate ${}^1r_{Q_1P} = [0.1, \quad 0.2, \quad -0.4]^T$ with respect to the x_1-y_1-z_1 coordinate system. Find the coordinate of the point in the x-y-z coordinate system.

6. Show that ${}^0A_1$ in equation (2.16) and ${}^0A_1$ in equation (2.17) are the same if the rotations are small. That is, α, β, and γ are such that $\cos\alpha = \cos\beta = \cos\gamma = 1$. Also, $\sin\alpha = \alpha$, $\sin\beta = \beta$, $\sin\gamma = \gamma$. Moreover, terms involving products of α, β, and γ can be neglected.

7. Suppose the orientation of the rotating frame is described by ${}^0A_1$ in equation (2.16). Show that (i) $({}^0A_1)^T({}^0\dot{A}_1)$ is skew-symmetric, and (ii) verify that the components of the angular velocity vector satisfy equation (2.18).

8. Show that the inverse of equation (2.18) is given by equation (2.19).

9. Show that the direction cosine matrix associated with the Z_α-X_β-Z_γ Euler angles is given by

$$
{}^0A_1 = \begin{bmatrix} c_\alpha c_\gamma - s_\alpha c_\beta s_\gamma & -c_\alpha s_\gamma - s_\alpha c_\beta c_\gamma & s_\alpha s_\beta \\ s_\alpha c_\gamma + c_\alpha c_\beta s_\gamma & -s_\alpha s_\gamma + c_\alpha c_\beta c_\gamma & -c_\alpha s_\beta \\ s_\beta s_\gamma & s_\beta c_\gamma & c_\beta \end{bmatrix},
$$

where $c_\alpha = \cos\alpha$, $s_\alpha = \sin\alpha$, etc. Also, show that the angular velocities of the x_1, y_1 and z_1 axes are related to the Euler angles via the equation

$$
\begin{bmatrix} \omega_1 \\ \omega_2 \\ \omega_3 \end{bmatrix} = \begin{bmatrix} s_\beta s_\gamma & c_\gamma & 0 \\ s_\beta c_\gamma & -s_\gamma & 0 \\ c_\beta & 0 & 1 \end{bmatrix} \begin{bmatrix} \dot{\alpha} \\ \dot{\beta} \\ \dot{\gamma} \end{bmatrix}.
$$

10. Consider a point P that is moving in the x_1-y_1-z_1 rectangular coordinate system as shown in Fig. 2.2. Suppose that the x_1-y_1-z_1 frame can translate and rotate with respect to the fixed rectangular coordinate system x-y-z. Then show that the acceleration of P with respect to the fixed frame is given by

$$
{}^0\bar{a} = {}^0\ddot{\bar{R}} + {}^1\ddot{\bar{r}} + {}^1\bar{\alpha} \times {}^1\bar{r} + {}^1\bar{\omega} \times ({}^1\bar{\omega} \times {}^1\bar{r}) + 2({}^1\bar{\omega} \times {}^1\dot{\bar{r}}),
$$

where ${}^0\ddot{\bar{R}} = \ddot{X}\,\hat{i} + \ddot{Y}\,\hat{j} + \ddot{Z}\,\hat{k}$ is the acceleration of Q_1 relative to Q, ${}^1\ddot{\bar{r}} = \ddot{x}_1\,\hat{i}_1 + \ddot{y}_1\,\hat{j}_1 + \ddot{z}_1\,\hat{k}_1$ is the acceleration of P relative to Q_1, ${}^1\bar{\alpha} = \alpha_1\,\hat{i}_1 + \alpha_2\,\hat{j}_1 + \alpha_3\,\hat{k}_1$ is the angular acceleration of the x_1-y_1-z_1 frame, and all other terms are defined in Section 2.1.4.

11. The diagram below shows a fixed rectangular coordinates system x-y-z with origin at Q. The link QB rotates about point Q with angular velocity ω_Z that is directed along the z-axis. Moreover, QB remains in the x-y plane. The link BC rotates about point B with angular velocity ω_B that is directed along a line perpendicular to QB. In addition, BC remains in the plane formed by the z-axis and the line QB. The thin disk shown

rotates about point C with angular velocity ω_C that is directed along the line BC. The radial direction of the disk is perpendicular to the line BC. The point P is a point on the edge of the disk.

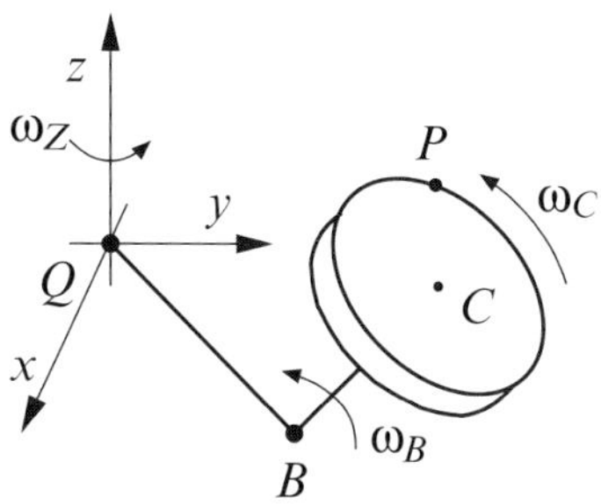

The length of QB is l_1, the length of BC is l_2, and the radius of the disk is ρ. Develop a set of equations to determine the position and velocity of point P starting from some suitable initial position.

12. For the mechanisms shown below determine; (i) the number and types of links, (ii) the number and types of joints, and (iii) the number of degrees of freedom.

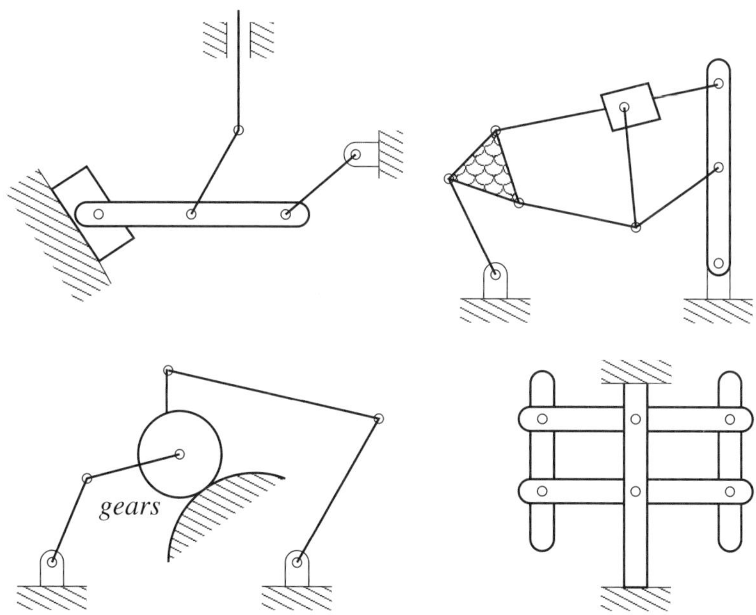

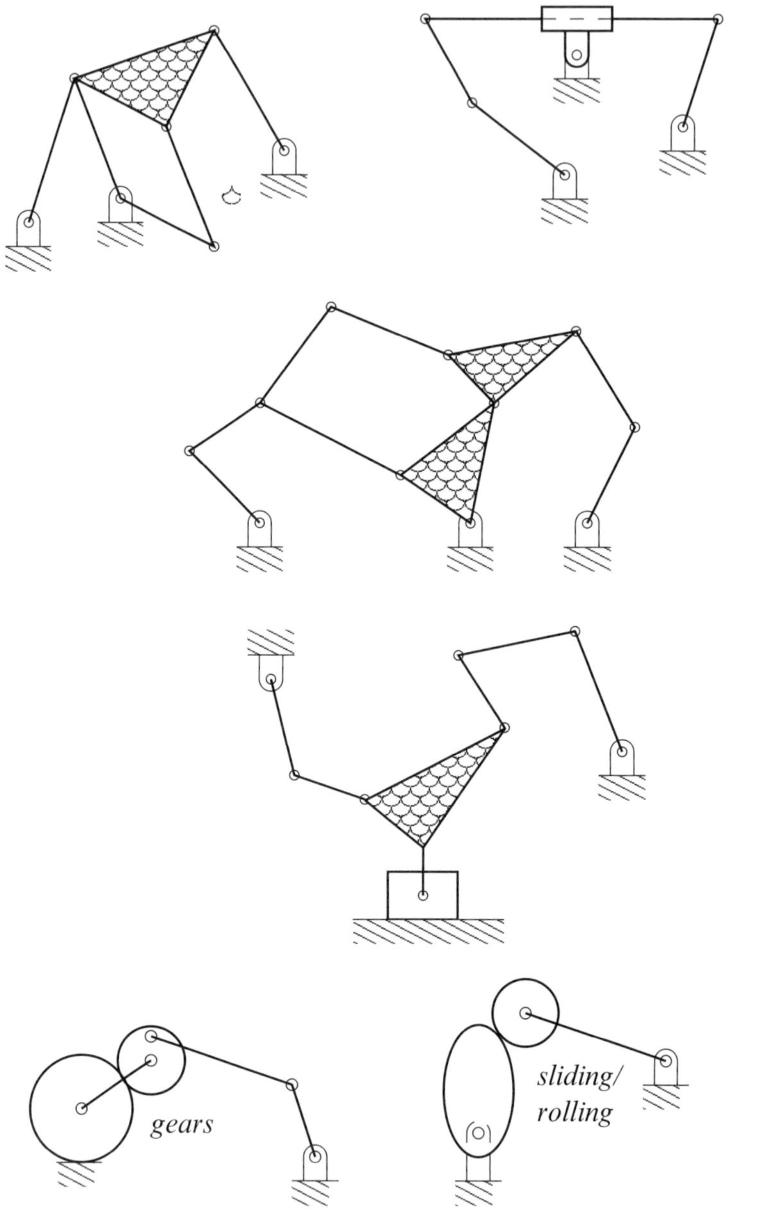
gears
sliding/
rolling

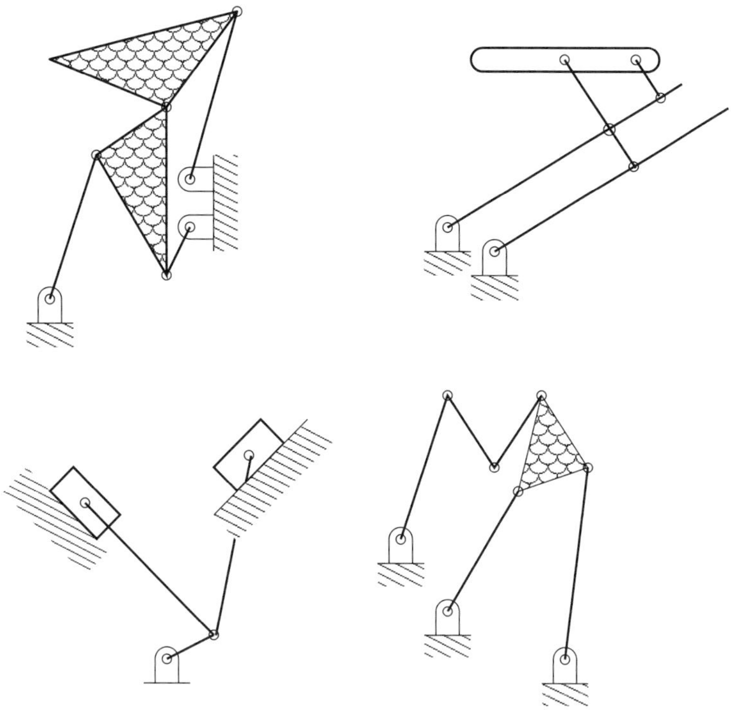

13. Find the number of degrees of freedom for the spatial mechanisms shown below.

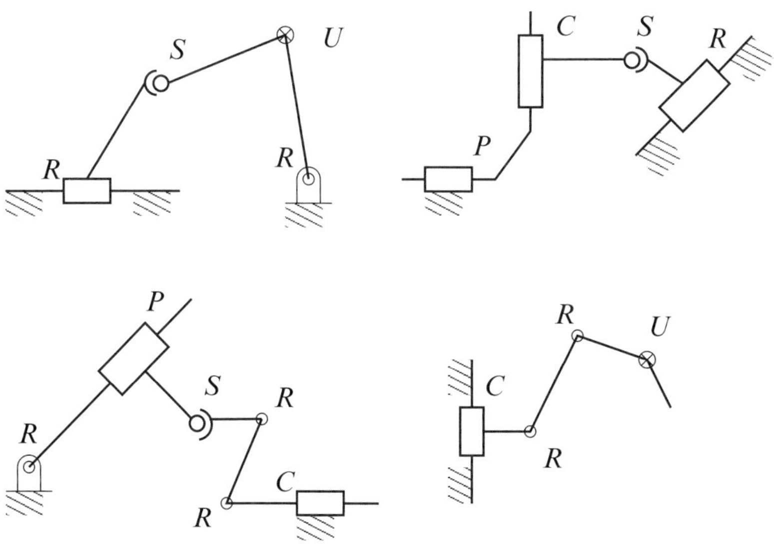

14. For the networks shown below determine; (i) assign nodes to the system, (ii) determine the number of branches, (iii) determine the number of

independent loops, (iv) assign flows to the branches, and (iv) write flow constraint equations for the nodes.

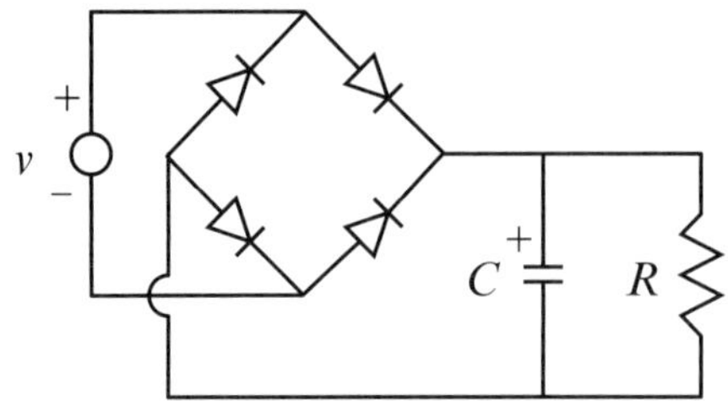

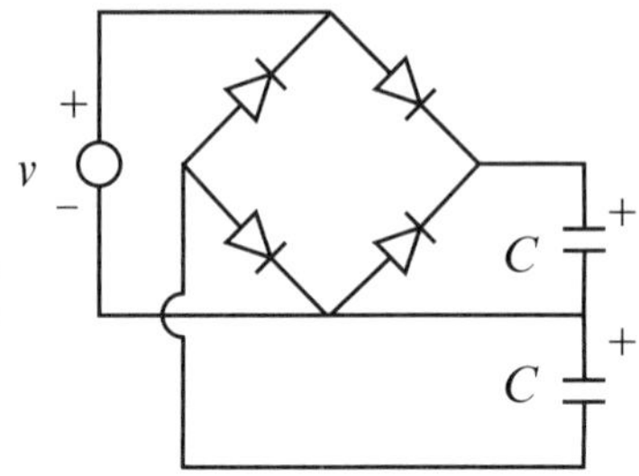

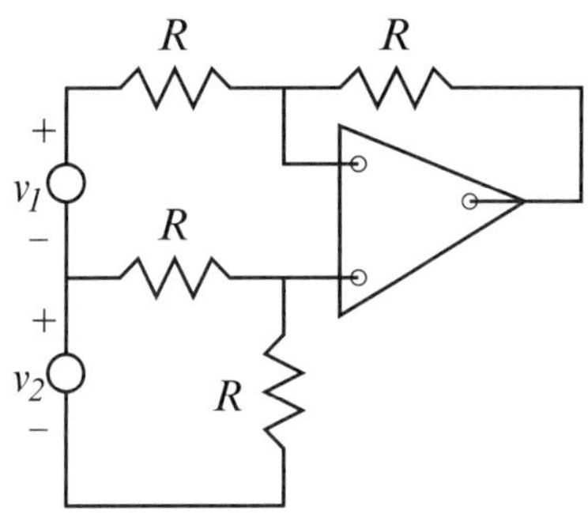

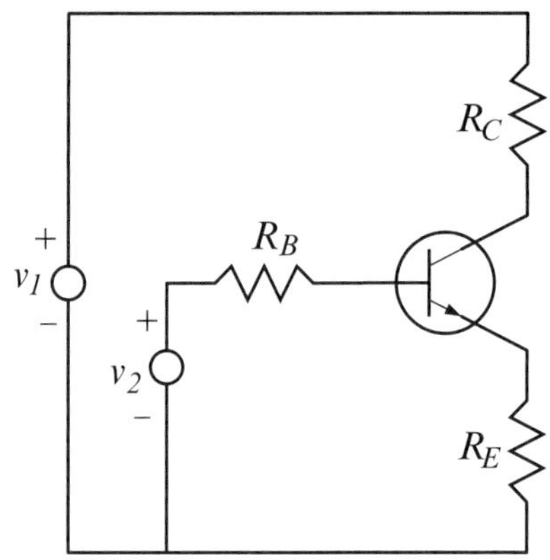

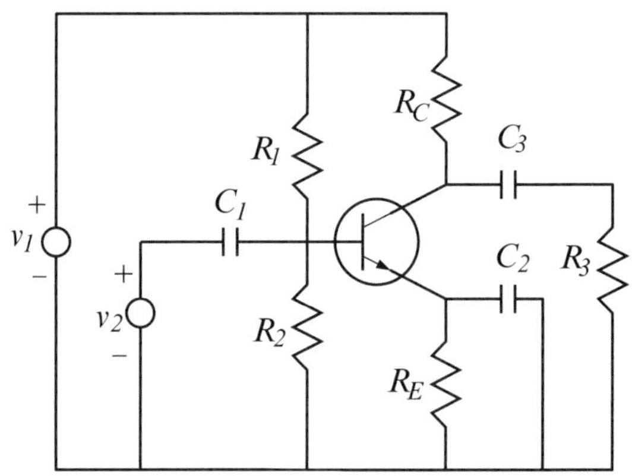

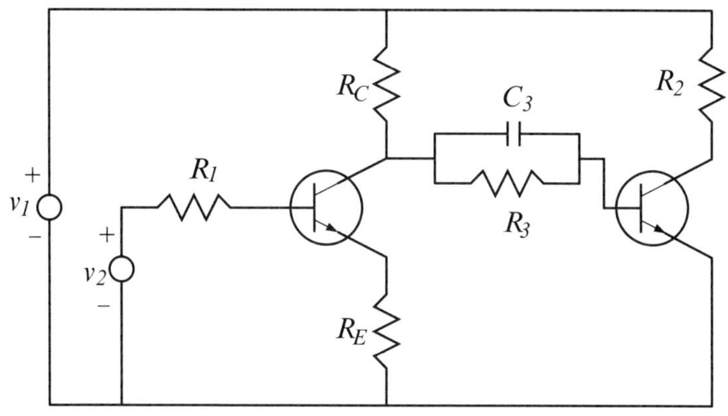

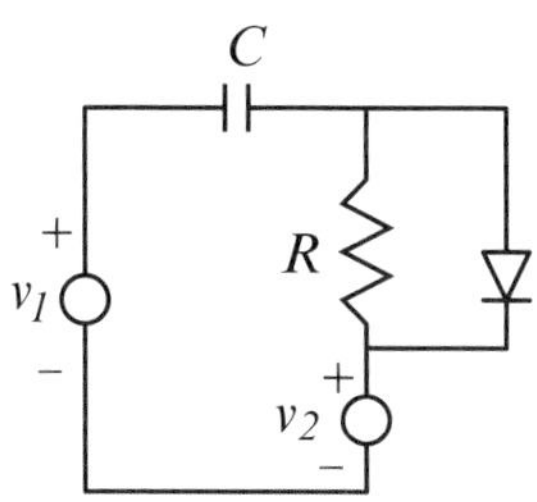

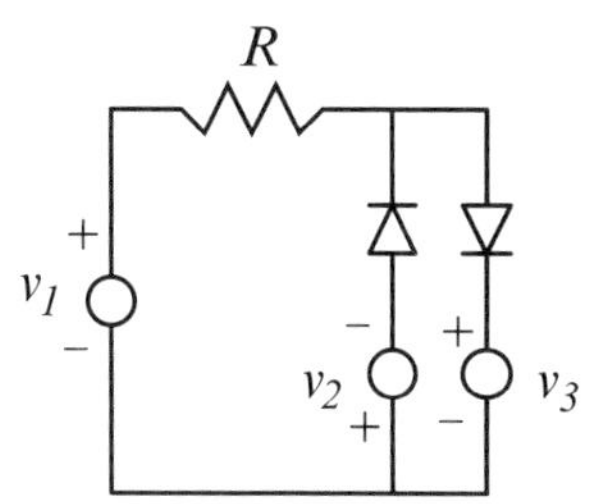

15. In the system shown below x-y-z represents the fixed rectangular coordinate system. The disk shown is attached to the shaft AB, and the shaft has a constant angular velocity, ω_z, about the fixed z-axis. The principal axes of inertia for the disk are aligned with the x_1-y_1-z_1 rectangular coordinate system, which is embedded in the disk. The z_1-axis make a constant angle α with the z-axis. Find an expression for the angular velocity of the disk. Specifically, determine $\bar{\omega} = \omega_1 \hat{i}_1 + \omega_2 \hat{j}_1 + \omega_3 \hat{k}_1$, where $\hat{i}_1$, $\hat{j}_1$ and $\hat{k}_1$ are the unit vectors in the x_1, y_1 and z_1 directions respectively.

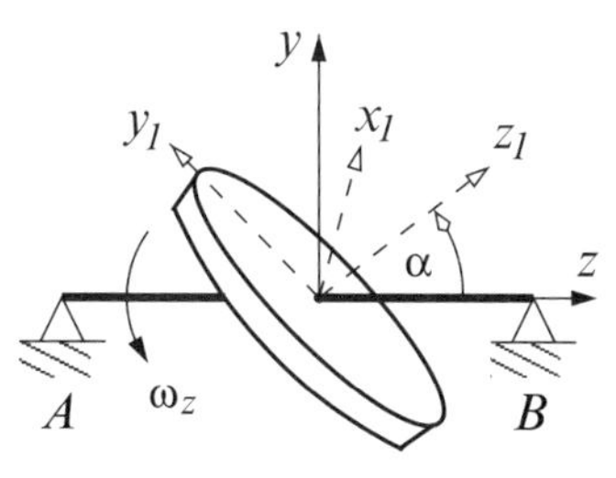

16. The disk shown below has radius r, and it rolls without slipping on the horizontal plane. The angular velocity of the disk is $\omega = 2\pi$ rad/s.

  a. Compute and plot the position, velocity and acceleration of a point P that is on the edge of the disk.
  b. What is the velocity and acceleration of P when P is in contact with the ground?

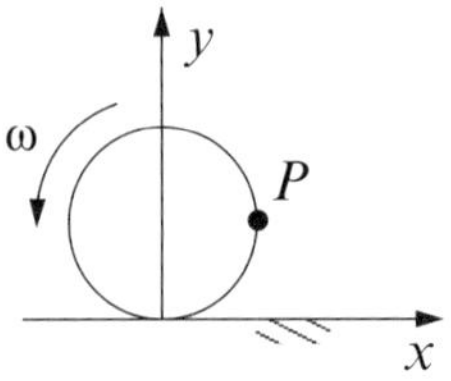

17. A disk with radius r rolls with out slipping on the inside of a cylinder that has a radius R. Determine the position and velocity the the point C on the edge of the disk, if the disk rotates around the cylinder with constant angular velocity $\dot{\theta} = 2\pi$ radians/second.

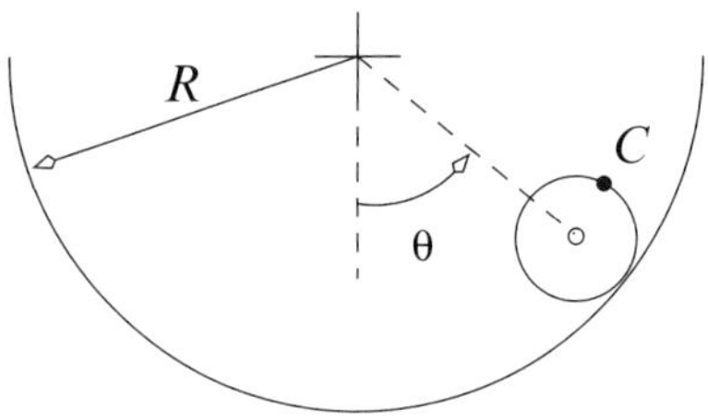

18. Develop the equations that determine the position and velocity of the point C on the robots shown below. The robot shown in (a) moves in a plane and has a revolute joint and a prismatic joint. The robot shown in (b) has a spherical joint at A an a revolute joint at B.

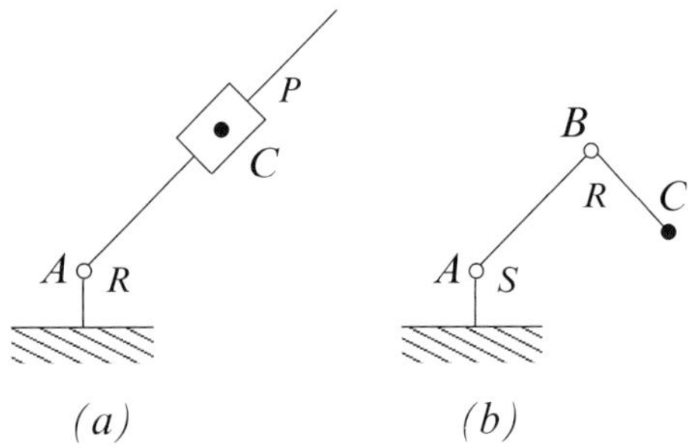

19. The figure below shows a planar R-P robot. The point A on the end effector moves from $(x = 3,\ y = -3)$ to $(x = 3,\ y = 3)$ along a straight line in 3 seconds. The motion is such that A starts at rest and ends at rest.

   a. Select an appropriate trajectory for the coordinates $x_A(t)$, and $y_A(t)$, for $0 \leq t \leq 3$.
   b. Given $x_A(t)$ and $y_A(t)$, compute $r(t)$, $\dot{r}(t)$, $\ddot{r}(t)$, $\theta(t)$, $\dot{\theta}(t)$ and $\ddot{\theta}(t)$.
   c. Plot these variables versus time.

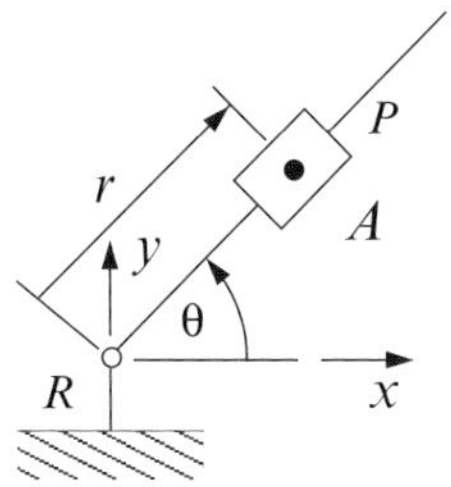

20. The fourbar mechanism shown below has the following proportions: $AD = 2.5$, $AB = 1$, $BC = 4$, $DC = 2$, $BE = 2$, and $EF = 0.75$. Note that EF is perpendicular to BC. The crank AB has an angular velocity of 2π radians/second counterclockwise. Find the equations required to compute the position and velocity of the point F. Plot the path of the point F, and its velocity.

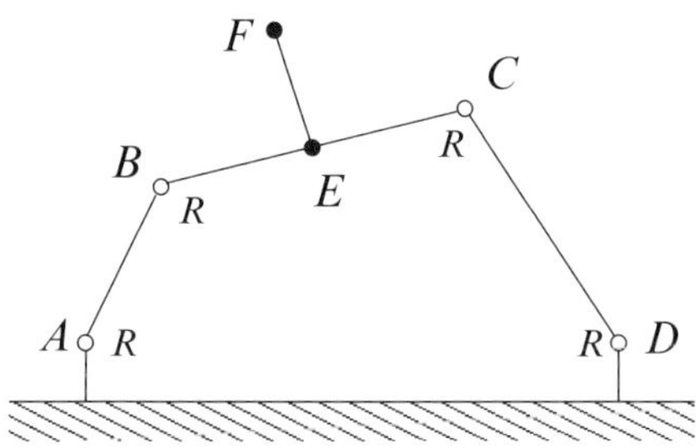

21. The quick return mechanism shown below has revolute joints at A, B, C, E, and F, and prismatic joints at D and G. The proportions of the mechanism are as follows: $AB = 4$ cm, $BC = 2$ cm, $AE = 8$ cm, and $EF = 3$ cm. If the crank, BC, has angular velocity 2π radians/second find the equations required to compute the the position and velocity of the point F as a function of the crank angle. Plot the configuration of the mechanism for crank angles $0, 0.2\pi, 0.4\pi, \cdots, 1.8\pi$ radians. Also, plot the velocity of F, the angular velocity of AE, and the angular velocity of EF.

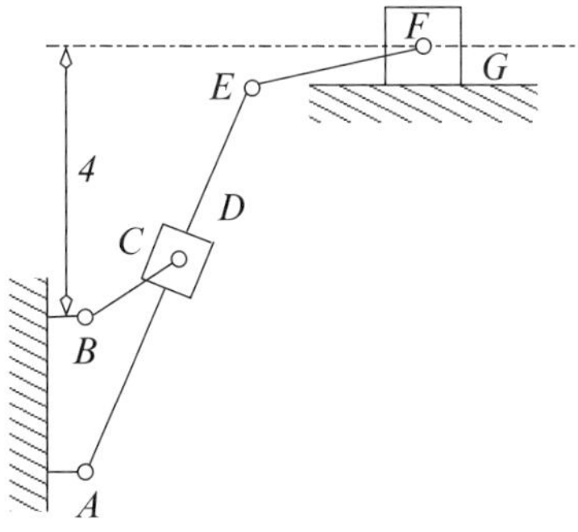

22. The mechanism shown below has revolute joints at A, B, and C, and the disk rolls on the ground without slipping at E. Find the equations necessary to determine the configuration of the mechanism at all times. If the crank has an angular velocity 2π radians/second, determine the position and velocity of the points C and D (the center of the disk). The mechanism has the following proportions: $QA = 0.5$, $AB = 1.0$, $BC = 5.0$, $CD = 2.5$, and $r = 3$.

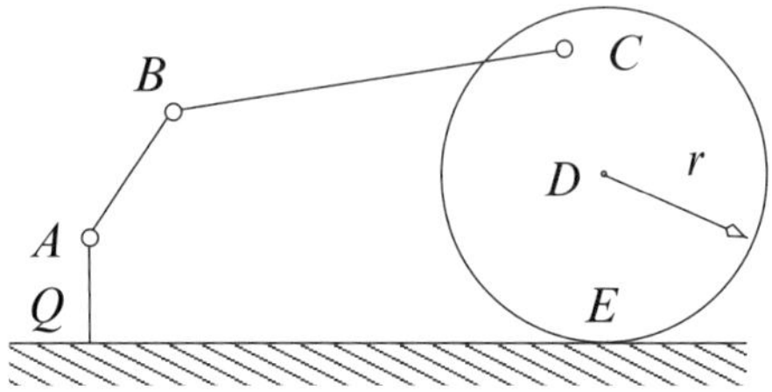

23. The mechanism shown below is an R-C robot. The link AD is connected to the ground at A via a revolute joint. Moreover, the rotation at A takes place about an axis that is perpendicular to the page. The link BE is connected to the link AB via a cylindrical joint at B. The link BE can translate along the line AD, and rotate about the line AD. Find a set of equations to determine the position and velocity of the point E.

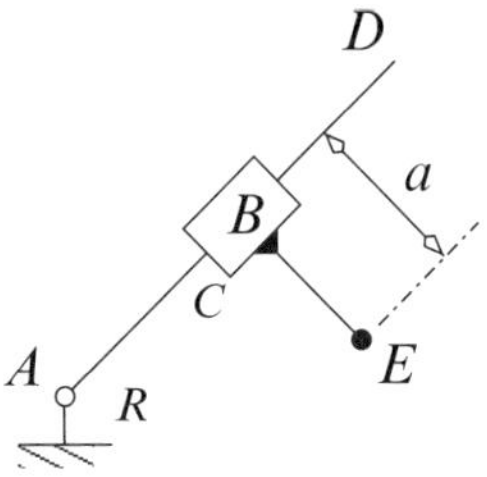

24. Determine the equations required to compute the position and velocity of the point D on the mechanism shown here. The slider A stay in contact with the wall while the disk CB rolls without slipping. The link AB is attached to the slider via a revolute joint at A, and it is attached to the disk via a revolute joint at B. Use the following dimensions: $AB = l$, $AD = d$, and $CB = r$ is the radius of the disk. The angle α is fixed.

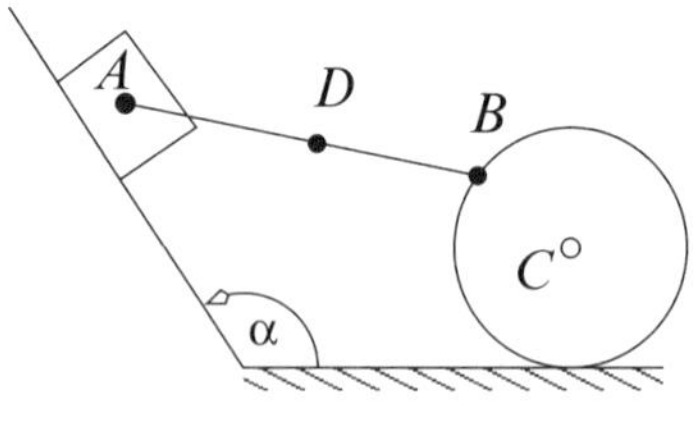

Chapter 3
Lagrange's Equation of Motion

This chapter develops Lagrange's equation of motion for a class of multidiscipline dynamic systems. To derive Lagrange's equation we utilize some concepts from analytical dynamics, and the first law of thermodynamics. By carrying out the development using the fundamental variables it is clear that the results obtained are applicable to all the engineering disciplines described in Chapter 1.

Section 3.1 introduces the concepts of generalized displacement, virtual displacement and virtual work. In Section 3.2 Lagrange's equation of motion is derived starting from the first law of thermodynamics. In Section 3.3 various examples are used to illustrate the application of Lagrange's equation to mechanical, electrical, fluid, and multidiscipline systems.

3.1 Analytical Dynamics

As discussed in Chapter 1 dynamic systems can be described as an assemblage of inductors, capacitors, resistors, constraint elements and sources. Associated with each of these elements is the set of fundamental variables displacement, q, flow, f, effort, e, and momentum, p. From Paynter's diagram (Fig. 1.4) it can be observed that all four variables are not all required to determine the state of the system at any time. In fact only two of the fundamental variables are required to determine the states of the system since, the other two variables can be determined using the system properties and the differential/integral relationship between the variables.

In Lagrangian dynamics the displacement and flow variables are used to describe the system behavior. In the Hamiltonian description of dynamic systems the momentum and displacement variables are used to describe the system. The bond graph and linear graph description of dynamic systems use the effort and flow variables to describe the system behavior.

B. Fabien, *Analytical System Dynamics: Modeling and Simulation*,
DOI 10.1007/978-0-387-85605-6_3,

3.1.1 Generalized variables

The number of displacement and flow variables that can be assigned to a dynamic system can be quite large. However, for each system there is a minimum number of displacement variables that are required to uniquely determine the state or configuration of the system. This minimum set of displacement variables are called the *generalized displacements*. The corresponding set of flow variables are called the *generalized flows*. The number of generalized displacement variables used to the describe the system is equal to the number of *degrees of freedom*. (See Section 2.2 and Section 2.3.)

Example 3.1.

Consider the spring, mass damper system shown in Figure 3.1a. Figure 3.1b shows the displacement coordinates assigned to the system components. Assigned to the mass is the displacement, x_1. The left and right end of the

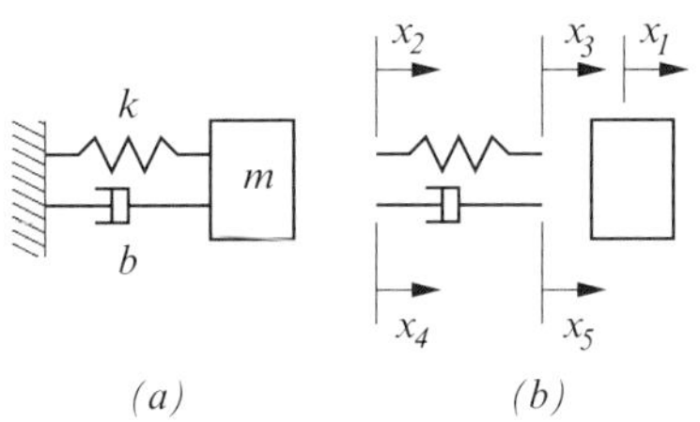

Fig. 3.1 Spring-mass-damper

spring are assigned displacement variables x_2, and x_3, respectively. The left and right end of the damper are assigned displacement variables x_4, and x_5, respectively Since the left end of the spring and damper are fixed, it is clear that $x_2 = x_4 = 0$. Also, since the right ends of the spring and damper are attached to the mass we have $x_5 = x_3 = x_1$. Thus, the system can be described using the single displacement variable x_1, and the the corresponding flow variable v_1. This system has 1 degree of freedom with x_1 being the the generalized displacement and v_1 the corresponding generalized flow.

Example 3.2.

Figure 3.2a shows a resistor, R, an inductor, L, and a capacitor, C, in series with a voltage source, v. Figure 3.2b shows the flow variables assigned to the system components. The current i_1 is assigned to the voltage source, i_2 to the resistor, i_3 to the inductor, and i_4 to the capacitor. However, Kirchhoff's

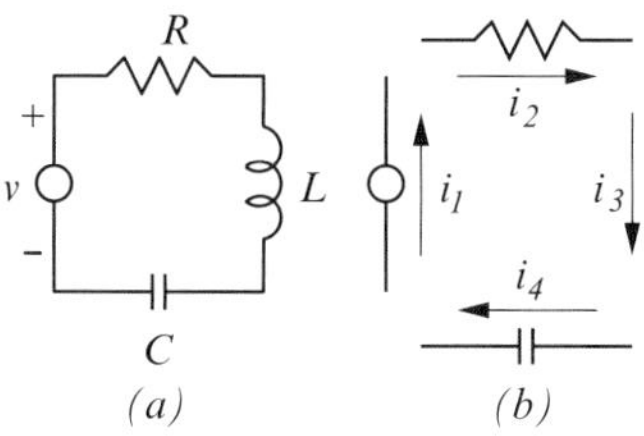

Fig. 3.2 Resistor-inductor-capacitor

current law states that the currents flowing into a node must sum to zero. As a result, $i_1 = i_2$, $i_2 = i_3$, $i_3 = i_4$, and $i_4 = i_1$. Thus, the system has 1 degree of freedom (and only one flow variable is required). We can select i_1 as the generalized flow variable, with $q_1 = \int i_1 \, dt$ as the generalized displacement.

Example 3.3.

Figure 3.3a shows a simple pendulum. In this system the mass, m, is attached to an inertialess rod of length l. The rod rotates freely about point O.

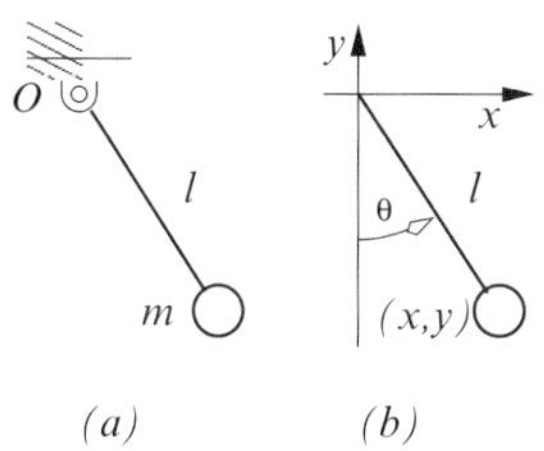

Fig. 3.3 Simple pendulum

The configuration of the system can be described using the displacement variables x and y. However, these variables are not independent. Since the rod is rigid we must satisfy the displacement constraint

$$\phi = x^2 + y^2 - l^2 = 0. \qquad (a)$$

This displacement constraint can be used to eliminate x or y as follows:

$$x = \pm\sqrt{l^2 - y^2} \quad (b), \qquad y = \pm\sqrt{l^2 - x^2} \quad (c).$$

Unfortunately, the transformation (b) gives an ambiguous result when $y = 0$, and the transformation (c) gives an ambiguous result when $x = 0$.

Alternatively, the angle θ can be used to establish that

$$\phi_1 = x - l\sin\theta = 0 \quad (d), \qquad \phi_2 = y + l\cos\theta = 0 \quad (e).$$

Thus, if θ is known we can unambiguously determine x and y using the constraints ϕ_1 and ϕ_2.

Clearly, this system has 1 degree of freedom, and we can select θ as the generalized displacement, with $\dot{\theta}$ as the generalized flow.

Example 3.4.

Figure 3.4a shows a hydraulic system consisting of a pump that delivers fluid at a desired pressure to a long pipe. The pipe is connected to a storage tank which is in turn connected to a valve. A schematic of a model for the system is shown in Fig. 3.4b. The model has an effort source P for the pump, the inertia of the fluid in the long pipe is model using L_f, the capacitance of the tank is included in C_f, and the resistance of the valve is modeled using R_f. The flow variables that describe the configuration of the system are Q_1

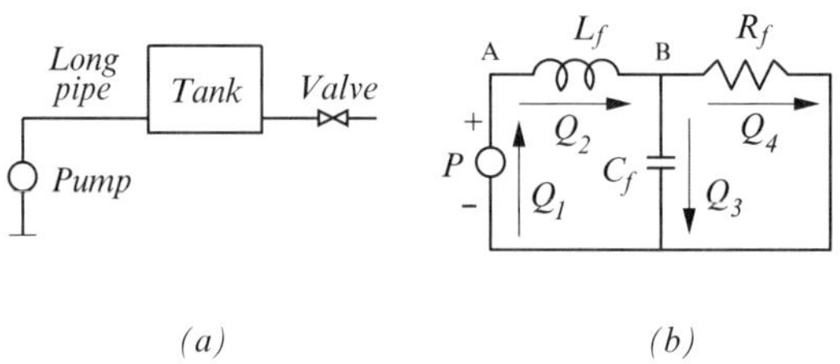

Fig. 3.4 Hydraulic system

the volume flow through the pump, Q_2 the volume flow through the pipe, Q_3 the volume flow stored in the tank, and Q_4 the volume flow through the valve.

These four flow variables are not independent. Since there are no leaks in the system, the continuity of flow requires that at node A,

$$\psi_1 = Q_1 - Q_2 = 0, \tag{a}$$

and at node B,

$$\psi_2 = Q_2 - Q_3 - Q_4 = 0. \tag{b}$$

Equations (a) and (b) represent flow constraints that must be satisfied. These two equations can be used to eliminate two of the flow variables. Thus, the system has 2 degrees of freedom. (Only two of the four variables are independent.) Suppose we select Q_1 and Q_4 as the generalized flow variables. Then, from (a) and (b) we get $Q_2 = Q_1$ and $Q_3 = Q_1 - Q_4$, respectively.

From the examples above it can be observed that a system can be described by N configuration coordinates say, $u_1, u_2, \cdots, u_N$, that are different form the n generalized displacements $q_1, q_2, \cdots, q_n$. Here, n is the number of degrees of freedom, and $N \geq n$. Moreover, the n generalized displacements are all independent.

3.1.2 Virtual displacements

A virtual displacement is a small displacement of the system that is (i) consistent with the constraints, and (ii) takes place contemporaneously. It is the second requirement that distinguishes virtual displacements from differential displacements. (Here, a differential displacement a small displacement of the system that is consistent with the constraints.)

To clarify the difference between differential displacements and virtual displacements consider the following examples.

Example 3.5.

Figure 3.5a shows a simple pendulum with configuration displacements x and y, and generalized displacement θ. From Example 3.3 we know that the system satisfies the two displacement constraints

$$\phi_1 = x - l \sin\theta = 0, \qquad \phi_2 = y + l\cos\theta = 0.$$

Now, suppose the system undergoes a small displacement $d\theta$ that takes place

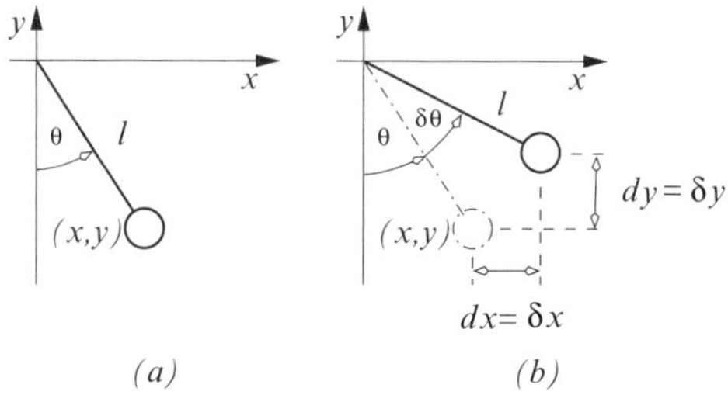

Fig. 3.5 Virtual displacement: simple pendulum

over the time interval dt as shown in Fig. 3.5b. Then, the differential of the constraints gives

$$d\phi_1 = dx - (l\cos\theta)d\theta = 0, \qquad d\phi_2 = dy - (l\sin\theta)d\theta = 0.$$

Hence, for a differential displacement $d\theta$, the configuration displacements that are consistent with the constraints are $dx = (l\cos\theta)d\theta$ and $dy = (l\sin\theta)d\theta$.

A virtual displacement is the same as a differential displacement but with time held fixed, i.e., $dt = 0$. Let $\delta\theta$ denote a virtual displacement of the generalized displacement θ, δx denote a virtual displacement of x, and δy denote a virtual displacement of y. Then the *variation* of the constraints gives

$$\delta\phi_1 = \delta x - (l\cos\theta)\delta\theta = 0, \qquad \delta\phi_2 = \delta y - (l\sin\theta)\delta\theta = 0.$$

Thus, in this case the virtual displacements and the differential displacements are the same. Also, note that a variation is similar to a differential but with time held fixed.

Example 3.6.

Figure 3.6a shows a particle that is constrained to move along a ramp that has a fixed incline α. The ramp is free to move along the x-axis, and its position is given by $X = A\sin\omega t$, where A and ω are constants, and t is the time. The configuration displacements for the system are x and y, and the generalized displacement is r. From the geometry of the system we get the

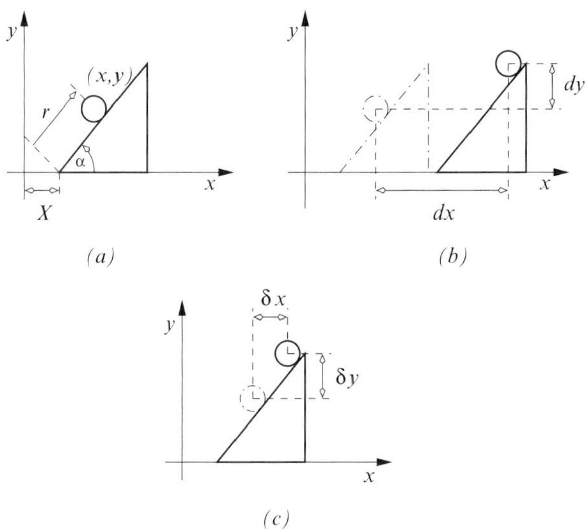

Fig. 3.6 Virtual displacement: moving ramp

two displacement constraints

$$\phi_1 = x - A\sin\omega t - r\cos\alpha = 0,$$
$$\phi_2 = y - r\sin\alpha = 0.$$

Now, suppose the system undergoes a small displacement, dr, that takes place over the time interval dt as shown in Fig. 3.6b. Then, the differential of the constraints gives

$$\begin{aligned} d\phi_1 &= dx - (A\omega \cos \omega t)dt - dr \cos \alpha = 0, \\ d\phi_2 &= dy - dr \sin \alpha = 0. \end{aligned}$$

On the other hand if the system undergoes a virtual displacement, δr, then, the variation of the constraints gives

$$\begin{aligned} \delta\phi_1 &= \delta x - \delta r \cos \alpha = 0, \\ \delta\phi_2 &= \delta y - \delta r \sin \alpha = 0. \end{aligned}$$

This variation is illustrated in Fig. 3.6c. In this case the differential displacement dx, and the virtual displacement δx are not the same. This is due to the fact that the constraint surface is moving, i.e., the constraint depends explicitly on time. Note also that the variation of the constraints can be obtained from the differential of the constraints by setting $dt = 0$.

3.1.3 Virtual work

The *virtual work* is the work done by the efforts in carrying the system through a virtual displacement, i.e.,

$$\delta \mathcal{W} = \sum_{i=1}^{n} e_i \, \delta q_i. \tag{3.1}$$

Here, $\delta \mathcal{W}$ is the virtual work, δq_i are the virtual displacements, e_i are the applied efforts, and $i = 1, 2, \cdots, n$ with n being the number of degrees of freedom. Note that q_i are the generalized displacements, and e_i are the corresponding *generalized efforts*. The virtual work is also equal to the work done by the flows in carrying the system through a virtual momentum, i.e.,

$$\delta \mathcal{W} = \sum_{i=1}^{n} f_i \, \delta p_i, \tag{3.2}$$

where δp_i are the virtual momentum, and f_i are the flows. Equations (3.1) and (3.2) follow from the definitions of the increment in work, i.e., equations (1.3) and (1.4), respectively.

An important consequence of the definition of virtual work is that constraint efforts do not contribute to the virtual work. This is most easily illustrated via an example.

Example 3.7.

Figure 3.7a shows a mass, m, that is free to slide on a smooth ramp under the influence of the acceleration due to gravity, g. The ramp is fixed and has a fixed slope with angle α. A free body diagram for the mass is

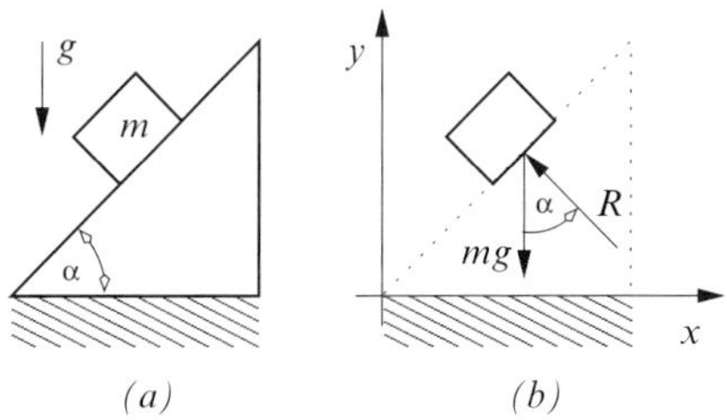

Fig. 3.7 Virtual work: constraint effort

shown in Fig. 3.7b. Here, the weight acts downward, and the reaction force (i.e., the constraint effort), R, acts perpendicular to the ramp as shown. The configuration displacements for the system are x and y.

The virtual work done by all the efforts acting on the mass is

$$\delta\mathcal{W} = -mg\,\delta y + R\cos\alpha\,\delta y - R\sin\alpha\,\delta x. \tag{a}$$

The mass is constrained to move along the ramp. Thus,

$$\phi = y - x\tan\alpha, \tag{b}$$

is a displacement constraint that must be satisfied. The variation of this constraint gives

$$\delta\phi = \delta y - \delta x\tan\alpha = 0. \tag{c}$$

Using (c) in (a) gives

$$\delta\mathcal{W} = \left[-mg + R\cos\alpha - R\frac{\sin\alpha}{\tan\alpha}\right]\delta y$$
$$\delta\mathcal{W} = -mg\,\delta y$$

Therefore, the constraint force, R, does no virtual work on the system.

The result obtained in Example 3.7 can be shown to be true in general. In fact the constraint efforts are all perpendicular to the constraint surface, while the virtual displacements take place tangent to the constraint surface. Hence, the constraint efforts will make no contribution to the virtual work.

In many cases the applied efforts are given in terms of the configuration displacements instead of the generalized displacements. Such situations require that we use the relationship between the configuration and generalized displacements to find expressions for the generalized efforts. In particular, suppose u_i are the configuration displacements, and E_i are the corresponding configuration efforts, for $i = 1, 2, \cdots, N$. Let q_i be the generalized displacements, and e_i be the corresponding generalized efforts. Then, there are transformation equations of the form

$$u_i = U_i(q_1, q_2, \cdots, q_n) \tag{3.3}$$

that relate the configuration and generalized displacements (see the examples above).

The virtual work done by the applied efforts satisfies

$$\delta\mathcal{W} = \sum_{i=1}^{N} E_i\,\delta u_i = \sum_{i=1}^{n} e_i\,\delta q_i.$$

But from (3.3) it can be seen that the virtual displacements δu_i satisfy

$$\delta u_i = \sum_{j=1}^{n} \frac{\partial U_i}{\partial q_j}\,\delta q_j.$$

Hence,

$$E_i \delta u_i = E_i \sum_{j=1}^{n} \frac{\partial U_i}{\partial q_j}\,\delta q_j,$$

which yields

$$e_i = \sum_{j=1}^{N} E_j \frac{\partial U_j}{\partial q_i}. \tag{3.4}$$

We will make use of this result in several of the examples to follow.

3.2 Lagrange's Equations of Motion

In this section we develop Lagrange's equation of motion for unconstrained systems. By an unconstrained system we mean that the displacements used to describe the system are independent hence, they represent a set of generalized displacements. Constrained systems will be discussed in Chapter 4. [1]

The first law of thermodynamics can be written as

[1] An alternate derivation of Lagrange's equation of motion, for multidiscipline systems, can be developed by invoking Hamilton's principle of least action. See for example, Wellstead (1979); White and Woodson (1959); and Williams (1996).

$$dE = \check{d}\mathcal{W} + \check{d}\mathcal{Q}, \tag{3.5}$$

where dE is the change in energy of the system, $\check{d}\mathcal{W}$ is the increment in work done on the system, and $\check{d}\mathcal{Q}$ is the heat added to the system. In (3.5) the system undergoes a differential displacement that takes place over the time interval dt. If we consider a virtual displacement instead of a differential displacement, equation (3.5) becomes

$$\delta E = \delta\mathcal{W} + \delta\mathcal{Q}. \tag{3.6}$$

The virtual work done by the efforts (and flows) for various energy discipline is given in Table 3.1. Here, it can be seen that the virtual work done by the temperature in thermal systems is $T\,\delta S$ which is equal to the heat added to the system, i.e., $\delta\mathcal{Q} = T\,\delta S$. Thus, by using the fundamental variables described in Chapter 1, equation (3.6) can be rewritten as

$$\delta E = \delta\mathcal{W}, \tag{3.7}$$

where we assume that $\delta\mathcal{Q} = T\,\delta S$ is already included in $\delta\mathcal{W}$.

Discipline	Virtual Work	
	$e\,\delta q$	$f\,\delta p$
Mechanical Translation	$F\,\delta x$	$v\,\delta p$
Mechanical Rotation	$\tau\,\delta\theta$	$\omega\,\delta H$
Electrical	$v\,\delta q$	$i\,\delta\lambda$
Fluid	$P\,\delta V$	$Q\,\delta\Gamma$
Thermal	$T\,\delta S$	–

Table 3.1 Virtual Work

Define the following system variables:

$$\begin{aligned}
q &= [q_1, q_2, \cdots, q_n]^T, \text{ the generalized displacement,} \\
f &= [f_1, f_2, \cdots, f_n]^T, \text{ the generalized flow,} \\
p &= [p_1, p_2, \cdots, p_n]^T, \text{ the generalized momentum, and} \\
e &= [e_1, e_2, \cdots, e_n]^T, \text{ the generalized effort,}
\end{aligned}$$

where n is the number of degrees of freedom.

The total energy of the system is

$$E = T(p, q, t) + V(q), \tag{a}$$

where the kinetic energy, T, may depend explicitly on the momentum, p, the displacement, q, and the time, t (see Problem 5 in Chapter 1). The potential energy, V, is a function of the displacement only.

The generalized efforts can be grouped as follows:

$$e = e^{\phi} + e^{R} + e^{s}, \tag{b}$$

where e^ϕ are the constraint efforts, e^R are the efforts due to ideal resistors, and e^s are the efforts of the sources.

Using (a) and (b) equation (3.7) can be rewritten as

$$\delta(T(p,q,t) + V(q)) = \sum_{i=1}^{n} (e_i^\phi + e_i^R + e_i^s)\delta q_i. \tag{3.8}$$

Next we will examine each of the terms in (3.8) with the aim of finding an expression that must hold for all arbitrary virtual displacements, δq.

The variation of the kinetic energy, δT.

Since $T = T(p,q,t)$ the variation is

$$\delta T = \sum_{i=1}^{n} \frac{\partial T}{\partial p_i}\delta p_i + \sum_{i=1}^{n} \frac{\partial T}{\partial q_i}\delta q_i.$$

However, from the definition of the kinetic energy we have $f_i = \partial T/\partial p_i$. Therefore, the first term in δT is

$$\begin{aligned}\sum_{i=1}^{n} \frac{\partial T}{\partial p_i}\delta p_i &= \sum_{i=1}^{n} f_i \delta p_i \\ &= \sum_{i=1}^{n} e_i \delta q_i \\ &= \sum_{i=1}^{n} \frac{dp_i}{dt}\delta q_i \\ &= \sum_{i=1}^{n} \frac{d}{dt}\frac{\partial T^*}{\partial f_i}\delta q_i.\end{aligned}$$

The last equation is obtained using the fact that $p_i = \partial T^*/\partial f_i$, where $T^* = T^*(f,q,t)$ is the kinetic coenergy.

Equation (1.9) can be written as

$$T(p,q,t) + T^*(f,q,t) - \sum_{i=1}^{n} p_i f_i = 0.$$

The variation of this equation yields, for the i-th variable,

$$\left(\frac{\partial T}{\partial p_i} - f_i\right)\delta p_i + \left(\frac{\partial T^*}{\partial f_i} - p_i\right)\delta f_i + \left(\frac{\partial T}{\partial q_i} + \frac{\partial T^*}{\partial q_i}\right)\delta q_i = 0.$$

However, $f_i = \partial T/\partial p_i$, $p_i = \partial T^*/\partial f_i$, and since δq_i can be varied arbitrarily, we conclude that the second term in δT contains

$$\frac{\partial T}{\partial q_i} = -\frac{\partial T^*}{\partial q_i}.$$

Thus,

$$\delta T = \sum_{i=1}^{n} \left(\frac{d}{dt}\frac{\partial T^*}{\partial f_i} - \frac{\partial T^*}{\partial q_i} \right) \delta q_i. \tag{c}$$

The variation of the potential energy δV.

The variation of the potential energy is simply

$$\delta V = \sum_{i=1}^{n} \frac{\partial V}{\partial q_i} \delta q_i \tag{d}$$

The virtual work $\delta \mathcal{W}$.

The virtual work is given by

$$\delta \mathcal{W} = \sum_{i=1}^{n} (e_i^{\phi} + e_i^{R} + e_i^{s}) \delta q_i.$$

The first term in the summation is zero because the constraint efforts do no virtual work, i.e.,

$$\sum_{i=1}^{n} e_i^{\phi} \delta q_i = 0. \tag{e}$$

From the definition of an ideal resistor (equation (1.16)), the second term can be rewritten in terms of the dissipation function to get

$$\sum_{i=1}^{n} e_i^{R} \delta q_i = -\sum_{i=1}^{n} \frac{\partial D}{\partial f_i} \delta q_i. \tag{f}$$

Using (c), (d), (e) and (f) in (3.8) gives

$$\sum_{i=1}^{n} \left[\frac{d}{dt}\frac{\partial T^*}{\partial f_i} - \frac{\partial T^*}{\partial q_i} + \frac{\partial D}{\partial f_i} + \frac{\partial V}{\partial q_i} - e_i^{s} \right] \delta q_i = 0. \tag{g}$$

Each virtual displacement, δq_i, can be varied arbitrarily. So, the only way that (g) can be satisfied is if the terms in square brackets all vanish simultaneously, i.e.,

$$\frac{d}{dt}\frac{\partial T^*}{\partial \dot{q}_i} - \frac{\partial T^*}{\partial q_i} + \frac{\partial D}{\partial \dot{q}_i} + \frac{\partial V}{\partial q_i} = e_i^{s}, \quad i = 1, 2, \cdots, n. \tag{3.9}$$

Equation (3.9) is known as *Lagrange's equation of motion*. Note that Lagrange's equation can be used with any of the engineering disciplines discussed in Chapter 1. These are a set of second order, ordinary differential equations involving the generalized displacement variables for the system. Given an appropriate set of initial conditions these differential equations can be solved to find the trajectory of the system.

The application of equation (3.9) requires:

1. The selection of a set of the n generalized displacements $q = [q_1, q_2, \cdots, q_n]^T$, and the corresponding generalized flows $f = [f_1, f_2, \cdots, f_n]^T$.
2. The formulation of the kinetic coenergy, $T^*(f, q, t)$, the potential energy, $V(q)$, and the dissipation function, $D(f)$, in terms of the generalized displacement and flow variables.
3. The determination of the virtual work done by the applied efforts $\delta\mathcal{W} = \sum_{i=1}^{n} e_i^s \, \delta q_i$, which yields the generalized efforts due to effort sources, e_i^s, $i = 1, 2, \cdots, n$.

Note that the generalized efforts do not include the applied efforts due to capacitors and resistors, since these applied efforts are already accounted for by $V(q)$ and $D(f)$, respectively.

The Lagrangian approach to constructing the equation of motion does not require the determination of the constraint efforts. (This is because the constraint efforts do no virtual work.) The exclusion of the constraint efforts greatly simplifies the modeling process since, the resolution of the constraint efforts is often a cumbersome task.

The significance of the terms in (3.9) can be illustrated by rearranging the equation as follows. Let $e_i^C = -\partial V/\partial q_i$ be the efforts applied to the system by the ideal capacitors, $e_i^R = -\partial D/\partial \dot{q}_i$ be the efforts applied to the system by the ideal resistors, and $e_i^I = d/dt(\partial T^*/\partial \dot{q}_i) - \partial T^*/\partial q_i$ be the inertia efforts. Then, (3.9) can be rewritten as

$$\frac{d}{dt}\frac{\partial T^*}{\partial \dot{q}_i} - \frac{\partial T^*}{\partial q_i} = -\frac{\partial D}{\partial \dot{q}_i} - \frac{\partial V}{\partial q_i} + e_i^s,$$

$$e_i^I = e_i^R + e_i^C + e_i^s = e_i, \quad i = 1, 2, \cdots, n.$$

where e_i are the applied efforts due to the capacitors, resistors and sources. Reinserting the last equation into (g) gives

$$\sum_{i=1}^{n} (e_i^I - e_i)\, \delta q_i = 0. \tag{3.10}$$

In analytical dynamics equation (3.10) is known as *D'Alembert's Principle.* Which is an application of Bernoulli's *Principle of Virtual Work* to dynamic systems.

The principle of virtual work states that; A system is in equilibrium if and only if the virtual work done by the applied efforts is zero, i.e., $\delta\mathcal{W} = \sum_{i=1}^{n} e_i \, \delta q_i = 0$.

D'Alembert's principle states that; The system trajectory is such that the inertia efforts e_i^I and the applied efforts e_i are in equilibrium. Thus, the virtual work done by the effective effort, $e_i^I - e_i$, is zero, i.e., $\delta\mathcal{W} = \sum_{i=1}^{n}(e_i^I - e_i)\,\delta q_i = 0$.

3.3 Applications

3.3.1 Mechanical systems

The application of Lagrange's equation to mechanical systems can be accomplished using the following procedure.

1. *Kinematic analysis.*

   A kinematic analysis is required to determine the velocity of the center of mass for each body in the system. These velocities are most easily determined using rectangular coordinates to define the configuration displacements. The transformation equations between the configuration displacements and the generalized displacements are then used write the configuration velocities in terms of the generalized displacements and flows.

2. *Applied effort analysis.*

   Typically, the effort sources are given in terms of the configuration displacements, not the generalized displacements. The coordinate transformation relating the configuration displacements to the generalized displacement is used in the expression for the virtual work to determine the generalized efforts.

This basic approach is used to determine the equations of motion for mechanical systems described in the examples below.

Example 3.8.

The figure below shows a mass, m, a linear spring with stiffness k, a damper with damping coefficient b, and an applied force F.

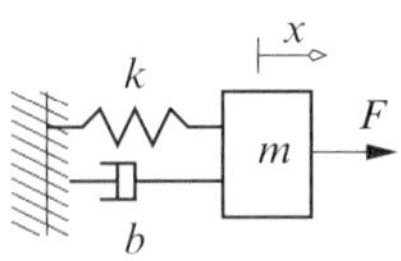

Kinematic analysis:

This system has 1 degree of freedom, i.e., only one displacement variable is require to determine the position of the mass. We will use x as the generalized displacement, and $\dot{x}$ as the corresponding generalized flow.

Applied effort analysis:

The virtual work done by the applied force F is $\delta\mathcal{W} = F\,\delta x = e^s_x\,\delta x$, where $e^s_x = F$ is the generalized effort for the system.

Lagrange's equations:

The kinetic coenergy, potential energy and the dissipation function are $T^* = m\dot{x}^2/2$, $V = kx^2/2$, and $D = b\dot{x}^2/2$, respectively. Therefore, Lagrange's equation for this system is

$$\frac{d}{dt}\frac{\partial T^*}{\partial \dot{x}} - \frac{\partial T^*}{\partial x} + \frac{\partial D}{\partial \dot{x}} + \frac{\partial V}{\partial x} = e^s_x,$$
$$m\ddot{x} + b\dot{x} + kx = F. \qquad (i)$$

To obtain (i) we have use the fact that;

$$\frac{\partial T^*}{\partial \dot{x}} = m\dot{x}, \quad \frac{d}{dt}\frac{\partial T^*}{\partial \dot{x}} = m\ddot{x}, \quad \frac{\partial T^*}{\partial x} = 0, \quad \frac{\partial D}{\partial \dot{x}} = b\dot{x}, \quad \text{and} \quad \frac{\partial V}{\partial x} - kx.$$

Given initial conditions for the displacement, x, and the flow, $\dot{x}$, the second-order ordinary differential equation (i) can be solved to find the trajectory of the system.

Example 3.9.

A simple pendulum is shown in the figure below. The system consists of a massless rod of length l that is free to rotate about point O. The mass m is attached to the end of the rod. The acceleration due to gravity acts downward, and a counterclockwise torque, τ, is applied to the rod as shown.

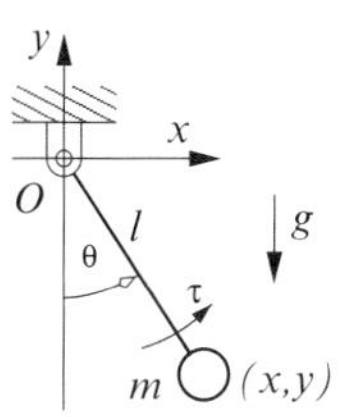

Kinematic analysis:

The position of the mass is given by the configuration displacements x and y. However, we will select θ as the generalized displacement for this 1 degree of freedom system, and $\dot{\theta}$ is the corresponding flow.

The goal of the kinematic analysis that follows is to find the velocity of the mass in terms of the generalized displacement and the generalized flow. From the figure above it can be seen that

$$x = l\sin\theta, \qquad y = -l\cos\theta. \tag{i}$$

The velocity of the point mass is $\bar{v} = \dot{x}\,\hat{\imath} + \dot{y}\,\hat{\jmath}$, where $\hat{\imath}$ is the unit vector along the x-axis, $\hat{\jmath}$ is the unit vector along the y-axis, $\dot{x} = l\dot{\theta}\cos\theta$, and $\dot{y} = l\dot{\theta}\sin\theta$. Therefore, the velocity of the mass, m, is

$$\bar{v} = l\dot{\theta}\cos\theta\,\hat{\imath} + l\dot{\theta}\sin\theta\,\hat{\jmath}.$$

Applied effort analysis:

The virtual work done by the torque and weight is $\delta\mathcal{W} = \tau\,\delta\theta - mg\,\delta y$. But, from (i) it can be seen that $\delta y = (l\sin\theta)\,\delta\theta$. Therefore,

$$\delta\mathcal{W} = \tau\,\delta\theta - mg\,\delta y = (\tau - mgl\sin\theta)\,\delta\theta = e^s_\theta\,\delta\theta,$$

where $e^s_\theta = \tau - mgl\sin\theta$ is the generalized effort.

Note that in this example, and the others to follow, the weight is included in the virtual work expression. In the Problems section the reader is asked to derive the equations of motion by including the weight in the potential energy function (see Section 1.2.2) instead of the virtual work.

Lagrange's equation:

The kinetic coenergy, potential energy and the dissipation function are

$$T^* = \frac{1}{2}m\bar{v}\cdot\bar{v} = \frac{1}{2}m(\dot{x}^2 + \dot{y}^2) = \frac{1}{2}ml^2\dot{\theta}^2,$$

$V = 0$, and $D = 0$, respectively.

Therefore, Lagrange's equation for this system is

$$\frac{d}{dt}\frac{\partial T^*}{\partial\dot{\theta}} - \frac{\partial T^*}{\partial\theta} + \frac{\partial D}{\partial\dot{\theta}} + \frac{\partial V}{\partial\theta} = e^s_\theta,$$

$$ml^2\ddot{\theta} = \tau - mgl\sin\theta. \tag{ii}$$

To arrive at (ii) we have used the following expressions:

$$\frac{\partial T^*}{\partial\dot{\theta}} = ml^2\dot{\theta}, \quad \frac{d}{dt}\frac{\partial T^*}{\partial\dot{\theta}} = ml^2\ddot{\theta}, \quad \frac{\partial T^*}{\partial\theta} = 0, \quad \frac{\partial D}{\partial\dot{\theta}} = 0, \quad \text{and} \quad \frac{\partial V}{\partial\theta} = 0.$$

Given initial conditions for the displacement, θ, and the flow, $\dot{\theta}$, the second-order ordinary differential equation (ii) can be solved to find the trajectory of the system.

Example 3.10.

The figure below shows an inverted pendulum system which consists of a mass m_1, a massless rod of length l, and a mass m_2 that is connected to the rod. The mass m_1 is constrained to move along the x-axis as shown, while the rod can rotate freely about the revolute joint at O. A force, F, is applied to mass m_1 along the x-axis, and the acceleration due to gravity acts downward as shown.

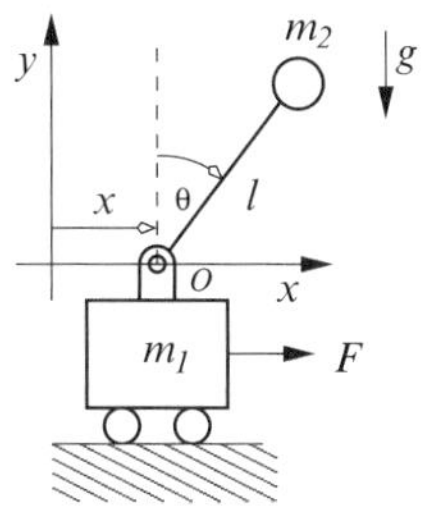

Kinematic analysis:

This system has 2 degrees of freedom. (We need one one displacement variable to locate m_1, and a second displacement variable to locate m_2.) The generalized displacements for the system are selected as x and θ as shown in the figure above. The corresponding generalized flows are $\dot{x}$ and $\dot{\theta}$. The kinematic analysis that follows aims to find the velocity of m_1 and m_2 in terms of the generalized displacements and the generalized flows.

Let $\hat{\imath}$ be the unit vector directed along the x-axis, and $\hat{\jmath}$ be the unit vector directed along the y-axis. Then, the position and velocity of mass m_1 is given by $\bar{r}_1 = x\,\hat{\imath}$, and $\dot{\bar{r}}_1 = \dot{x}\,\hat{\imath}$, respectively. The position and velocity of mass m_2 is given by $\bar{r}_2 = (x + l\sin\theta)\,\hat{\imath} + l\cos\theta\,\hat{\jmath}$, and $\dot{\bar{r}}_2 = (\dot{x} + l\dot{\theta}\cos\theta)\,\hat{\imath} - l\dot{\theta}\sin\theta\,\hat{\jmath}$, respectively.

Applied effort analysis:

The virtual work done by the force F, and the weight m_2g is $\delta\mathcal{W} = F\,\delta x - m_2g\,\delta y_2$, where $y_2 = l\cos\theta$ is the vertical position of mass m_2. Therefore, $\delta y_2 = -(l\sin\theta)\,\delta\theta$, and the virtual work becomes

$$\delta\mathcal{W} = F\,\delta x + m_2gl\sin\theta\,\delta\theta = e_x^s\,\delta x + e_\theta^s\,\delta\theta.$$

Thus, the generalized efforts are $e_x^s = F$, and $e_\theta^s = m_2 gl \sin\theta$. Here, e_x^s is the generalized effort associated with the displacement x, and e_θ^s is the generalized effort associated with the displacement θ.

Lagrange's equations:

The kinetic coenergy, potential energy and the dissipation function are

$$\begin{aligned} T^* &= \frac{1}{2} m_1 \dot{\bar{r}}_1 \cdot \dot{\bar{r}}_1 + \frac{1}{2} m_2 \dot{\bar{r}}_2 \cdot \dot{\bar{r}}_2 \\ &= \frac{1}{2} m_1 \dot{x}^2 + \frac{1}{2} m_2 (\dot{x}^2 + l^2 \dot{\theta}^2 + 2l\dot{x}\dot{\theta}\cos\theta), \end{aligned}$$

$V = 0$, and $D = 0$, respectively.

Lagrange's equations for this system are

$$\frac{d}{dt}\frac{\partial T^*}{\partial \dot{x}} - \frac{\partial T^*}{\partial x} + \frac{\partial D}{\partial \dot{x}} + \frac{\partial V}{\partial x} = e_x^s, \qquad (a)$$

$$\frac{d}{dt}\frac{\partial T^*}{\partial \dot{\theta}} - \frac{\partial T^*}{\partial \theta} + \frac{\partial D}{\partial \dot{\theta}} + \frac{\partial V}{\partial \theta} = e_\theta^s. \qquad (b)$$

Using T^*, V, and D we can see that

$$\begin{aligned} \frac{\partial T^*}{\partial \dot{x}} &= (m_1 + m_2)\dot{x} + m_2 l\dot{\theta}\cos\theta, \\ \frac{d}{dt}\frac{\partial T^*}{\partial \dot{x}} &= (m_1 + m_2)\ddot{x} + m_2 l\ddot{\theta}\cos\theta - m_2 l\dot{\theta}^2\sin\theta, \\ \frac{\partial T^*}{\partial x} &= 0, \quad \frac{\partial D}{\partial \dot{x}} = 0, \quad \frac{\partial V}{\partial x} = 0, \\ \frac{\partial T^*}{\partial \dot{\theta}} &= m_2(l^2\dot{\theta} + l\dot{x}\cos\theta), \\ \frac{d}{dt}\frac{\partial T^*}{\partial \dot{\theta}} &= m_2 l^2\ddot{\theta} + m_2 l\ddot{x} - m_2 l\dot{x}\dot{\theta}\sin\theta, \\ \frac{\partial T^*}{\partial \theta} &= -m_2 l\dot{x}\dot{\theta}\sin\theta, \quad \frac{\partial D}{\partial \dot{\theta}} = 0, \quad \frac{\partial V}{\partial \theta} = 0. \end{aligned}$$

Thus, (a) and (b) are

$$\begin{aligned} (a) &\Rightarrow (m_1 + m_2)\ddot{x} + m_2 l(\ddot{\theta}\cos\theta - \dot{\theta}^2\sin\theta) = F, \\ (b) &\Rightarrow m_2(l^2\ddot{\theta} + l\ddot{x}\cos\theta) = m_2 gl\sin\theta. \end{aligned}$$

These are the coupled nonlinear ordinary differential equations that determine the behavior of the inverted pendulum model described above.

3.3.1.1 Kinetic coenergy of rigid bodies

Heretofore we have consider mechanical system containing point masses. In order to extend the results from the previous sections to mechanical systems in that are in spatial motion, we must develop expressions for the kinetic coenergy of rigid bodies.

Figure 3.8 shows an idealized representation of a rigid body. Here, the body is assumed to consist of N (possible infinite) point masses, m_i, $i = 1, 2, \cdots, N$. The rectangular coordinate system x-y-z, with origin at Q, represents a fixed

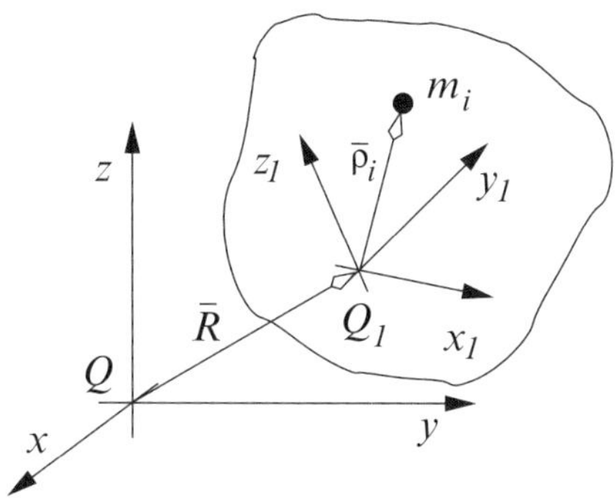

Fig. 3.8 A rigid body model

reference frame. The rectangular coordinate system x_1-y_1-z_1, with origin at Q_1, is attached to the body. Thus, the x_1-y_1-z_1 frame moves with the body, and is often referred to as a 'body-fixed' coordinate system.

To develop an expression for the kinetic coenergy of the body we must determine the velocity of each point mass. Let the vector $\bar{r}_i = X_i\,\hat{\imath} + Y_i\,\hat{\jmath} + Z_i\,\hat{k}$ be the position of the i-th point mass with respect to the fixed frame. Here, the unit vectors $\hat{\imath}$, $\hat{\jmath}$ and $\hat{k}$ are parallel to the x-axis, y-axis, and z-axis, respectively. Also, let the vector $\bar{R} = x_{Q_1}\,\hat{\imath} + y_{Q_1}\,\hat{\jmath} + z_{Q_1}\,\hat{k}$ be the position of Q_1 relative to Q. Finally, let the vector $\bar{\rho}_i = x_i\,\hat{\imath}_1 + y_i\,\hat{\jmath}_1 + z_i\,\hat{k}_1$ be the position of the point mass m_i relative to Q_1. The unit vectors $\hat{\imath}_1$, $\hat{\jmath}_1$ and $\hat{k}_1$ are parallel to the x_1-axis, y_1-axis, and z_1-axis, respectively.

Using Fig. 3.8 we see that the position of the point mass m_i satisfies

$$\bar{r}_i = \bar{R} + \bar{\rho}_i. \tag{a}$$

With equation (a), and the results from Section 2.1.4, it can be shown that the velocity of point mass m_i with respect to the fixed frame is given by

$$\dot{\bar{r}}_i = \dot{\bar{R}} + \bar{\omega} \times \bar{\rho}_i, \tag{b}$$

where

$$\dot{\bar{r}}_i = \dot{X}_i\,\hat{\imath} + \dot{Y}_i\,\hat{\jmath} + \dot{Z}_i\,\hat{k},$$
$$\dot{\bar{R}} = \dot{x}_{Q_1}\,\hat{\imath} + \dot{y}_{Q_1}\,\hat{\jmath} + \dot{z}_{Q_1}\,\hat{k},$$
$$\bar{\omega} = \omega_x\,\hat{\imath}_1 + \omega_y\,\hat{\jmath}_1 + \omega_z\,\hat{k}_1.$$

Where ω_x, ω_y and ω_z are the angular velocities of the body-fixed axes x_1, y_1 and z_1, respectively. In addition, we have used the fact that $\frac{d}{dt}(\bar{\rho}_i) = 0$, i.e., since the body is rigid, the point mass does not move relative to the x_1-y_1-z_1 coordinate system.

The kinetic coenergy of the rigid body can thus be written as

$$\begin{aligned}
T^* &= \frac{1}{2}\sum_{i=1}^{N} m_i\dot{\bar{r}}_i\cdot\dot{\bar{r}}_i,\\
&= \frac{1}{2}\sum_{i=1}^{N} m_i(\dot{\bar{R}} + \bar{\omega}\times\bar{\rho}_i)\cdot(\dot{\bar{R}} + \bar{\omega}\times\bar{\rho}_i),\\
&= \frac{1}{2}\sum_{i=1}^{N} m_i\dot{\bar{R}}\cdot\dot{\bar{R}} + \dot{\bar{R}}\cdot\left[\bar{\omega}\times\sum_{i=1}^{N} m_i\bar{\rho}_i\right]\\
&\quad + \frac{1}{2}\sum_{i=1}^{N} m_i(\bar{\omega}\times\bar{\rho}_i)\cdot(\bar{\omega}\times\bar{\rho}_i),\\
&= \frac{1}{2}\sum_{i=1}^{N} m_i\dot{\bar{R}}\cdot\dot{\bar{R}} + \dot{\bar{R}}\cdot\left[\bar{\omega}\times\sum_{i=1}^{N} m_i\bar{\rho}_i\right]\\
&\quad + \frac{1}{2}\bar{\omega}\cdot\left[\sum_{i=1}^{N} m_i\bar{\rho}_i\times(\bar{\omega}\times\bar{\rho}_i)\right].
\end{aligned} \tag{3.11}$$

This result can be simplified for two important configurations of the system. The first being rigid bodies where the point Q_1 coincides with the *center of mass*, and the second being rigid bodies where the point Q_1 is fixed.

Center of mass of a rigid body

The center of mass of the rigid body is a location $\bar{\rho}_c$ in the x_1-y_1-z_1 coordinate system where the moments due to all the point masses vanish. That is,

$$\sum_{i=1}^{N}(m_i\bar{\rho}_c - m_i\bar{\rho}_i) = 0. \tag{c}$$

Let m denote the total mass of the body then,

$$m = \sum_{i=1}^{N} m_i. \tag{d}$$

Using equation (d) in equation (c) gives the location of the center of mass as

$$\bar{\rho}_c = \frac{\sum_{i=1}^{N} m_i \bar{\rho}_i}{m} \tag{e}$$

Kinetic coenergy with respect to the center of mass

If Q_1 coincides with the center of mass for the rigid body then, $\bar{\rho}_c = 0$, and as a result $\sum_{i=1}^{N} m_i \bar{\rho}_i = 0$. Also, $\bar{R} = x_c \hat{i} + y_c \hat{j} + z_c \hat{k}$, defines the position of the center of mass, and $\dot{\bar{R}} = \dot{x}_c \hat{i} + \dot{y}_c \hat{j} + \dot{z}_c \hat{k}$ defines the velocity of the center of mass. Hence, the kinetic coenergy given by equation (3.11) becomes

$$T^* = \frac{1}{2} m \dot{\bar{R}} \cdot \dot{\bar{R}} + \frac{1}{2} \bar{\omega} \cdot \left[\sum_{i=1}^{N} m_i \bar{\rho}_i \times (\bar{\omega} \times \bar{\rho}_i) \right]. \tag{f}$$

Expanding the term in the square brackets from equation (f) gives

$$\begin{aligned}
\sum_{i=1}^{N} m_i \bar{\rho}_i \times (\bar{\omega} \times \bar{\rho}_i) &= \sum_{i=1}^{N} m_i \left\{ (\bar{\rho}_i \cdot \bar{\rho}_i)\bar{\omega} - (\bar{\rho}_i \cdot \bar{\omega})\bar{\rho}_i \right\}, \\
&= \sum_{i=1}^{N} m_i \Big\{ (\lambda \omega_x - \mu x_i)\, \hat{i}_1 + (\lambda \omega_y - \mu y_i)\, \hat{j}_1 \\
&\qquad + (\lambda \omega_z - \mu z_i)\, \hat{k}_1 \Big\}, \\
&= H_x\, \hat{i}_1 + H_y\, \hat{j}_1 + H_z\, \hat{k}_1 \\
&= \bar{H}.
\end{aligned}$$

Here,

$$\begin{aligned}
\lambda &= x_i^2 + y_i^2 + z_i^2, \quad \mu = x_i \omega_x + y_i \omega_y + z_i \omega_z, \\
H_x &= I_{xx}\omega_x + I_{xy}\omega_y + I_{xz}\omega_z, \\
H_y &= I_{xy}\omega_x + I_{yy}\omega_y + I_{yz}\omega_z, \\
H_z &= I_{xz}\omega_x + I_{yz}\omega_y + I_{zz}\omega_z, \\
I_{xx} &= \sum_{i=1}^{N} m_i (y_i^2 + z_i^2), \quad I_{xy} = -\sum_{i=1}^{N} m_i x_i^2 y_i^2, \quad I_{xz} = -\sum_{i=1}^{N} m_i x_i^2 z_i^2, \\
I_{yy} &= \sum_{i=1}^{N} m_i (x_i^2 + z_i^2), \quad I_{yz} = -\sum_{i=1}^{N} m_i y_i^2 z_i^2, \quad I_{zz} = \sum_{i=1}^{N} m_i (x_i^2 + y_i^2).
\end{aligned}$$

The vector $\bar{H} = H_x\, \hat{i}_1 + H_y\, \hat{j}_1 + H_z\, \hat{k}_1$ is the angular momentum of the rigid body. The terms I_{xx}, I_{yy}, and I_{zz} are called the moments of inertia, and the terms I_{xy}, I_{xz}, and I_{yz} are called the products of inertia.

If we define the matrices $v_c = [\dot{x}_c, \quad \dot{y}_c, \quad \dot{z}_c]^T$, $\omega = [\omega_x, \quad \omega_y, \quad \omega_z]^T$, and the *inertia matrix*

$$I_c = \begin{bmatrix} I_{xx} & I_{xy} & I_{xz} \\ I_{xy} & I_{yy} & I_{yz} \\ I_{xz} & I_{yz} & I_{zz} \end{bmatrix} \tag{g}$$

then, the kinetic coenergy, (equation (f)), becomes

$$T^* = \frac{1}{2} m v_c^T v_c + \frac{1}{2} \omega^T I_c \omega. \tag{3.12}$$

The first term in equation (3.12) is due to the translation of the body while the second term is due to the rotation of the body. Recall that v_c is the velocity of the center of mass with respect to the fixed coordinate system, and ω is the angular velocity of the x_1-y_1-z_1 coordinate system that is fixed to the body. It must be emphasized however, that the components of ω are directed along the x_1-axis, y_1-axis, and z_1-axis not the x-axis, y-axis, and z-axis.

For any rigid body we can find an orientation of the x_1-y_1-z_1 coordinate system such that the products of inertia vanish, i.e., $I_{xy} = I_{xz} = I_{yz} = 0$. In such orientations the x_1-y_1-z_1 axes are called the *Principal Axes of Inertia*, and the inertia matrix, I_c becomes diagonal.

Kinetic coenergy with respect to a fixed point

If the point Q_1 is a fixed point then, $\dot{\bar{R}} = 0$, and the kinetic coenergy given by equation (3.11) becomes,

$$\begin{aligned} T^* &= \frac{1}{2} \bar{\omega} \cdot \left[\sum_{i=1}^{N} m_i \bar{\rho}_i \times (\bar{\omega} \times \bar{\rho}_i) \right], \\ &= \frac{1}{2} \omega^T I_{Q_1} \omega. \end{aligned} \tag{3.13}$$

where $\omega = [\omega_x, \quad \omega_y, \quad \omega_z]^T$, and the inertia matrix, I_{Q_1}, is computed with respect to the fixed point Q_1.

Plane motion of a rigid body

If the body is constrained to move in the x-y plane then the kinetic coenergy given in equation (3.12) becomes

$$T^* = \frac{1}{2} m (\dot{x}_c^2 + \dot{y}_c^2) + \frac{1}{2} I_{zz} \omega_z^2, \tag{3.14}$$

where the moment of inertia, I_{zz}, is computed relative to the center of mass.

If the body is constrained to rotate in the x-y plane, about a fixed point then the kinetic coenergy given in equation (3.13) becomes

$$T^* = \frac{1}{2}\hat{I}_{zz}\omega_z^2, \tag{3.15}$$

where the moment of inertia, $\hat{I}_{zz}$, is computed relative to the fixed point. The examples that follow will illustrate the use of these kinetic coenergy expressions.

Example 3.11.

The center of mass of the disk shown below is at point O. The disk can rotate about O in a plane as shown. The moment of inertia of the disk, about the point O is I_O. A linear spring with stiffness k is attached to the disk at radius a, and a mass, m, is attached to the disk via a chord at a radius b. The acceleration due to gravity acts downward as shown. The dynamic equations of motion for this system can be determined as follows.

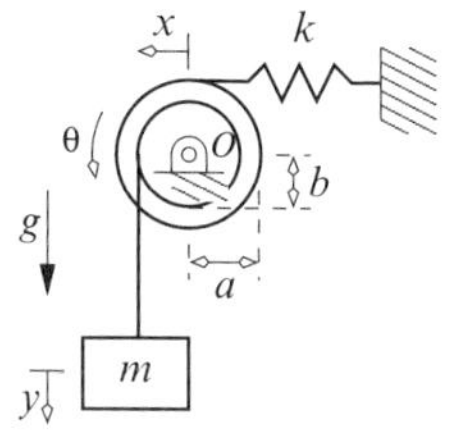

Kinematic analysis:

The angular displacement of the disk is denoted by the angle θ. The variable y gives the vertical displacement of the mass, and the variable x measures the linear displacement of the spring. These three variables are not all independent. In fact as the disk undergoes an angular displacement, θ, the geometry of the system requires that x and y satisfy the two constraints:

$$\begin{aligned}\phi_1 &= x - a\theta = 0,\\ \phi_2 &= y - b\theta = 0.\end{aligned}$$

Thus, the system has only 1 degree of freedom. Here we select θ as the generalized displacement, and $\dot{\theta}$ as the corresponding generalized flow.

To complete the kinematic analysis we note that the disk rotates about a fixed point with angular velocity $\dot{\theta}$, and the mass m has velocity $\dot{y} = \frac{d}{dt}(b\theta)$ $= b\dot{\theta}$. Therefore, we have determined the velocity of all the inertia elements in terms of the generalized displacement and flow.

Applied effort analysis:

The virtual work done by the weight is

$$\delta\mathcal{W} = mg\,\delta y.$$

However, from the constraint ϕ_2 it can be seen that $\delta y = b\,\delta\theta$. Thus, in terms of the generalized displacement, θ, the virtual work becomes

$$\delta\mathcal{W} = mgb\,\delta\theta = e_\theta\,\delta\theta,$$

where $e_\theta^s = mgb$ is the generalized effort associated with the generalized displacement θ.

Lagrange's equation:

The kinetic coenergy for the system is

$$\begin{aligned} T^* &= \frac{1}{2}I_O\dot{\theta}^2 + \frac{1}{2}m\dot{y}^2, \\ &= \frac{1}{2}I_O\dot{\theta}^2 + \frac{1}{2}mb^2\dot{\theta}^2, \\ &= \frac{1}{2}(I_O + mb^2)\dot{\theta}^2. \end{aligned}$$

The term $I_O\dot{\theta}^2/2$ follows from equation (3.15) since, the disk is a rigid body rotating about a fixed point in a plane. The term $m\dot{y}^2/2$ is due to the vertical motion of the mass, and is rewritten in terms of the generalized displacement to obtain the final result for the kinetic coenergy.

The potential energy due to the spring is

$$V = \frac{1}{2}kx^2 = \frac{1}{2}ka^2\theta^2,$$

where we have used the fact that $x = a\theta$, and the dissipation function is $D = 0$.

Using T^*, V, and D it can be seen that

$$\frac{\partial T^*}{\partial\dot{\theta}} = (I_O + mb^2)\dot{\theta}, \quad \frac{d}{dt}\frac{\partial T^*}{\partial\dot{\theta}} = (I_O + mb^2)\ddot{\theta}, \quad \frac{\partial T^*}{\partial\theta} = 0,$$

$$\frac{\partial D}{\partial\dot{\theta}} = 0, \quad \text{and} \quad \frac{\partial V}{\partial\theta} = ka^2\theta.$$

Hence, Lagrange's equation of motion for the system is

$$\begin{aligned} \frac{d}{dt}\frac{\partial T^*}{\partial\dot{\theta}} - \frac{\partial T^*}{\partial\theta} + \frac{\partial D}{\partial\dot{\theta}} + \frac{\partial V}{\partial\theta} &= e_\theta^s, \\ (I_O + mb^2)\ddot{\theta} + ka^2\theta &= mgb. \end{aligned}$$

Example 3.12.

The system shown in the figure below, consists of a disk with radius r that is attached to a linear spring. The disk has mass m and moment of inertia I about its center of mass. The disk rolls on the incline without slipping. The linear spring has stiffness k, and is attached to the center of the disk. The incline has a fixed angle α, and the acceleration due to gravity acts downward as shown. The dynamic equations of motion for this system can be developed as follows.

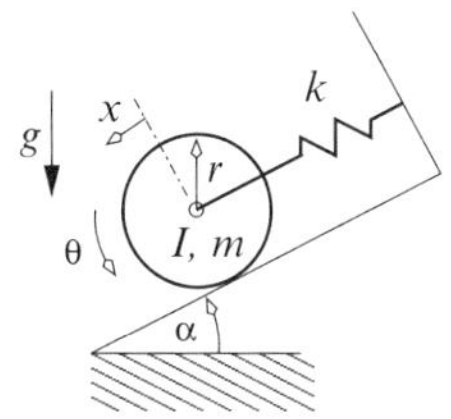

Kinematic analysis:

Since the disk rolls without slipping the system has 1 degree of freedom. Here, we select θ and the generalized displacement, and $\dot{\theta}$ as the generalized flow. We also note that the linear displacement x and the angular displacement θ are related by the constraint equation

$$\phi = x - r\theta = 0.$$

Applied effort analysis:

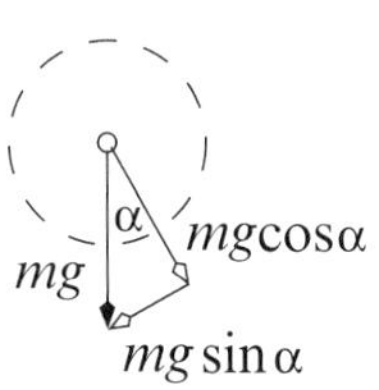

The diagram on the left shows the weight of the disk (a vector) resolved into components that are parallel and perpendicular to the incline. The component of the weight that is perpendicular to the incline does no virtual work. Hence, the virtual work done by the weight of the disk is

$$\delta\mathcal{W} = mg\sin\alpha\,\delta x = mgr\sin\alpha\,\delta\theta = e_\theta^s\,\delta\theta.$$

Here, we have used the fact that $x = r\theta$ to get $\delta x = r\delta\theta$.

Lagrange's equation:

The disk in this system is a rigid body undergoing general plane motion. In which case, the kinetic coenergy is given by equation (3.14), i.e.,

$$T^* = \frac{1}{2}m\dot{x}^2 + \frac{1}{2}I\dot{\theta}^2.$$

The first term in T^* is due to the translation of the center of mass of the disk, and the second term is due to the rotation of the disk. Using the fact that $\dot{x} = r\dot{\theta}$ gives

$$T^* = \frac{1}{2}(I + mr^2)\dot{\theta}^2.$$

The potential energy due to the spring is

$$V = \frac{1}{2}kx^2 = \frac{1}{2}kr^2\theta^2,$$

and the dissipation function is $D = 0$.

Using T^*, V, and D we obtain

$$\frac{\partial T^*}{\partial \dot{\theta}} = (I + mr^2)\dot{\theta}, \quad \frac{d}{dt}\frac{\partial T^*}{\partial \dot{\theta}} = (I + mr^2)\ddot{\theta}, \quad \frac{\partial T^*}{\partial \theta} = 0,$$

$$\frac{\partial D}{\partial \dot{\theta}} = 0, \quad \text{and} \quad \frac{\partial V}{\partial \theta} = kr^2\theta.$$

Therefore, Lagrange's equation in terms of the generalized displacement θ is

$$\frac{d}{dt}\frac{\partial T^*}{\partial \dot{\theta}} - \frac{\partial T^*}{\partial \theta} + \frac{\partial D}{\partial \dot{\theta}} + \frac{\partial V}{\partial \theta} = e^s_\theta,$$

$$(I + mr^2)\ddot{\theta} + kr^2\theta = mgr\sin\alpha.$$

Example 3.13.

A thin disk with radius r and mass m rolls without slipping on the inner surface of a cylinder that has radius R, as shown in the figure below. The acceleration due to gravity acts downward. The equations of motion for this system can be determined as follows.

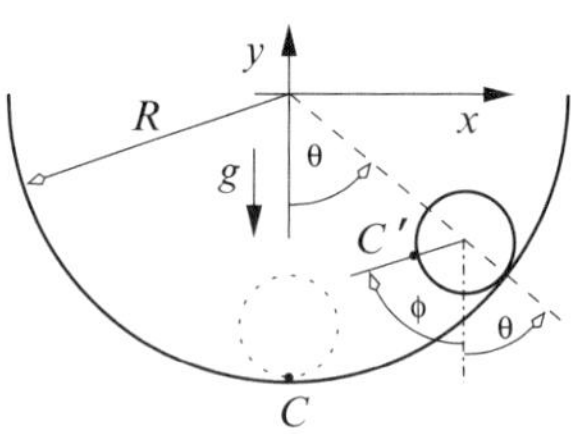

Kinematic analysis:

Since the disk rolls without slipping we have 1 degree of freedom. The displacement variable θ measures the angle that the center of the disk makes the y-axis. As the disk rolls on the cylinder the point C moves to C' as shown in the figure above. The displacement ϕ measures the rotation of the disk relative to the y-axis, i.e., ϕ is the angular displacement from C to C'.

Let $\hat{\imath}$ be the unit vector directed along the x-axis, and $\hat{\jmath}$ be the unit vector directed along the y-axis. Then the position and velocity of the center of the disk is given by

$$\begin{aligned}\bar{s} &= (R-r)\sin\theta\,\hat{\imath} - (R-r)\cos\theta\,\hat{\jmath},\\ \dot{\bar{s}} &= (R-r)\dot{\theta}\cos\theta\,\hat{\imath} + (R-r)\dot{\theta}\sin\theta\,\hat{\jmath}.\end{aligned}$$

Since the disk rolls without slipping we have the relationship $R\theta = r(\theta + \phi)$. Hence, the angular velocity of the disk is given by $\dot{\phi} = ((R-r)/r)\dot{\theta}$. In the analysis that follows we will use θ as the generalized displacement, and $\dot{\theta}$ as the generalized flow.

Applied effort analysis:

The virtual work done by the weight mg is $\delta\mathcal{W} = -mg\,\delta y$, where $y = -(R-r)\cos\theta$ is the vertical position of the center of the disk. Therefore, $\delta y = ((R-r)\sin\theta)\,\delta\theta$, and the virtual work becomes $\delta\mathcal{W} = -mg(R-r)\sin\theta\,\delta\theta = e^s_\theta\,\delta\theta$. Hence, the generalized effort is $e^s_\theta = -mg(R-r)\sin\theta$.

Lagrange's equation:

The kinetic coenergy, potential energy and the dissipation function are

$$\begin{aligned}T^* &= \frac{1}{2}m\dot{\bar{s}}\cdot\dot{\bar{s}} + \frac{1}{2}I\dot{\phi}^2,\\ &= \frac{1}{2}m(R-r)^2\dot{\theta}^2 + \frac{1}{2}\frac{mr^2}{2}\left[\left(\frac{R}{r}-1\right)\dot{\theta}\right]^2,\\ &= \frac{3}{4}m(R-r)^2\dot{\theta}^2,\end{aligned}$$

$V = 0$, and $D = 0$, respectively. (In T^* we have used the fact that the moment of inertia for a thin disk is $I = mr^2/2$.)

Using T^*, V, and D we obtain

$$\frac{\partial T^*}{\partial\dot{\theta}} = \frac{3}{2}m(R-r)^2\dot{\theta}, \quad \frac{d}{dt}\frac{\partial T^*}{\partial\dot{\theta}} = \frac{3}{2}m(R-r)^2\ddot{\theta}, \quad \frac{\partial T^*}{\partial\theta} = 0,$$

$$\frac{\partial D}{\partial\dot{\theta}} = 0, \quad \text{and} \quad \frac{\partial V}{\partial\theta} = 0.$$

Therefore, Lagrange's equation for this system is

$$\begin{aligned}\frac{d}{dt}\frac{\partial T^*}{\partial\dot{\theta}} - \frac{\partial T^*}{\partial\theta} + \frac{\partial D}{\partial\dot{\theta}} + \frac{\partial V}{\partial\theta} &= e^s_\theta,\\ \frac{3}{2}m(R-r)^2\ddot{\theta} &= -mg(R-r)\sin\theta.\end{aligned}$$

Example 3.14.

A model of a planar R-R robot is shown in the figure below. The link AB can rotate about the revolute joint at A in the x-y plane, and the link BC can rotate about the revolute joint at B in the x-y plane. The link AB has length l_1, mass m_1, and moment of inertia I_1 about its center of mass, which is in the geometric center of the link. The link BC has length l_2, mass m_2, and moment of inertia I_2 about its center of mass, which is in the geometric center of the link. The torques τ_1 and τ_2 are applied to AB and BC, respectively. The dynamic equations of motion for this system can be developed as follows.

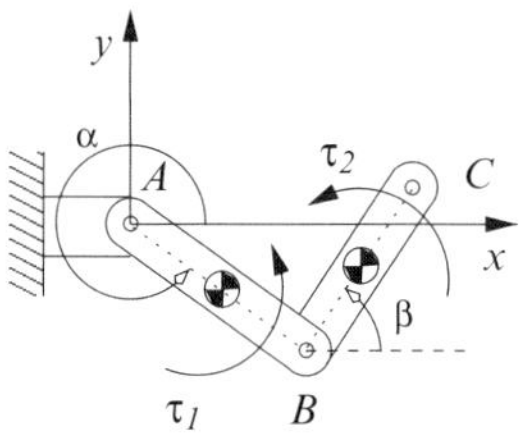

Kinematic analysis:

The rectangular coordinate system x-y represents the fixed or inertial reference frame for the system. The angle α is the angular displacement of the link AB with respect to the x-y coordinate system, and the angle β is the angular displacement of the link BC with respect to the x-y coordinate system. Using the results from Section 2.2 it is clear that this robot has 2 degrees of freedom. Here, the angles α and β are selected as the generalized displacements, and $\dot{\alpha}$ and $\dot{\beta}$ as the corresponding generalized flows.

To determine the kinetic coenergy for the system we require the velocities of the center of mass of each link. Let $\hat{\imath}$ be the unit vector directed along the x-axis, and let $\hat{\jmath}$ be the unit vector directed along the y-axis. Then, the position of the center of mass for link AB, with respect to the fixed frame, is given by

$$\bar{r}_1 = \frac{l_1}{2}\cos\alpha\,\hat{\imath} + \frac{l_1}{2}\sin\alpha\,\hat{\jmath},$$

and the position of the center of mass of link BC, with respect to the fixed frame, is given by

$$\bar{r}_2 = (l_1\cos\alpha + \frac{l_2}{2}\cos\beta)\,\hat{\imath} + (l_1\sin\alpha + \frac{l_2}{2}\sin\beta)\,\hat{\jmath}.$$

The velocity of the center of mass of link AB is thus,

$$\bar{v}_1 = \frac{d\bar{r}_1}{dt} = -\frac{l_1}{2}\dot{\alpha}\sin\alpha\,\hat{\imath} + \frac{l_1}{2}\dot{\alpha}\cos\alpha\,\hat{\jmath}, \qquad (a)$$

and the velocity of the center of mass of link BC is

$$\bar{v}_2 = \frac{d\bar{r}_2}{dt} = (-l_1\dot{\alpha}\sin\alpha - \frac{l_2}{2}\dot{\beta}\sin\beta)\,\hat{\imath} + (l_1\dot{\alpha}\cos\alpha + \frac{l_2}{2}\dot{\beta}\cos\beta)\,\hat{\jmath}. \qquad (b)$$

Applied effort analysis:

The virtual work done by the applied torques is

$$\delta\mathcal{W} = \tau_1\,\delta\alpha + \tau_2\,\delta\beta = e^s_\alpha\,\delta\alpha + e^s_\beta\,\delta\beta.$$

Hence, the generalized efforts are $e^s_\alpha = \tau_1$ and $e^s_\beta = \tau_2$.

Lagrange's equation:

Using equation (3.14) it can be seen that the kinetic coenergy of the system is given by

$$T^* = \frac{1}{2}m_1\bar{v}_1\cdot\bar{v}_1 + \frac{1}{2}I_1\dot{\alpha}^2 + \frac{1}{2}m_2\bar{v}_2\cdot\bar{v}_2 + \frac{1}{2}I_2\dot{\beta}^2,$$

The first two terms represent the kinetic coenergy contribution due to link AB, and the second two terms represent the kinetic coenergy contribution due to link BC. Using equations (a) and (b) the kinetic coenergy for the system becomes

$$T^* = \frac{1}{2}\left[I_1 + \frac{m_1 l_1^2}{4}\right]\dot{\alpha}^2 + \frac{1}{2}m_2\left[l_1^2\dot{\alpha}^2 + \frac{l_2^2}{4}\dot{\beta}^2 + l_1 l_2\dot{\alpha}\dot{\beta}\cos(\alpha-\beta)\right] + \frac{1}{2}I_2\dot{\beta}^2. \qquad (c)$$

For this system the potential energy is $V = 0$, and the dissipation function is $D = 0$. Lagrange's equations for the system are,

$$\frac{d}{dt}\frac{\partial T^*}{\partial\dot{\alpha}} - \frac{\partial T^*}{\partial\alpha} + \frac{\partial D}{\partial\dot{\alpha}} + \frac{\partial V}{\partial\alpha} = e^s_\alpha, \qquad (d)$$

$$\frac{d}{dt}\frac{\partial T^*}{\partial\dot{\beta}} - \frac{\partial T^*}{\partial\beta} + \frac{\partial D}{\partial\dot{\beta}} + \frac{\partial V}{\partial\beta} = e^s_\beta. \qquad (e)$$

Using equation (c) it can be shown that equations (d) and (e) lead to the coupled nonlinear differential equations

$$\left[I_1 + m_1\frac{l_1^2}{4} + m_2 l_1^2\right]\ddot{\alpha} + \frac{m_2 l_1 l_2}{2}\ddot{\beta}\cos(\alpha-\beta) + \frac{m_2 l_1 l_2}{2}\dot{\beta}^2\sin(\alpha-\beta) = \tau_1,$$

$$\left[I_2 + m_2\frac{l_2^2}{4}\right]\ddot{\beta} + \frac{m_2 l_1 l_2}{2}\ddot{\alpha}\cos(\alpha-\beta) - \frac{m_2 l_1 l_2}{2}\dot{\alpha}^2\sin(\alpha-\beta) = \tau_2.$$

Before leaving this example it will prove instructive to reexamine the kinetic coenergy of the link AB. In the analysis above the link AB is treated as a rigid body in general plane motion. Hence, equation (3.14) is used to determine the kinetic coenergy of the link as

$$T^*_{AB} = \frac{1}{2}m_1\bar{v}_1 \cdot \bar{v}_1 + \frac{1}{2}I_1\dot{\alpha}^2 = \frac{1}{2}\left[I_1 + \frac{m_1 l_1^2}{4}\right]\dot{\alpha}^2, \tag{f}$$

where m_1 is the mass of the link, $\bar{v}_1$ is the velocity of the center of mass of the link, I_1 is the moment of inertia about the center of mass, and $\dot{\alpha}$ is the angular velocity of the link.

However, the link AB can also be treated as a rigid body rotating about a fixed point, i.e., link AB rotates about the fixed point A. In which case the kinetic coenergy of the link AB is given by equation (3.15) as

$$T^*_{AB} = \frac{1}{2}I_{1_A}\dot{\alpha}^2, \tag{g}$$

where I_{1_A} is the moment of inertia of the link AB about point A. Comparing equations (f) and (g) it is evident that

$$I_{1_A} = I_1 + m_1\frac{l_1^2}{4}.$$

Which is in fact the result one would obtain by applying the parallel axis theorem. It is then clear that we can replace equation (f) with with (g) in T^*, and obtain the same differential equations of motion.

Example 3.15.

A model of a symmetric top is shown in the figure below. Here, the top is free to rotate about a fixed point Q. The rectangular coordinate system x-y-z represents the fixed reference frame for the system. The rectangular coordinate system x_1-y_1-z_1 is attached to the top at the point Q such that the z_1-axis is along the line from Q to C. The point C is the center of mass for the top, and l is the distance from Q to C. The mass of the top is m. The moments of inertia about the point Q are I_{12}, I_{12}, and I_3 in the x_1, y_1, and z_1 directions, respectively. (Note that the moments of inertia are given with respect to the fixed point Q, not the center of mass.) The acceleration due to gravity acts downward as shown. The dynamic equations of motion for this system can be developed as follows.

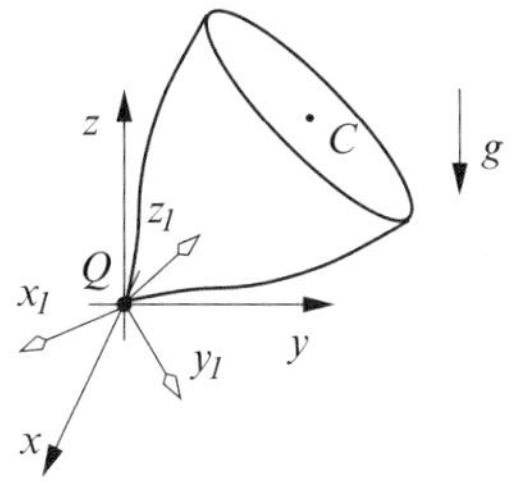

Kinematic analysis:

Since the top is free to rotate in about the point Q the system has 3 degrees of freedom, i.e., the variables that determine the direction cosine matrix of the x_1-y_1-z_1 coordinate system. Here, the orientation of the x_1-y_1-z_1 coordinate system with respect to the fixed frame will be described using the Z_α-X_β-Z_γ Euler angles (see Section 2.1.5). A point $r = [x,\ y,\ z]^T$ in the x-y-z coordinate system is then related to a point ${}^1r = [x_1,\ y_1,\ z_1]^T$ in the x_1-y_1-z_1 coordinate system via the equation

$$r = {}^0A_1\,{}^1r, \tag{a}$$

where the direction cosine matrix is

$${}^0A_1 = \begin{bmatrix} c_\alpha c_\gamma - s_\alpha c_\beta s_\gamma & -c_\alpha s_\gamma - s_\alpha c_\beta c_\gamma & s_\alpha s_\beta \\ s_\alpha c_\gamma + c_\alpha c_\beta s_\gamma & -s_\alpha s_\gamma + c_\alpha c_\beta c_\gamma & -c_\alpha s_\beta \\ s_\beta s_\gamma & s_\beta c_\gamma & c_\beta \end{bmatrix}.$$

Here, $c_\alpha = \cos\alpha$, $s_\alpha = \sin\alpha$, etc. The angles α, β and γ are taken as the generalized displacements, and $\dot\alpha$, $\dot\beta$ and $\dot\gamma$ are the corresponding generalized flows.

Let ω_1, ω_2 and ω_3 be the angular velocities of the x_1-axis, y_1-axis, and z_1-axis, respectively. Then, ω_1, ω_2 and ω_3 are related to the Euler angles via the equation

$$\begin{bmatrix} \omega_1 \\ \omega_2 \\ \omega_3 \end{bmatrix} = \begin{bmatrix} s_\beta s_\gamma & c_\gamma & 0 \\ s_\beta c_\gamma & -s_\gamma & 0 \\ c_\beta & 0 & 1 \end{bmatrix} \begin{bmatrix} \dot\alpha \\ \dot\beta \\ \dot\gamma \end{bmatrix}. \tag{b}$$

(See Problem 9 in Chapter 2.)

Applied effort analysis:

The virtual work done by the weight of the top is $\delta\mathcal{W} = -mg\,\delta z_C$ where δz_C is the variation in the position of the center of mass, C, along the z-axis. The coordinate of C in the x_1-y_1-z_1 system is ${}^1r_C = [0,\ 0,\ l]^T$. Using equation (a) it can be seen that the coordinate of C in the x-y-z system is

$$r_C = \begin{bmatrix} x_C \\ y_C \\ z_C \end{bmatrix} = {}^0A_1{}^1r_c = \begin{bmatrix} l s_\alpha s_\beta \\ -l c_\alpha s_\beta \\ l c_\beta \end{bmatrix}.$$

Therefore, $\delta z_C = -l s_\beta \, \delta\beta$, and the virtual work due to the weight is

$$\delta\mathcal{W} = mgls_\beta \, \delta\beta.$$

Which gives the generalized efforts as $e_\alpha^s = 0$, $e_\beta^s = mgls_\beta$, and $e_\gamma^s = 0$.

Lagrange's equation:

The kinetic coenergy of the top is

$$T^* = \frac{1}{2}(I_{12}{\omega_1}^2 + I_{12}{\omega_2}^2 + I_3{\omega_3}^2) = \frac{1}{2}I_{12}\left[s_\beta^2\dot{\alpha}^2 + \dot{\beta}^2\right] + \frac{1}{2}I_3\left(c_\beta\dot{\alpha} + \dot{\gamma}\right)^2.$$

For this system the potential energy is $V = 0$, and the dissipation function is $D = 0$.

Lagrange's equations for the system are

$$\frac{d}{dt}\frac{\partial T^*}{\partial\dot{\alpha}} - \frac{\partial T^*}{\partial\alpha} + \frac{\partial D}{\partial\dot{\alpha}} + \frac{\partial V}{\partial\alpha} = e_\alpha^s,$$

$$\frac{d}{dt}\frac{\partial T^*}{\partial\dot{\beta}} - \frac{\partial T^*}{\partial\beta} + \frac{\partial D}{\partial\dot{\beta}} + \frac{\partial V}{\partial\beta} = e_\beta^s,$$

$$\frac{d}{dt}\frac{\partial T^*}{\partial\dot{\gamma}} - \frac{\partial T^*}{\partial\gamma} + \frac{\partial D}{\partial\dot{\gamma}} + \frac{\partial V}{\partial\gamma} = e_\gamma^s.$$

Using T^*, V, and D it can be shown that these yield the differential equations

$$\frac{d}{dt}\left[(I_{12}s_\beta^2 + I_3c_\beta^2)\dot{\alpha} + I_3c_\beta\dot{\gamma}\right] = 0,$$

$$I_{12}\ddot{\beta} - \left[I_{12} - I_3\right]s_\beta c_\beta\dot{\alpha}^2 + I_3s_\beta\dot{\alpha}\dot{\gamma} = mgls_\beta,$$

$$\frac{d}{dt}\left[I_3(c_\beta\dot{\alpha} + \dot{\gamma})\right] = 0.$$

3.3.2 Electrical systems

The application of Lagrange's equation to electrical systems can be accomplished using the following procedure.

1. *Kinematic analysis.*

The electrical systems considered here form a network of elements, i.e., ideal inductors, capacitors and resistors. Each system element forms a branch of the network, and the interconnection between the elements define the nodes of the network. A simple approach to modeling these systems is to assign a flow variable (i.e., a current) to each branch of the network. This will usually result in a large number of variables that are not all independent. In fact, if B is the number of branches in the network, and N is the number of nodes in the network, then the number of independent flow variables will be $n = B - N + 1$. Equations relating the dependent variables to the independent variables can be obtained by applying Kirchhoff's current law at the nodes of the network.

2. *Applied effort analysis.*

The virtual work done by all effort (i.e., voltage) sources must be defined in terms of the generalized displacements. Here, the virtual work done by an effort source is positive if the current flows from the negative terminal to the positive terminal of the source.

Example 3.16.

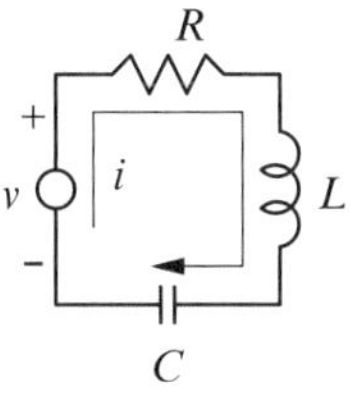

RLC circuit

The RLC (resistor R, inductor L, capacitor C) shown here has an applied voltage v. The dynamic analysis of this system using Lagrange's equation proceeds as follows. The current, $i = dq/dt = \dot{q}$, is selected as the generalized flow variable, and q is the corresponding charge (i.e., the generalized displacement).

Applied effort analysis:

The virtual work done by the voltage is $\delta\mathcal{W} = v\,\delta q = e_q^s\,\delta q$. Therefore, the generalized effort is $e_q^s = v$.

Lagrange's equation:

The kinetic coenergy, potential energy and the dissipation function are $T^* = L\dot{q}^2/2$, $V = q^2/(2C)$, and $D = R\dot{q}^2/2$, respectively. Using these functions we have the following results:

$$\frac{\partial T^*}{\partial \dot{q}} = L\dot{q}, \quad \frac{d}{dt}\frac{\partial T^*}{\partial \dot{q}} = L\ddot{q}, \quad \frac{\partial D}{\partial \dot{q}} = R\dot{q}, \quad \text{and} \quad \frac{\partial V}{\partial q} = \frac{q}{C}.$$

Hence, Lagrange's equation for this system is

$$\frac{d}{dt}\frac{\partial T^*}{\partial \dot{q}} - \frac{\partial T^*}{\partial q} + \frac{\partial D}{\partial \dot{q}} + \frac{\partial V}{\partial q} = e_q^s,$$

$$L\ddot{q} + R\dot{q} + \frac{q}{C} = v.$$

This second order ordinary differential equation describes the behavior of the system in terms of the charge q. This equation can be rewritten in terms of the current as

$$L\frac{di}{dt} + Ri + \frac{1}{C}\int i\,dt = v.$$

Example 3.17.

A model of an electrical lead filter with applied voltage v is shown in the figure below. The dynamic equations of motion for this system can be determined as follows.

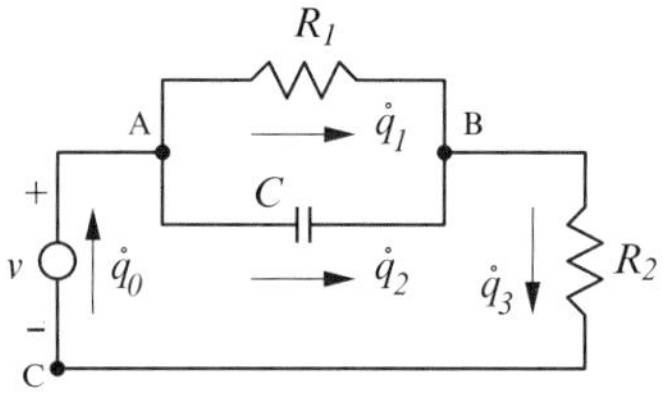

Kinematic analysis:

The circuit has four branches and three nodes (i.e., A, B and C). A current is assigned to each branch of the network. The system has $n = B - N + 1 = 4 - 3 + 1 = 2$ degrees of freedom. (Recall that B is the number of branches and N is the number of nodes in the network.) Therefore, only two of the four flows $\dot{q}_0$, $\dot{q}_1$, $\dot{q}_2$, $\dot{q}_3$ are independent. Here, we select q_1 and q_3 as the independent (generalized) displacements. The corresponding generalized flows are $\dot{q}_1$ and $\dot{q}_3$.

Applying Kirchhoff's current law to nodes B and C gives the relationships $\dot{q}_2 = \dot{q}_3 - \dot{q}_1$, and $\dot{q}_0 = \dot{q}_3$, respectively. Hence, we have obtained explicit expressions for the dependent flows in terms of the independent flows. Moreover, integrating the first of these equations leads to $q_2 = q_3 - q_1 + q_{20}$, where q_{20} is the initial charge in the capacitor.

Applied effort analysis:

The virtual work done by the voltage source is $\delta\mathcal{W} = v\,\delta q_0 = v\,\delta q_3 = e^s_{q_1}\,\delta q_1 + e^s_{q_3}\,\delta q_3$. Therefore, the generalized efforts are $e^s_{q_1} = 0$, and $e^s_{q_3} = v$.

Lagrange's equations:

The kinetic coenergy, potential energy and the dissipation function are $T^* = 0$, $V = (q_3 - q_1 + q_{20})^2/(2C)$, and $D = (R_1\dot{q}_1^2 + R_2\dot{q}_3^2)/2$, respectively.

Lagrange's equations for this system are

$$\frac{d}{dt}\frac{\partial T^*}{\partial \dot{q}_1} - \frac{\partial T^*}{\partial q_1} + \frac{\partial D}{\partial \dot{q}_1} + \frac{\partial V}{\partial q_1} = e^s_{q_1} \qquad (a)$$

$$\frac{d}{dt}\frac{\partial T^*}{\partial \dot{q}_3} - \frac{\partial T^*}{\partial q_3} + \frac{\partial D}{\partial \dot{q}_3} + \frac{\partial V}{\partial q_3} = e^s_{q_3}. \qquad (b)$$

Using the energy relationships and the generalized efforts in (a) and (b) yields the coupled differential equations

$$(a) \Rightarrow R_1\dot{q}_1 - \frac{1}{C}(q_3 - q_1 - q_{20}) = 0,$$

$$(b) \Rightarrow R_2\dot{q}_3 + \frac{1}{C}(q_3 - q_1 - q_{20}) = v.$$

Example 3.18.

The circuit shown in the figure below is called a half-wave rectifier. The network consists of a voltage source, v, a diode, a capacitor, C, and a resistor, R.

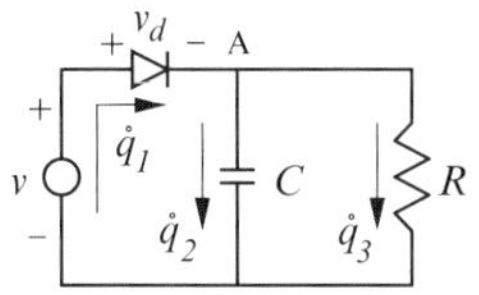

Kinematic analysis:

Three flow variables ($\dot{q}_1, \dot{q}_2$, and $\dot{q}_3$) have been assigned to the network. However, it is easy to verify that the system only has 2 degrees of freedom. Here, we select q_1 and q_3 to be the generalized displacements, and $\dot{q}_1$ and $\dot{q}_3$ as the corresponding generalized flows.

To eliminate the excess displacement variable, q_2, we apply Kirchhoff's current law to node A. This gives $\dot{q}_2 = \dot{q}_1 - \dot{q}_3$. Integrating this equation leads to $q_2 = q_1 - q_3 + q_{20}$, where q_{20} is the initial charge in the capacitor. We therefore have an explicit expression for q_2 in terms of the generalized displacements.

Applied effort analysis:

Here we consider the diode to be an effort regulated flow source. (See Section 1.2.5.) In particular, the diode voltage v_d, and the current $\dot{q}_1$ satisfy the equation

$$\dot{q}_1 = I_s(e^{\alpha v_d} - 1), \qquad (a)$$

where $I_s > 0$, and $\alpha > 0$ are constants. Using (a) it can be seen that

$$v_d = \frac{1}{\alpha} \ln\left(\frac{\dot{q}_1}{I_s} + 1\right).$$

Our analysis must account for the virtual work done by the diode voltage, v_d.

The virtual work done by the voltage source, v, and the diode voltage v_d is $\delta\mathcal{W} = (v - v_d)\,\delta q_1 = e^s_{q_1}\,\delta q_1 + e^s_{q_3}\,\delta q_3$. (Recall that the work done by a voltage source is positive if the current flows from the negative terminal to the positive terminal. Hence, the virtual work done by the source v is $v\,\delta q_1$, and the virtual work done by the diode voltage, v_d, is $-v_d\,\delta q_1$). The net result is that the generalized efforts are $e^s_{q_1} = v - v_d$, and $e^s_{q_3} = 0$.

Lagrange's equations:

The kinetic coenergy, potential energy and the dissipation function are, $T^* = 0$, $V = (q_1 - q_3 + q_{20})^2/(2C)$, and $D = R\dot{q}_3^2/2$, respectively.

Lagrange's equations for this system are

$$\frac{d}{dt}\frac{\partial T^*}{\partial \dot{q}_1} - \frac{\partial T^*}{\partial q_1} + \frac{\partial D}{\partial \dot{q}_1} + \frac{\partial V}{\partial q_1} = e^s_{q_1}, \tag{b}$$

$$\frac{d}{dt}\frac{\partial T^*}{\partial \dot{q}_3} - \frac{\partial T^*}{\partial q_3} + \frac{\partial D}{\partial \dot{q}_3} + \frac{\partial V}{\partial q_3} = e^s_{q_3}. \tag{c}$$

Using the energy relationships and the generalized efforts gives

$$(b) \Rightarrow \frac{1}{C}(q_1 - q_3 + q_{20}) = v - \frac{1}{\alpha}\ln\left(\frac{\dot{q}_1}{I_s} + 1\right),$$

$$(c) \Rightarrow R\dot{q}_3 - \frac{1}{C}(q_1 - q_3 + q_{20}) = 0.$$

Example 3.19.

The schematic in (a) shows an electrical network that includes an operational amplifier. We would like to find a relationship between the applied voltage, v_a, and the voltage at terminal 3 of the operational amplifier, i.e., v_3.

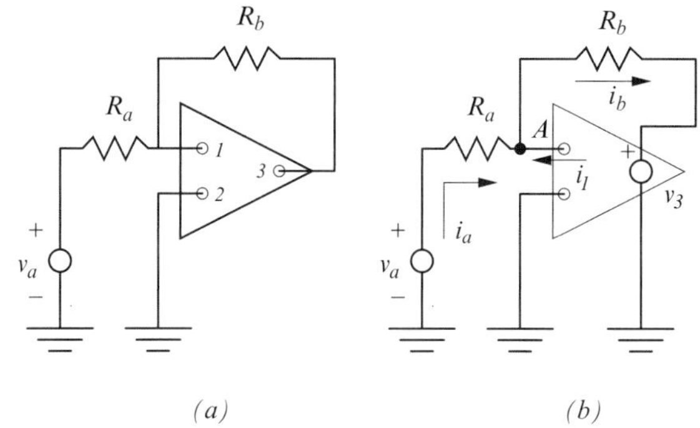

Kinematic analysis:

An equivalent network model of the system is shown in (b). Here, we have assigned the current $i_a = \dot{q}_a$ to the branch that includes the voltage source, v_a, and the resistor R_a. The current $i_b = \dot{q}_b$ is assigned to the resistor R_b, and $i_1 = \dot{q}_1$ is assigned to terminal 1 of the operational amplifier.

Applying Kirchhoff's current law to node A gives the flow constraint equation

$$\dot{q}_a + \dot{q}_1 - \dot{q}_b = 0.$$

However, the input resistance to the operational amplifier is very large and as a result $\dot{q}_1 \approx 0$. Hence, we get $\dot{q}_a = \dot{q}_b$ is the independent flow variable for the network.

Applied effort analysis:

Using the equivalent network model we can see that the virtual work done by the voltage source, v_a, and the output voltage v_3 is $\delta\mathcal{W} = (v_a - v_3)\,\delta q_a = e^s_{q_a}\,\delta q_a$. Therefore, the generalized effort is $e^s_{q_a} = v_a - v_3$.

Lagrange's equations:

The kinetic coenergy, potential energy and the dissipation function are $T^* = 0$, $V = 0$, and $D = (R_a + R_b)\dot{q}_a^2/2$, respectively.

Lagrange's equation for this system is

$$\frac{d}{dt}\frac{\partial T^*}{\partial \dot{q}_a} - \frac{\partial T^*}{\partial q_a} + \frac{\partial D}{\partial \dot{q}_a} + \frac{\partial V}{\partial q_a} = e^s_{q_a}.$$

Which yields

$$(R_a + R_b)\dot{q}_a = v_a - v_3.$$

To eliminate $\dot{q}_a$ from the last equation we use the fact that the operational amplifier behaves according to the rule

$$v_3 = k_g(v_2 - v_1),$$

where k_g is very large constant, v_1 is the voltage at terminal 1, and v_2 is the voltage at terminal 2.

Using the equivalent network model we can see that $v_1 = v_a - R_a\dot{q}_a$, and $v_2 = 0$. Therefore, $v_3 = k_g(R_a\dot{q}_a - v_a)$, from which we obtain

$$\dot{q}_a = \frac{v_3 + k_g v_a}{k_g R_a} \approx \frac{v_a}{R_a}.$$

Putting this in the equation of motion produces

$$\begin{aligned}(R_a + R_b)\frac{v_a}{R_a} &= v_a - v_3 \\ v_3 &= -\frac{R_b}{R_a}v_a.\end{aligned}$$

We can thus conclude that the network represents an inverting amplifier with gain R_b/R_a.

Example 3.20.

A model of a DC motor is shown in the figure below. The motor is excited by a voltage source, v. The resistance and inductance of the motor coils are modeled using R and L, respectively. The rotor has moment of inertia I, and is subjected to torsional damping with damping coefficient b.

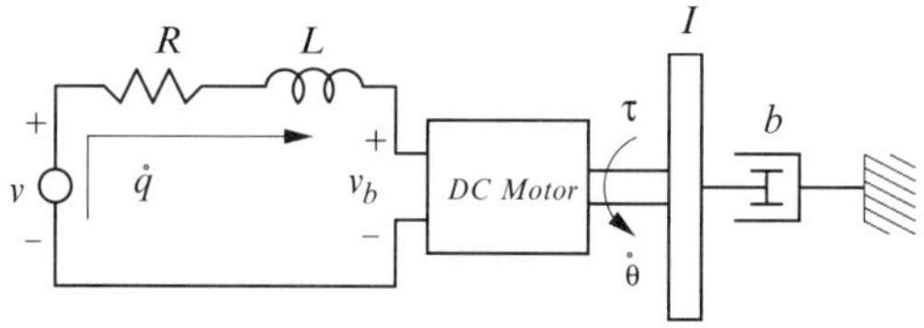

Kinematic analysis:

The generalized displacements for the system are q, the charge in the electrical circuit, and θ the angular position of the rotor. The corresponding generalized flows are $\dot{q}$, the current in the circuit, and $\dot{\theta}$, the angular velocity of the rotor.

Applied effort analysis:

The effort sources in this model are the applied voltage, v, the motor back emf v_b, and the motor torque, τ. The virtual work done by these efforts is $\delta\mathcal{W} = (v - v_b)\,\delta q + \tau\,\delta\theta = e_q^s\,\delta q + e_\theta^s\,\delta\theta$. Therefore, the generalized efforts are

$e_q^s = v - v_b$, and $e_\theta^s = \tau$. For a DC motor the back emf and the torque are related to the generalized velocities by

$$v_b = K_b \dot{\theta}, \tag{a}$$

$$\tau = K_t \dot{q}, \tag{b}$$

where K_b is the back emf constant and K_t is the torque constant.

Lagrange's equations:

The kinetic coenergy, potential energy and the dissipation function are $T^* = L\dot{q}^2/2 + I\dot{\theta}^2/2$, $V = 0$, and $D = R\dot{q}^2/2 + b\dot{\theta}^2/2$, respectively.

Lagrange's equation for this system are

$$\frac{d}{dt}\frac{\partial T^*}{\partial \dot{q}} - \frac{\partial T^*}{\partial q} + \frac{\partial D}{\partial \dot{q}} + \frac{\partial V}{\partial q} = e_q^s, \tag{c}$$

$$\frac{d}{dt}\frac{\partial T^*}{\partial \dot{\theta}} - \frac{\partial T^*}{\partial \theta} + \frac{\partial D}{\partial \dot{\theta}} + \frac{\partial V}{\partial \theta} = e_\theta^s. \tag{d}$$

Using the energy relationships, the generalized efforts gives, and (a) and (b), we get

$$(c) \Rightarrow L\ddot{q} + R\dot{q} = v - K_b\dot{\theta},$$
$$(d) \Rightarrow I\ddot{\theta} + b\dot{\theta} = K_t\dot{q}.$$

3.3.3 Fluid systems

A simple approach to analyzing fluid systems is to construct an equivalent electrical circuit model for the system then, apply the electrical systems modeling techniques to this equivalent circuit. In the fluid system long pipes can be modeled as ideal inductors, valves and nozzles can be modeled as ideal resistors, and tanks can be modeled as ideal capacitors.

Example 3.21.

Consider the fluid system shown in Fig. 3.4a and its corresponding equivalent circuit model Fig. 3.4b. In this system a pump delivers fluid to a long pipe which is connected to a tank. The fluid exiting the tank flows through a valve. The pump is modeled as an effort source, P, the pipe is modeled as fluid inertia, L_f, the tank is modeled as a fluid capacitance, C_f, and the valve is modeled as a fluid resistance, R_f.

Kinematic analysis:

As shown in Fig. 3.4b a flow variable is assigned to each element of the system. However, there are only two independent variables ($n = B - N + 1 = 4 - 3 + 1 = 2$.) Here, we select Q_2 and Q_4 as the generalized flows.

To eliminate the flows Q_1 and Q_3 we note that the continuity of flow at nodes A and B gives, $Q_1 = Q_2$, and $Q_3 = Q_2 - Q_4$, respectively. Integrating these equations leads to

$$V_1 = V_2 + V_{10}, \tag{a}$$

$$V_3 = V_2 - V_4 + V_{30}. \tag{b}$$

Here, V_i is the volume (displacement) that corresponds to the flow Q_i, for $i = 1, 2, 3, 4$. (Note that $Q_i = \dot{V}_i$.) Also, V_{10} and V_{30} represent initial volumes.

Applied effort analysis:

The virtual work done by the pressure source, P, is $\delta\mathcal{W} = P\,\delta V_1$. However, from (a) we get that $\delta V_1 = \delta V_2$. Hence, $\delta\mathcal{W} = P\,\delta V_2 = e^s_{V_2}\,\delta V_2 + e^s_{V_4}\,\delta V_4$. Therefore, the generalized efforts are $e^s_{V_2} = P$, and $e^s_{V_4} = 0$.

Lagrange's equations:

The kinetic coenergy, potential energy and the dissipation function are $T^* = L_f\dot{V}_2^2/2$, $V = (V_2 - V_4 + V_{30})^2/(2C_f)$, and $D = R_f\dot{V}_4^2/2$, respectively.

Lagrange's equations for this system are

$$\frac{d}{dt}\frac{\partial T^*}{\partial \dot{V}_2} - \frac{\partial T^*}{\partial V_2} + \frac{\partial D}{\partial \dot{V}_2} + \frac{\partial V}{\partial V_2} = e^s_{V_2}, \tag{c}$$

$$\frac{d}{dt}\frac{\partial T^*}{\partial \dot{V}_4} - \frac{\partial T^*}{\partial V_4} + \frac{\partial D}{\partial \dot{V}_4} + \frac{\partial V}{\partial V_4} = e^s_{V_4}. \tag{d}$$

Using the energy relationships and the generalized efforts gives

$$(c) \Rightarrow L_f\ddot{V}_2 + \frac{1}{C_f}(V_2 - V_4 + V_{30}) = P, \tag{e}$$

$$(d) \Rightarrow R_f\dot{V}_4 - \frac{1}{C_f}(V_2 - V_4 + V_{30}) = 0. \tag{f}$$

Equations (e) and (f) can be written in terms of the generalized flows instead of the generalized displacements. Specifically, put (f) into (e), and differentiate (f) to get

$$(e) \Rightarrow L_f\dot{Q}_2 + R_f\dot{Q}_4 = P,$$
$$(f) \Rightarrow R_f\dot{Q}_4 - \frac{1}{C_f}(Q_2 - Q_4) = 0.$$

Example 3.22.

Figure (a) below shows a schematic of a fluid system, and the Fig. (b) shows the equivalent network model for the system. The pump supplies fluid to the first tank with given pressure, P. The tanks are connected to each other via a nonlinear valve. Each tank is treated as an ideal capacitor with capacitance C_1 and C_2. The pressure drop across the nonlinear valve, P_v, satisfies the equation,

$$P_v = R_f|\dot{V}_v|\dot{V}_v, \tag{a}$$

where R_f is a constant and $\dot{V}_v$ is the volume flow rate through the valve.

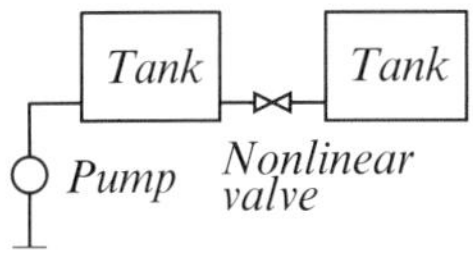

(a)

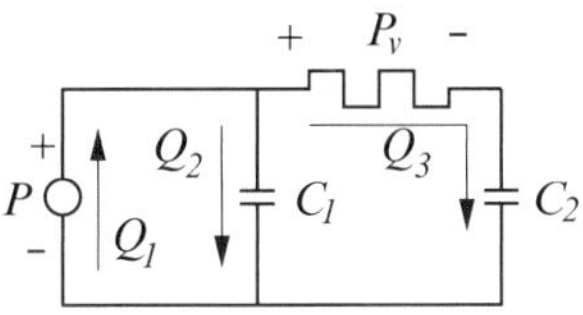

(b)

Kinematic analysis:

Three flow variables have been assigned to the system, however there are only two independent variables ($n = B - N + 1 = 3 - 2 + 1 = 2$.) Here, we select Q_1 and Q_3 as the generalized flows. The continuity of flow at the tank C_1 gives $Q_2 = Q_1 - Q_3$. Integrating this equation leads to

$$V_2 = V_1 - V_3 + V_{20}. \tag{b}$$

Here, V_i is the volume (displacement) that corresponds to the flow Q_i, for $i = 1, 2, 3$. (Note that $Q_i = \dot{V}_i$.) Also, V_{20} represents the initial volume in the tank C_1.

Applied effort analysis:

The virtual work done by the pressure source, P, and by the nonlinear valve is $\delta\mathcal{W} = P\,\delta V_1 - P_v\,\delta V_3 = e^s_{V_1}\,\delta V_1 + e^s_{V_3}\,\delta V_3$. Therefore, the generalized efforts are $e^s_{V_1} = P$, and $e^s_{V_3} = -P_v$. Note, that the virtual work done by the nonlinear valve is negative because the flow is directed from the positive terminal to the negative terminal.

Lagrange's equations:

The kinetic coenergy, potential energy and the dissipation function are $T^* = 0$, $V = (V_1 - V_3 + V_{20})^2/(2C_1) + V_3^2/(2C_2)$, and $D = 0$, respectively.

Lagrange's equations are

$$\frac{d}{dt}\frac{\partial T^*}{\partial \dot{V}_1} - \frac{\partial T^*}{\partial V_1} + \frac{\partial D}{\partial \dot{V}_1} + \frac{\partial V}{\partial V_1} = e^s_{V_1}, \tag{c}$$

$$\frac{d}{dt}\frac{\partial T^*}{\partial \dot{V}_3} - \frac{\partial T^*}{\partial V_3} + \frac{\partial D}{\partial \dot{V}_3} + \frac{\partial V}{\partial V_3} = e^s_{V_3}. \tag{d}$$

Using the energy relationships, the generalized efforts, and (a) gives

$$(c) \Rightarrow \frac{1}{C_1}(V_1 - V_3 + V_{20}) = P,$$

$$(d) \Rightarrow -\frac{1}{C_1}(V_1 - V_3 + V_{20}) + \frac{V_3}{C_2} = -R_f|\dot{V}_3|\dot{V}_3.$$

References

1. C. Lanczos, *The Variational Principles of Mechanics*, vol. 4, 4th ed. University of Toronto Press, 1970.
2. G. W. Ogar and J. J. D'Azzo, "A Unified Procedure for Deriving the Differential Equations of of Electrical and Mechanical Systems," IRE Transactions on Education, vol. E-5, pp. 18-26, 1962.
3. L. Meirovitch, *Methods of Analytical Dynamics*, McGraw-Hill, 1970.
4. L. A. Pars, *A Treatise on Analytical Dynamics*. Heinemann, 1965.
5. R. J. Smith, *Circuits devices and systems*, John Wiley and Sons, 1976.
6. G. J. Van Wylen and R. E. Sonntag, *Fundamentals of Classical Thermodynamics*, 3rd ed. Wiley & Sons, 1985.
7. P. E. Wellstead, *Introduction to Physical System Modelling*, Academic Press, 1979.
8. D. C. White, and H. H. Woodson, *Electromechanical Energy Conversion*, John Wiley & Sons, 1959.
9. J. H. Williams, Jr., *Fundamentals of Applied Dynamics*, John Wiley & Sons, 1996.

Problems

1. In Example 3.5, show that the reaction forces at O do not contribute to the virtual work.

2. Show that with the proper choice of units for the fundamental variables the virtual work for all the terms in Table 3.1 have the same units.
3. Derive Lagrange's equation for the viscoelastic system shown below.

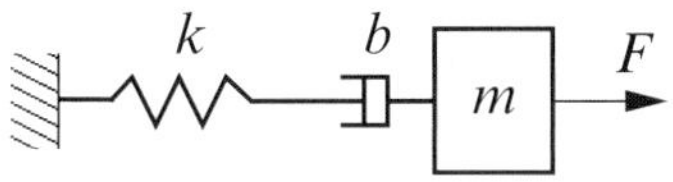

4. In the system shown here mass m_1 can move freely while, mass m_2 is subjected to static and kinetic (sliding) friction forces. The coefficient of static friction is μ_s, and the coefficient of kinetic friction is μ_k. Derive Lagrange's equation of motion for the system.

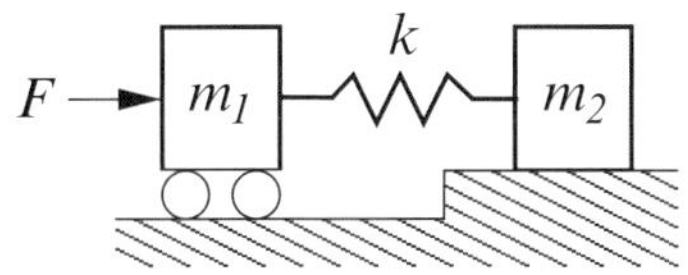

5. The pendulum on the left is connected to the pendulum on the right via nonlinear spring that behaves according to the law

$$f_s = ky^3,$$

where f_s is the force applied to the spring, k is a constant, and y is the net deflection of the spring. Derive Lagrange's equations of motion for this system. Assume small angular displacements for each pendulum, and that the acceleration due to gravity acts downward as shown.

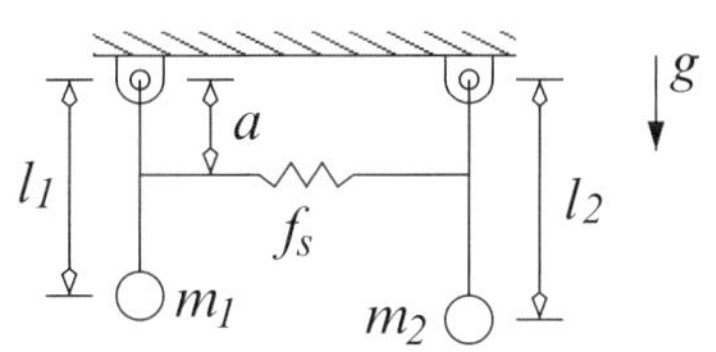

6. Derive Lagrange's equation for the (a) double pendulum, and (b) the spherical pendulum shown here. In the case of the double pendulum the rod l_1 can rotate about the pivot at O in the x-y plane. The rod l_2 can rotate about M_1 in the x-y plane. For the spherical pendulum the rod, l, is free to rotate about the x-axis and y-axis simultaneously. Derive Lagrange's equation of motion for each system.

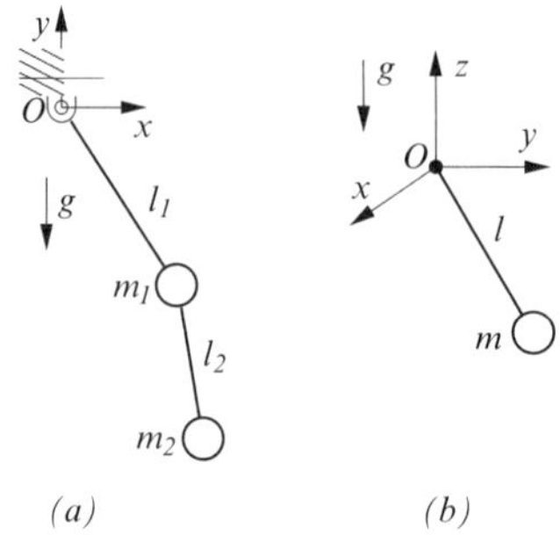

7. A $\frac{1}{4}$-car model is shown in the figure below. Here, $y(t)$ represents the profile of the road from a fixed reference. The effective spring stiffness of the tire is k_t, and m_t is the mass of the tire. The suspension has stiffness k, and damping coefficient b. The variable m denotes one quarter of the mass of the car. Derive Lagrange's equation for this system. (Neglect the horizontal motion of the system.)

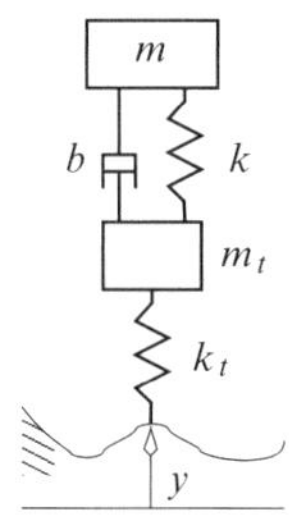

8. Derive Lagrange's equation of motion for the system shown below. Here, the rod has mass m and moment of inertia I_c about its center of mass. The linear spring has stiffness k, and the linear damper has damping coefficient c. (Assume small small angular displacements for the rod.)

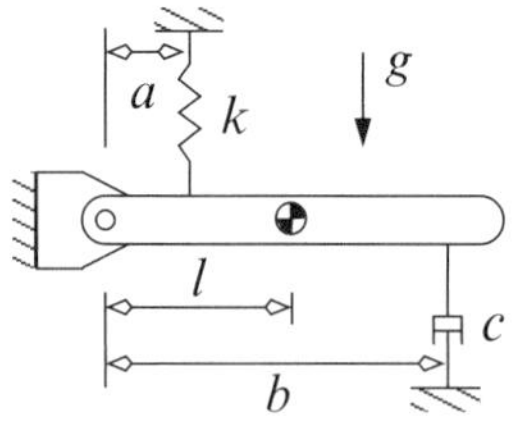

9. The figure below shows a $\frac{1}{2}$-car model. The road profile at the front of the car is $y_f(t)$, and the road profile at the rear of the car is $y_r(t)$. Both measure from the same fixed reference. The front tire has an effective spring stiffness k_{ft}, and mass m_{ft}. The rear tire has an effective spring stiffness k_{rt}, and mass m_{rt}. The front suspension has stiffness k_f, and damping coefficient b_f. The rear suspension has stiffness k_r, and damping coefficient b_r. The effective mass and moment of inertia for one half of the

car are m, and I, respectively. Derive Lagrange's equation for this system. (Neglect the horizontal motion of the system.) If the car has a constant horizontal velocity, v, how are $y_f(t)$ and $y_r(t)$ related?

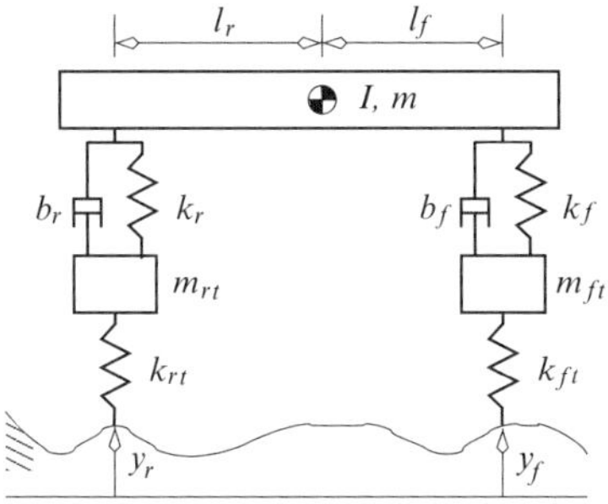

10. The figure below shows a model of an unbalanced disk (rotor). The disk is free to rotate about point O, while point O can move in the plane. The bearings at O have stiffness k_x in the x-direction, and k_y in the y direction. The rotor has mass m, and moment of inertia I about its center of mass. Moreover, the center of mass is offset from the geometric center of the rotor by a distance e. Derive Lagrange's equation for this system.

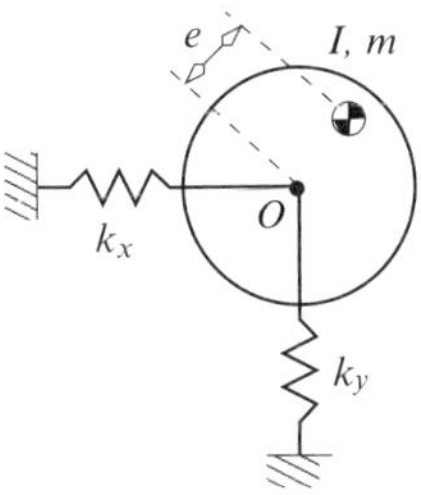

11. The block A has mass m_1 and is constrained to move in the vertical direction. Attached to the block is a linear spring with stiffness k and a linear damper with damping coefficient c. The mass m_2 is also attached to the block A via a massless rod of length e. The mass m_2 has a constant angular velocity ω. Derive Lagrange's equations of motion for this system. (Neglect gravity.)

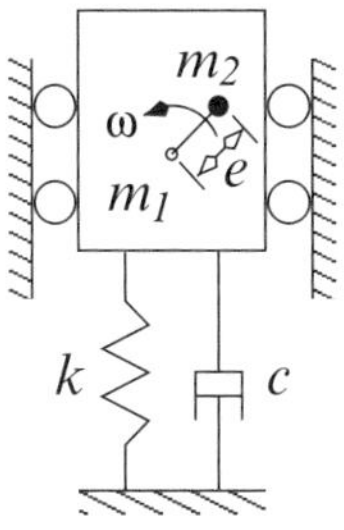

12. Using generalized displacements (i.e., independent displacements) derive Lagrange's equation of motion for the system shown here. The block A has mass m_1, the block B has mass m_2, and the rod AB has mass m_3 and moment of inertia I_3 about its center of mass. The uniform rod AB has length l_3, and its center of mass is at the geometric center of the rod. The rod is connected to the blocks via revolute joints.

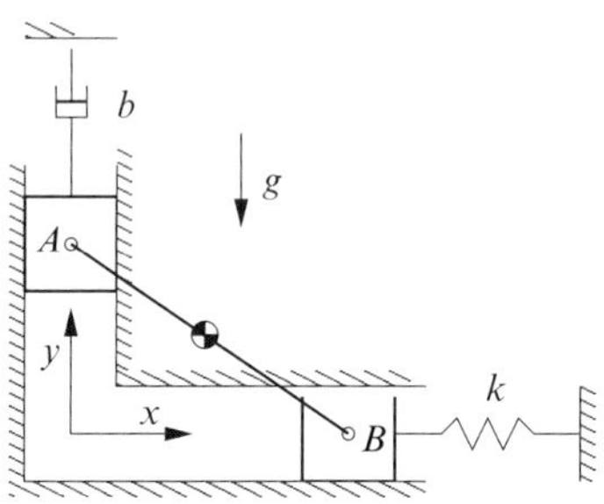

13. Verify the Lagrange's equations of motion in Example 3.14.
14. Verify the Lagrange's equations of motion in Example 3.15.
15. Derive Lagrange's equation of motion of the systems shown in Examples 3.9, 3.10, 3.11, 3.12, 3.13 and 3.15. In each case include the force(s) due to gravity in the potential energy instead of the virtual work.
16. Derive Lagrange's equation for the following electrical circuits.

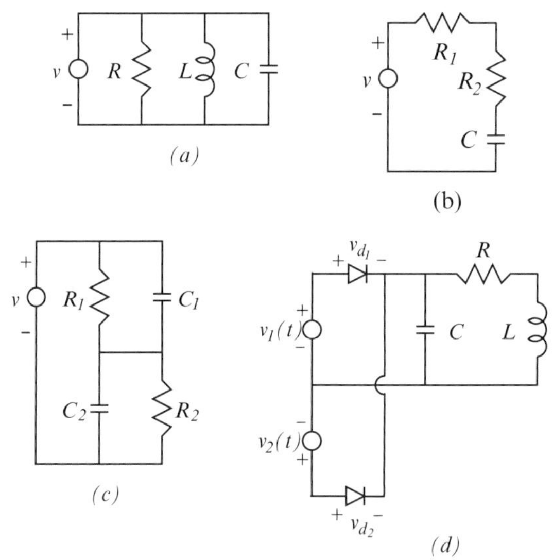

17. Derive Lagrange's equation for the electromagnetic suspension shown below. Here, the inductance of the coil is given by $L = \frac{\gamma_0}{\gamma_1 + x}$, where γ_0, and γ_1 are constants. The variable x is air gap distance between the suspended mass and the inductor. The acceleration due to gravity acts downward as shown.

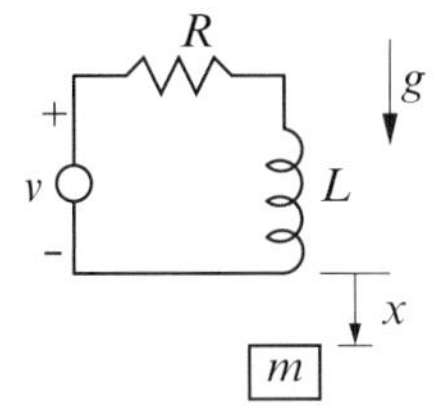

18. For the fluid systems shown below derive Lagrange's equation.

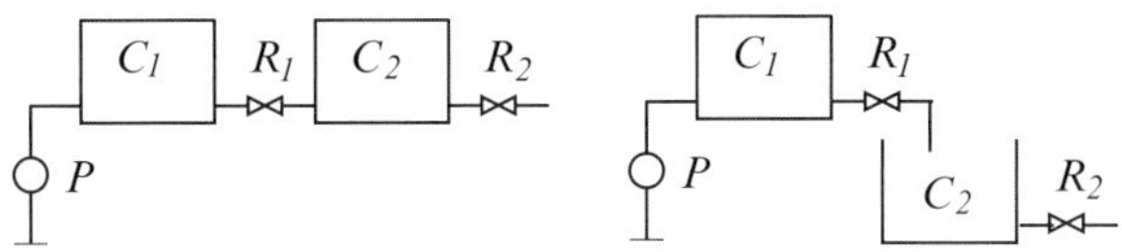

(a) (b)

19. Using Lagrange's equation of motion determine the differential equation relating v_a and v_3 for the operational amplifier networks shown below.

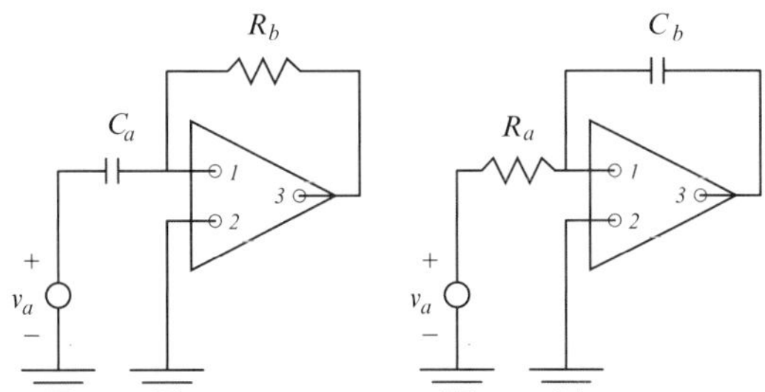

(a) (b)

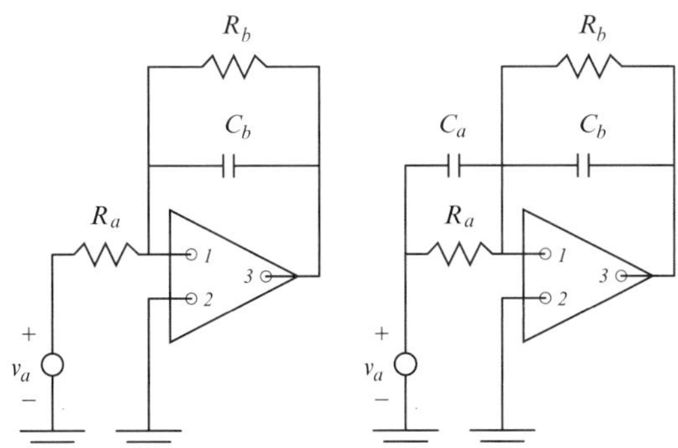

(c) (d)

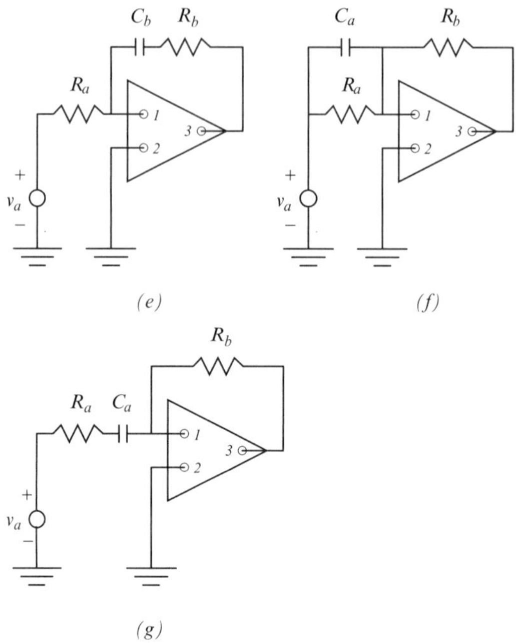

(e) (f) (g)

20. Derive Lagrange's equation of motion for the system shown below. In this system the mass, m, is attached to a cord that is wrapped around the cylinder that has a radius r. In the position shown the cord has length l_0. The cylinder is fixed an can not rotate. Also, assume that during the motion the cord is always in tension.

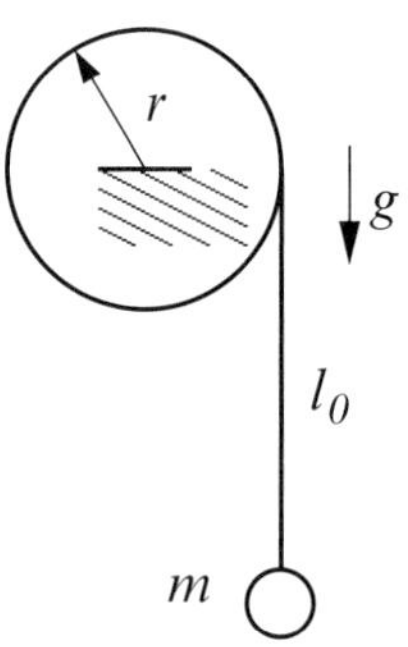

21. Derive Lagrange's equation of motion for the systems shown below. In (a) the rod has mass m_1 and moment of inertia I_1 about the center of mass. In (b) the uniform disk has radius r, mass m_1 and moment of inertia I_1 about the center of mass. In (c) each of the uniform disks have radius r and roll without slipping.

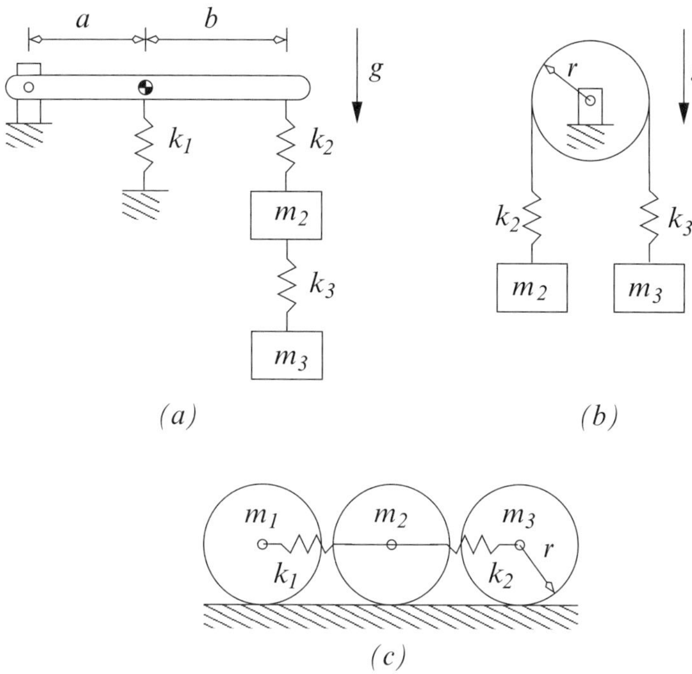

22. The mass m is attached to the flexible belt-pulley system as shown below. Portions of the belt has linear stiffness k_1 and k_2. The pulleys have moments of inertia I_1 and I_2. Each pulley has radius r. A torque τ is applied to the first pulley. In addition the pulley bearings are subject to linear torsional damping with damping coefficient b. Derive Lagrange's equation of motion for this system.

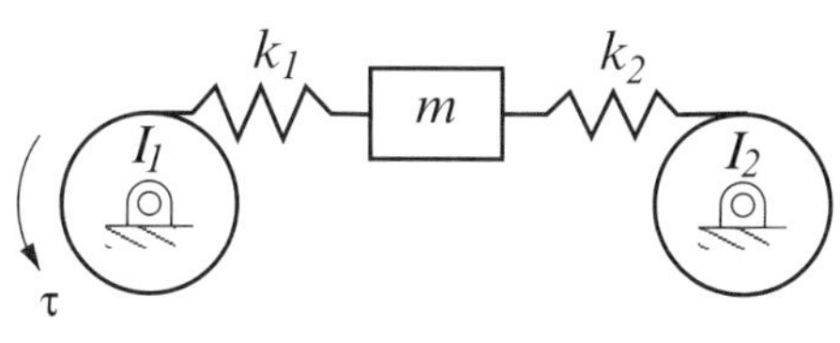

23. Derive Lagrange's equation of motion for the system shown below using θ as the generalized displacement. Here, the disk with mass m and moment of inertia I rolls without slipping on the incline. The linear spring, with stiffness k, is attached to the outer edge of the disk.

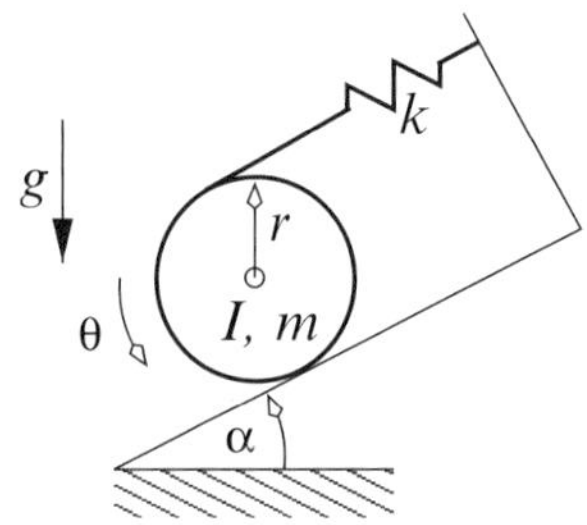

24. For the pulley system shown below, determine (i) the number of degrees of freedom, and (ii) derive Lagrange's equation of motion using the generalize coordinates. Note that each pulley has a mass and a moment of inertia.

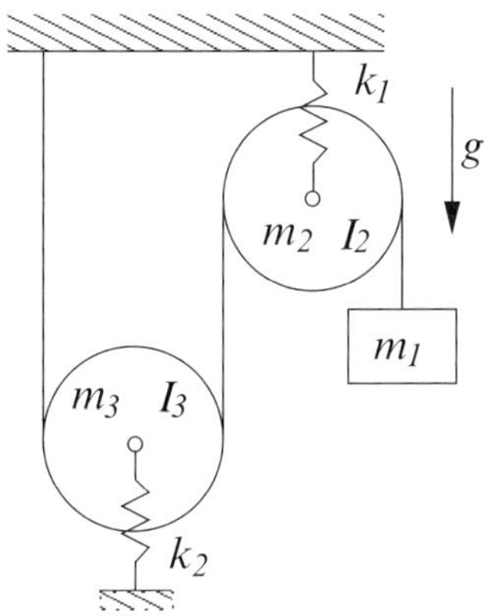

25. The rotordynamic system shown here consists of a uniform disk that is attached to a vertical shaft AB. At A the shaft can rotate in a frictionless pivot. At B the shaft is supported by linear springs with stiffness k_x and k_y along the fixed x-axis and y-axis, respectively. The disk rotates with angular velocity w_{z_1}. Derive Lagrange's equation of motion for this system neglecting the weight of the shaft. Assume that the shaft can only undergo small displacements in the x and y directions at B.

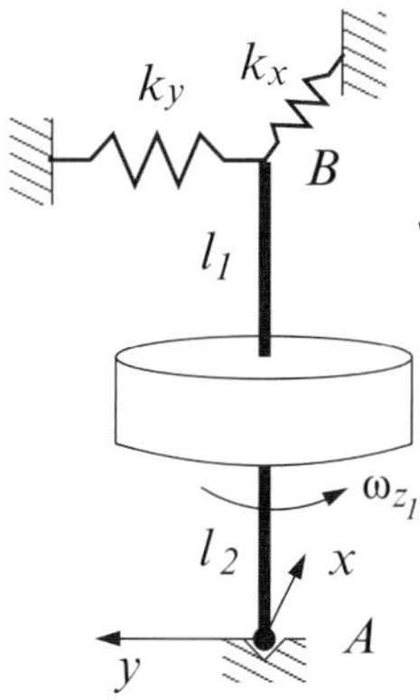

26. In the system shown below x-y-z represents the fixed rectangular coordinate system. (Note that the x-axis is directed into the page.) The disk shown is attached to the shaft AB. A torque τ is applied to the shaft and is directed along the positive z-axis. The principal axes of inertia for the disk are aligned with the x_1-y_1-z_1 rectangular coordinate system, which is embedded in the disk. Take the moments of inertia of the disk to be $I_1 = I_2 = I$ about the x_1 and y_1 axes, and I_3 about the z_1 axis. The z_1 axis make a constant angle α with the z-axis. Derive Lagrange's equation of motion for the system. Also, develop expressions for the reaction forces at the bearings A and B. Assume that the shaft has negligible mass and length l. (See Chapter 2, problem 15.)

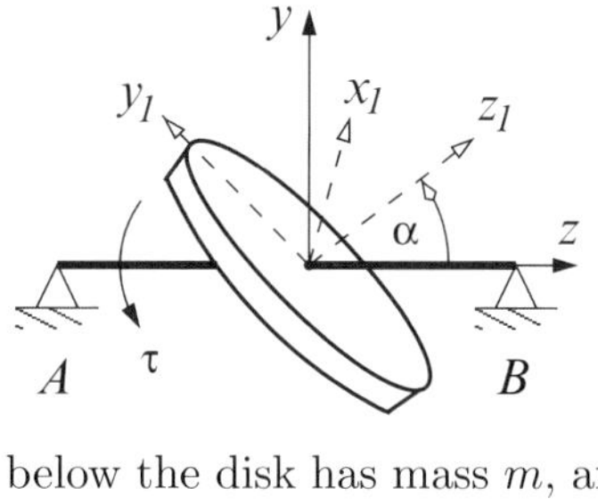

27. In the system shown below the disk has mass m, and moments of inertia $I_1 = I_2 = I$ about the x_1 and y_1 axes, and I_3 about the z_1-axis. The center of mass of the disk is offset from the shaft AB by a distance ϵ. Also, the principal axes of inertia are aligned with the x_1-y_1-z_1 rectangular coordinate system, where the z_1-axis make a constant angle α with the z-axis. Derive Lagrange's equation of motion for this system, an expressions for the reaction forces at the bearings A and B. Assume that the shaft has negligible mass and length l.

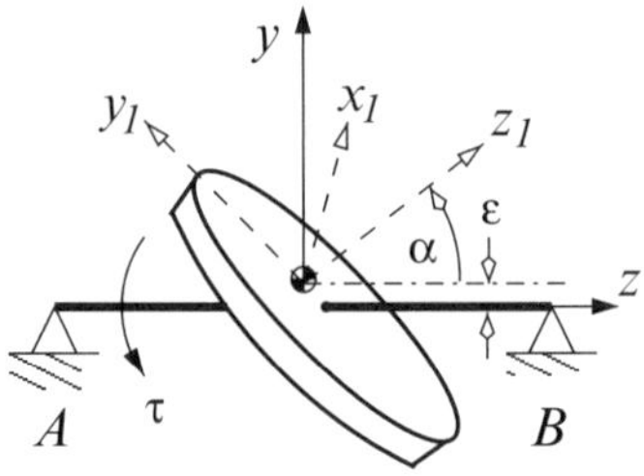

28. Repeat problem 27 for the case where the bearings are elastic. That is, at A the bearing has stiffness k_{Ax} along the x-axis, and k_{Ay} along the y-axis. Similarly, at B the bearing has stiffness k_{Bx} along the x-axis, and k_{By} along the y-axis. Assume that the shaft has negligible mass and length l.

Chapter 4
Constrained Systems

In the previous chapters it was shown that the configuration of the system can be described using N displacement variables, say, $q = [q_1, q_2, \cdots, q_N]^T$, and the associated flow, momentum and effort variables. Although the system can be described using N displacement variables there may only be n degrees of freedom, where $n \leq N$. Recall that in Chapter 3 Lagrange's equation was developed by selecting n independent displacement variables from the set $q = [q_1, q_2, \cdots, q_N]^T$. These n independent displacements are called the generalized displacements. The remaining $N - n$ dependent displacement variables are related to the generalized displacements via $N - n$ displacement and/or flow constraint equations. These constraint relationships are used to eliminate the dependent variables from the kinetic coenergy, the potential energy, the dissipation function, and the virtual work done by the applied efforts. As a result the dynamic equations of motion are formulated in terms of the generalized coordinates alone.

In this chapter we will develop versions of Lagrange's equation that will allow us to retain all N configuration variables if the formulation of the equations of motion. This approach to systems modeling is particularly useful for complex systems where the relationship between the independent and dependent variables are highly nonlinear. For such systems eliminating the $N - n$ dependent variables is a nontrivial task.

Section 4.1 classifies the types of constraint equations treated in this text. In Section 4.2 the Theorem of Lagrange multiplies is used to develop a version of Lagrange's equation where the dependent variables associated with displacement constraints are retained. In Section 4.3 Lagrange's equation is further extended to include systems with displacement and flow constraints. Finally, Section 4.4 considers dynamic systems that include effort and dynamic constraints.

B. Fabien, *Analytical System Dynamics: Modeling and Simulation*,
DOI 10.1007/978-0-387-85605-6_4,

4.1 Constraint Classification

4.1.1 Displacement and flow constraints

In Chapter 2 we explored the kinematic constraints that arise in the models of mechanical systems. There, is was shown that the spatial properties of the model can lead to displacement constraints of the form

$$\phi_j(q_1, q_2, \cdots, q_N) = 0, \quad j = 1, 2, \cdots, m, \qquad (a)$$

where $q = [q_1, q_2, \cdots, q_N]$ are the N configuration displacement coordinates that are used to describe the system, and $m = N - n$, with n being the number of degrees of freedom in the model.

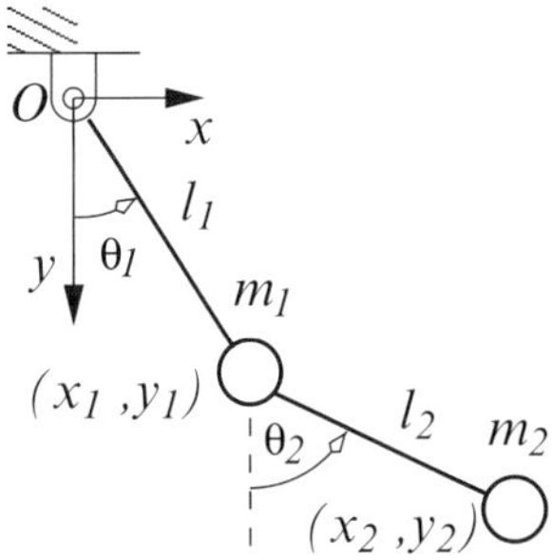

For example the double pendulum system shown here uses configuration coordinates x_1, y_1, θ_1, x_2, y_2, and θ_2, to describe the position of the masses m_1 and m_2. Here, (x_1, y_1) is the location of the mass m_1, θ_1 is the angle between l_1 and the y-axis, (x_2, y_2) is the location of the mass m_2, and θ_2 is the angle between l_2 and the y-axis. Since the system only has 2 degrees of freedom these 6 configuration coordinates are not all independent. If we select θ_1 and θ_2 as the independent coordinates (i.e., the generalized displacements) then we can determine 4 constraint equations that will determine the remaining 'excess' coordinates. Specifically, the constraint equations for the double pendulum are

$$\begin{aligned} \phi_1 &= x_1 - l_1 \sin\theta_1 = 0, \\ \phi_2 &= y_1 - l_1 \cos\theta_1 = 0, \\ \phi_3 &= x_2 - (l_1 \sin\theta_1 + l_2 \sin\theta_2) = 0, \\ \phi_4 &= y_2 - (l_1 \cos\theta_1 + l_2 \cos\theta_2) = 0. \end{aligned} \qquad (b)$$

Clearly these equations are in the form described by equation (a), and can be used to explicitly eliminate x_1, y_1, x_2, and y_2 from the formulation of the equations of motion.

It was also shown in Chapter 2 that network models give rise to flow constraints of the form

$$\psi_j(f_1, f_2, \cdots, f_N) = 0, \quad j = 1, 2, \cdots, m, \qquad (c)$$

where $f = [f_1, f_2, \cdots, f_N]^T$ are the N configuration flow coordinates associated with the system. Consider the network system shown here.

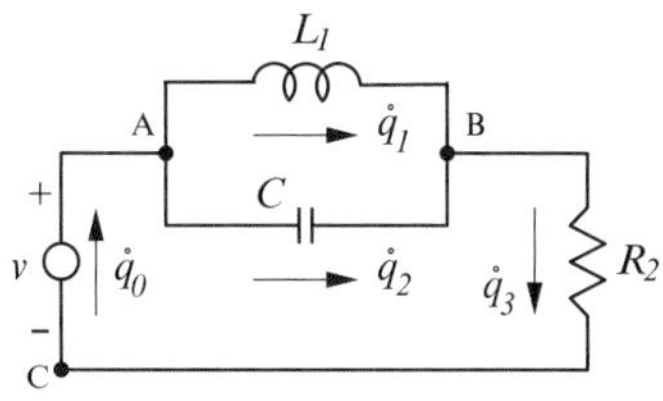

The system is assigned 4 flow variables ($\dot{q}_0$, $\dot{q}_1$, $\dot{q}_2$, and $\dot{q}_3$) but there are only 2 independent loops. Hence, 2 of the flow variables represent excess coordinates. If we select $\dot{q}_0$ and $\dot{q}_3$ to be the independent flow variables (i.e., the generalized flows) then the other two variables are determined from the flow constraints.

$$\begin{aligned}\psi_1 &= \dot{q}_0 - (\dot{q}_1 + \dot{q}_2) = 0,\\ \psi_2 &= \dot{q}_1 + \dot{q}_2 - \dot{q}_3 = 0.\end{aligned} \tag{c}$$

These equations are found by applying Kirchhoff's current law at nodes A and B.

A useful property of the flow constraints given by equation (c) is that they can be integrated to obtain displacement constraints as described by equation (a). For example, integrating the flow constraints given in equation (c) results in the displacement constraints

$$\begin{aligned}\phi_1 &= q_0 - (q_1 + q_2) + Q_0 = 0,\\ \phi_2 &= q_1 + q_2 - q_3 + Q_1 = 0,\end{aligned} \tag{d}$$

where Q_0 and Q_1 constants of integration that are selected to ensure that the displacement constraints are consistent at the initial time.

We also note that the displacements constraints given by equation (a) can be differentiated with respect to time to get flow constraints of the form (c). The resulting flow constraints can be integrated to recover the original displacement constraints. For example differentiating the displacement constraints (b) gives

$$\begin{aligned}\psi_1 &= \dot{x}_1 - l_1\dot{\theta}_1\cos\theta_1 = 0,\\ \psi_2 &= \dot{y}_1 + l_1\dot{\theta}_1\sin\theta_1 = 0,\\ \psi_3 &= \dot{x}_2 - (l_1\dot{\theta}_1\cos\theta_1 + l_2\dot{\theta}_2\cos\theta_2) = 0,\\ \psi_4 &= \dot{y}_2 + l_1\dot{\theta}_1\sin\theta_1 + l_2\dot{\theta}_2\sin\theta_2 = 0,\end{aligned} \tag{e}$$

which can be integrated to obtain (b). In fact, the equations in (e) are exact differentials of the displacement constraints, i.e.,

$$\begin{aligned}\psi_j\,dt = {}& \frac{d\phi_j}{dx_1}\,dx_1 + \frac{d\phi_j}{dy_1}\,dy_1 + \frac{d\phi_j}{d\theta_1}\,d\theta_1\\ &+ \frac{d\phi_j}{dx_2}\,dx_2 + \frac{d\phi_j}{dy_2}\,dy_2 + \frac{d\phi_j}{d\theta_2}\,d\theta_2 = 0, \quad j = 1,2,3,4\end{aligned}$$

In this text we will restrict our attention to flow constraints that have the general form

$$\psi_j(q, f, t)\, dt = \sum_{i=1}^{N} a_{ji}(q, t)\, dq_i + a_j(q, t)\, dt = 0, \quad j = 1, 2, \cdots, m, \qquad (f)$$

where $q = [q_1, q_2, \cdots, q_N]^T$ are the displacements, $f = [f_1, f_2, \cdots, f_N]^T$ are the flows, and the coefficients a_{ji} and a_j are functions of the displacements and the time, t. Equation (f) is called the *Pfaffian* form of the constraint. In terms of the flow variables, equation (f) can be written as

$$\psi_j(q, f, t) = \sum_{i=1}^{N} a_{ji}(q, t)\, f_i + a_j(q, t) = 0, \quad j = 1, 2, \cdots, m. \qquad (4.1)$$

If the coefficients a_{ji} and a_j are exact differentials of some function (say ϕ_j), then equation (4.1) can be integrated to obtain the displacement constraints

$$\phi_j(q, t) = 0, \quad j = 1, 2, \cdots, m. \qquad (4.2)$$

In analytical mechanics the flow constraints (4.1) are called *holonomic constraints*, if they can be integrated to obtain displacement constraints of the form (4.2). The dynamic systems associated with integrable flow constraints are called holonomic systems. The two examples given above, (equation (b) and equation (d)), describe holonomic constraints. If the time variable, t, appears explicitly in the displacement constraints (4.2) then, they are called *rheonomic* constraints. If the time variable does not appear explicitly in the displacement constraints (4.2) then, they are called *scleronomic* constraints.

If the constraints (4.1) are not integrable then, they are called *nonholonomic* constraints, and the associated dynamic system is called a nonholonomic system. Such a system is shown in the figure below.

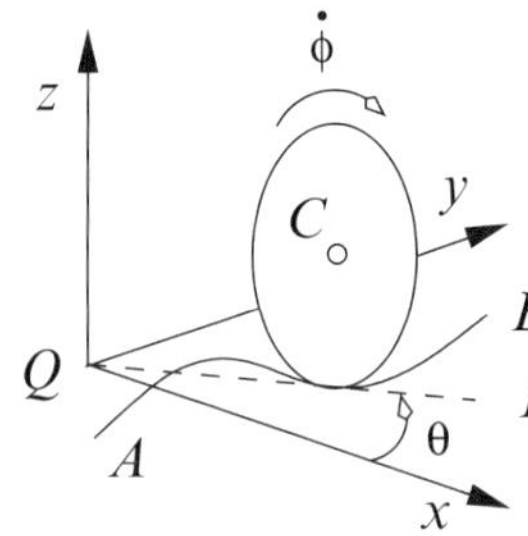

The figure shows a disk with radius r rolling without slipping along the path AB. At the instant shown the line QP is tangent to the path of the disk. The disk has angular velocity $\dot{\phi}$ that is directed about a line that passes though the center of the disk, C, and is perpendicular to QP. Hence, the x and y

components of velocity of the center of the disk satisfy the constraints

$$\dot{x} - r\dot{\phi}\cos\theta = 0,$$
$$\dot{y} - r\dot{\phi}\sin\theta = 0.$$

In differential form these constraints are

$$\begin{aligned} dx - r\cos\theta\, d\phi &= 0, \\ dy - r\sin\theta\, d\phi &= 0. \end{aligned} \tag{g}$$

Clearly, the constraints in equation (g) are in Pfaffian form where, x, y, ϕ and θ are all displacement variables. Moreover, there are no integrating factors that will reduce these equations that involve displacement variables alone. Thus, the constraints in (g) are nonholonomic constraints. Unlike holonomic constraints, the equations in (g) can not be used to eliminate two of the four displacement variables. Hence, all four displacement variables must be used to describe the system dynamics.

4.1.2 Dynamic constraints

Some of the regulated sources described in this text generate efforts that are related to the derivatives or integrals of the displacement and flow variables. To account for these relationships we introduce m_s dynamic variables $s = [s_1, s_2, \cdots, s_{m_s}]^T$, which will allows us to write the *dynamic constraints* as differential equations of the form

$$\dot{s}_j - \Sigma_j(q, f, e, s, t) = 0, \;\; j = 1, 2, \cdots, m_s. \tag{4.3}$$

Here, q=$[q_1, q_2, \cdots, q_N]^T$ are the displacement variables, f=$[f_1, f_2, \cdots, f_N]^T$ are the flow variables, $e = [e_1, e_2, \cdots, e_{m_s}]^T$ are the effort variables, t is the time, and Σ_j is a continuous function of the system variables. Some models of transistors can be put in the form of equation (4.3). Also, the behavior of control system elements can be described using dynamic constraints. In the systems modeling technique presented below the dynamic constraints are retained in the model formulation, and there is no need to explicitly eliminate the dynamic variables s.

4.1.3 Effort constraints

In addition to constraints involving displacement and flow variables, we consider in this text constraints that include effort variables. These *effort constraints* are algebraic equations of the form

$$\Gamma_j(q, f, e, s, t) = 0, \; j = 1, 2, \cdots, m_e, \tag{4.4}$$

where q=$[q_1, q_2, \cdots, q_N]^T$ are the displacement variables, f=$[f_1, f_2, \cdots, f_N]^T$ are the flow variables, $e = [e_1, e_2, \cdots, e_{m_e}]^T$ are the effort variables, $s = [s_1, s_2, \cdots, s_{m_s}]^T$ are the dynamic variables, t is the time, and Γ_j is a continuous function of the system variables. Constraints of this type arise from system elements that are regulated effort sources such as diodes, DC motors and Coulomb friction (see Section 1.2.5). It should be noted that the efforts in (4.4) are not necessarily the generalized efforts for the system. In principle the m_e algebraic equations, (4.4), can be used to solve for the m_e effort variables, e, explicitly. However, this is not always be desirable. In this chapter the equations of motion are constructed so that the effort constraints can remain in the form given by equation (4.4).

4.2 Lagrange's Equation with Displacement Constraints

In this section we develop a form of Lagrange's equation that can be applied to systems that are described in terms of the configuration coordinates and are subject to displacement constraints. Let $q = [q_1, q_2, \cdots, q_N]^T$, be the N configuration displacements for the systems, and let $f = [f_1, f_2, \cdots, f_N]^T$, be the corresponding flow variables. Here, the displacement variables are related by m_1 independent displacement constraints of the form

$$\phi_j(q, t) = 0, \quad j = 1, 2, \cdots, m_1, \tag{a}$$

where t denotes the time. In the formulation of Lagrange's equation presented in Section 3.2 the displacement constraints (a) are used to eliminate m_1 of the N configuration variables. Doing so leaves $n = N - m_1$ generalized coordinates which are used to determine the system energies and generalized efforts.

As we will see in the examples below, eliminating the m_1 excess coordinates can sometimes be very difficulty. For this reason we reconsider the derivation of Lagrange's equation with the aim of retaining all N configuration coordinates is the problem formulation. For the sake of simplicity we will assume that the displacements are selected such that the first m_1 displacements are dependent on the remaining $n = N - m_1$ displacements. Since the constraints are independent we can always rearrange the displacement variable to meet this condition.

To begin, let the total energy of the system be $E = T(q, p, t) + V(q)$, where $T(q, p, t)$ is the kinetic energy, $p = [p_1, p_2, \cdots, p_N]^T$ are the momentum variables, and $V(q)$ is the potential energy. The efforts applied to the system are $e = e^R + e^s$, where $e^R = [-dD(f)/df_1, -dD(f)/df_2, \cdots, -dD(f)/df_N]^T$ are the efforts due to ideal resistors, $D(f)$ is the dissipation function, and

$e^s = [e_1^s, e_2^s, \cdots, e_N^s]^T$ are the applied efforts. (Note the e^s excludes the efforts due to ideal capacitors and ideal resistors.)

The variational form of the first law of thermodynamics gives

$$\delta E = \delta(T(q,p,t) + V(q)) = \delta \mathcal{W} = \sum_{i=1}^{N} (e_i^R + e_i^s)\,\delta q_i. \tag{b}$$

Using the relationship between the kinetic energy, $T(q,p,t)$ and the kinetic coenergy, $T^*(q,f,t)$, it can be shown that equation (b) satisfies

$$\sum_{i=1}^{N} \left[\frac{d}{dt}\frac{\partial T^*}{\partial f_i} - \frac{\partial T^*}{\partial q_i} + \frac{\partial D}{\partial f_i} + \frac{\partial V}{\partial q_i} - e_i^s \right] \delta q_i = 0. \tag{c}$$

(See Section 3.2.) Here, the virtual displacements, δq_i, are not all independent. In fact, a variation of the constraints (a) yields

$$\sum_{i=1}^{N} \frac{\partial \phi_j(q,t)}{\partial q_i}\,\delta q_i = 0, \quad j = 1, 2, \cdots, m_1. \tag{d}$$

That is, the virtual displacements must be tangent to the constraint surface.

We will employ the method of Lagrange multipliers to ensure that the condition (d) is satisfied. In particular, multiply the j-th equation in (d) by λ_j to get $\sum_{i=1}^{N} \lambda_j \left(\partial \phi_j / \partial q_i\right) \delta q_i = 0$, $j = 1, 2, \cdots, m_1$. Adding this result to (c) gives

$$\sum_{i=1}^{N} \left[\frac{d}{dt}\frac{\partial T^*}{\partial f_i} - \frac{\partial T^*}{\partial q_i} + \frac{\partial D}{\partial f_i} + \frac{\partial V}{\partial q_i} + \sum_{j=1}^{m_1} \lambda_j \frac{\partial \phi_j}{\partial q_i} - e_i^s \right] \delta q_i = 0. \tag{e}$$

Here, λ_j, $j = 1, 2, \cdots, m_1$ are functions of the time, and are called the *Lagrange multipliers*. We can *choose* λ_j so that each of the first m_1 terms in (e) vanish, i.e., the Lagrange multipliers are selected so that

$$\frac{d}{dt}\frac{\partial T^*}{\partial f_i} - \frac{\partial T^*}{\partial q_i} + \frac{\partial D}{\partial f_i} + \frac{\partial V}{\partial q_i} + \sum_{j=1}^{m_1} \lambda_j \frac{\partial \phi_j}{\partial q_i} - e_i^s = 0, \quad i = 1, 2, \cdots, m_1. \tag{f}$$

Equation (e) then involves the summation of $N - m_1$ terms, i.e.,

$$\sum_{i=m_1+1}^{N} \left[\frac{d}{dt}\frac{\partial T^*}{\partial f_i} - \frac{\partial T^*}{\partial q_i} + \frac{\partial D}{\partial f_i} + \frac{\partial V}{\partial q_i} + \sum_{j=1}^{m_1} \lambda_j \frac{\partial \phi_j}{\partial q_i} - e_i^s \right] \delta q_i = 0. \tag{g}$$

However, we have assumed that the displacements are arranged so that q_i, $i = m_1 + 1, m_1 + 2, \cdots, N$ are all independent. In which case the virtual displacements, δq_i, in (g) can be varied arbitrarily, and the equation can only be satisfied if the terms in the square brackets all vanish simultaneously. Combining (a), (f) and (g) we see that the constrained dynamic system must satisfy the differential-algebraic equations

$$\frac{d}{dt}\frac{\partial T^*}{\partial f_i} - \frac{\partial T^*}{\partial q_i} + \frac{\partial D}{\partial f_i} + \frac{\partial V}{\partial q_i} + \sum_{j=1}^{m_1} \lambda_j \frac{\partial \phi_j}{\partial q_i} - e_i^s = 0, \quad i = 1, 2, \cdots, N,$$
$$\phi_i(q, t) = 0, \quad i = 1, 2, \cdots, m_1. \tag{4.5}$$

Given a set of consistent initial conditions the $N + m_1$ equations in (4.5) can be solved to determine the $N + m_1$ variables q and λ. In this formulation the term $\lambda_j\,(\partial\phi_j/\partial q_i)$ is an effort, and can be interpreted as the effort required to ensure that the j-th constraint is satisfied.

4.2.1 Application to displacement constrained systems

This section illustrates the application of equation (4.5) to dynamic systems that contain displacement constraints. Here our goal is simply to determine the equations of motion of the system, the analysis of the resultant differential-algebraic equations is considered in Chapter 5 and Chapter 6.

Example 4.1.

The mechanism shown in the figure below consists of two sliders (A and B) that are connected by a massless rod. The slider A has mass m_A and can move along the y-axis. The slider B has mass m_B and can move along the x-axis. A linear spring with stiffness k is attached to the slider B, and the rod AB has length l. Note that the rod is connected to the sliders via revolute joints at A and B. The acceleration due gravity acts downward as shown, and the effects of friction are neglected.

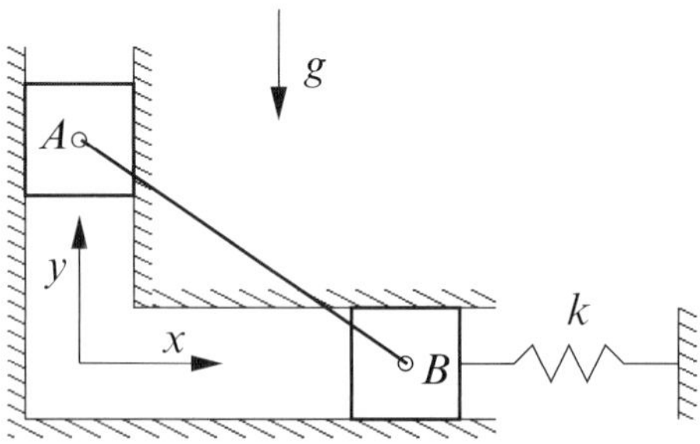

An analysis of this mechanism using Lagrange's equation proceeds as follows.

Kinematic analysis:

The configuration coordinates for the system are x and y which define the positions of B and A, respectively. A mobility analysis shows that this system has 1 degree of freedom thus, there is a constraint relationship between the two variables x and y. In fact it is easy to see that at any instant the distance between A and B is l, i.e., the system has the displacement constraint

$$\phi = x^2 + y^2 - l^2 = 0.$$

This constraint equation is sufficiently simple for us to eliminate one of the two displacement variables and proceed with the analysis using a generalized (independent) coordinate. Here however, we will retain both coordinates in the system description.

Applied effort analysis:

The virtual work done by the weight of slider A is equal to the work done by the applied efforts, i.e., $\delta\mathcal{W} = -m_A g\,\delta y = e_x^s\,\delta x + e_y^s\,\delta y$. Therefore, $e_x^s = 0$ and $e_y^s = -m_A g$. (Note that the applied efforts do not include the efforts due to capacitors and resistors, since those efforts are accounted for in the potential energy and the dissipation function.)

Lagrange's equation:

The kinetic coenergy for the system is $T^* = (m_A\dot{y}^2 + m_B\dot{x}^2)/2$, the potential energy for the system is $V = (kx^2)/2$, and the dissipation function is $D = 0$.

Using equation (4.5) Lagrange's equations of motion for this system can be stated as

$$\frac{d}{dt}\frac{\partial T^*}{\partial \dot{x}} - \frac{\partial T^*}{\partial x} + \frac{\partial D}{\partial \dot{x}} + \frac{\partial V}{\partial x} + \lambda\frac{\partial \phi}{\partial x} = e_x^s, \qquad (a)$$

$$\frac{d}{dt}\frac{\partial T^*}{\partial \dot{y}} - \frac{\partial T^*}{\partial y} + \frac{\partial D}{\partial \dot{y}} + \frac{\partial V}{\partial y} + \lambda\frac{\partial \phi}{\partial y} = e_y^s, \qquad (b)$$

$$\phi(x, y) = 0. \qquad (c)$$

Using T^*, V, D and ϕ, given above, equations (a), (b) and (c) become

$$(a) \Rightarrow m_B\ddot{x} + kx + 2\lambda x = 0,$$
$$(b) \Rightarrow m_A\ddot{y} + 2\lambda y = -m_A g,$$
$$(c) \Rightarrow x^2 + y^2 - l^2 = 0.$$

These are a set of differential-algebraic equations that determine the behavior of the system. There are two differential equations, (a) and (b), and one algebraic equation, (c), that are used to determine the displacements x and y, and the Lagrange multiplier λ. Given a consistent set of initial conditions we can solve these differential-algebraic equations to determine the trajectory of the system. Numerical techniques for the solution of these differential-algebraic equations will be discussed in Chapter 5 and Chapter 6.

Example 4.2.

The figure below shows a planar slider crank mechanism. The crank AB has mass m_1 and moment of inertia I_1 about the center of mass. The length of the crank is l_1, and the center of mass is at distance $l_1/2$ from A. The connecting rod BC has mass m_2 and moment of inertia I_2 about the center of mass. The length of the connecting rod is l_2, and the center of mass is a distance $l_2/2$ from B. The slider D has mass m_3. A torque, τ, is applied to the crank. An analysis of this device proceeds as follows.

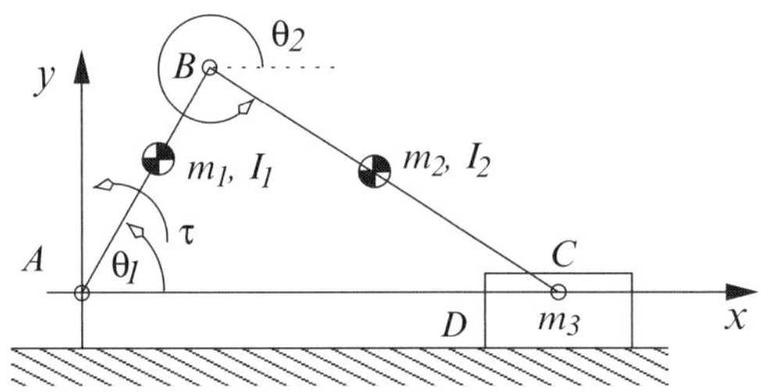

Kinematic analysis:

Let, (x_1, y_1) denote the coordinates is the center of mass of the crank AB, with respect to the fixed rectangular coordinate system x-y. Similarly, let (x_2, y_2) be the coordinate of the center of mass of the connecting rod BC, and let x_3 be the coordinate of the slider D. The angle θ_1 gives the orientation of the crank with respect to the x-axis, and the angle θ_2 gives the orientation of the connecting rod with respect to the x-axis. Thus, the 7 configuration displacement variables for the system are x_1, y_1, θ_1, x_2, y_2, θ_2, and x_3.

A mobility analysis of the system shows that there is 1 degree of freedom hence, there are 6 constraint equations that relate the 7 configuration displacements described above. These constraints are determined as follows.

The mass centers of the crank and connecting rod are determined via the 4 displacement constraints

$$\begin{aligned}
\phi_1 &= x_1 - \frac{l_1}{2}\cos\theta_1 = 0, \\
\phi_2 &= y_1 - \frac{l_1}{2}\sin\theta_1 = 0, \\
\phi_3 &= x_2 - (l_1\cos\theta_1 + \frac{l_2}{2}\cos\theta_2) = 0, \\
\phi_4 &= y_2 - (l_1\sin\theta_1 + \frac{l_2}{2}\sin\theta_2) = 0.
\end{aligned}$$

The vector loop closure equation $\bar{AB} + \bar{BC} = \bar{AC}$ yields the 2 constraints

$$\begin{aligned}
\phi_5 &= l_1\cos\theta_1 + l_2\cos\theta_2 - x_3 = 0, \\
\phi_6 &= l_1\sin\theta_1 + l_2\sin\theta_2 = 0.
\end{aligned}$$

These 6 displacement constraints could be used to eliminate 6 of the configuration variables from the problem description. Here however, all 7 displacement variables are retained in the problem formulation.

Applied effort analysis:

The virtual work done by the applied torque, τ, is equal to the work done by the applied efforts, i.e., $\delta\mathcal{W} = \tau\,\delta\theta_1 = e^s_{x_1}\,\delta x_1 + e^s_{y_1}\,\delta y_1 + e^s_{\theta_1}\,\delta\theta_1 + e^s_{x_2}\,\delta x_2 + e^s_{y_2}\,\delta y_2 + e^s_{\theta_2}\,\delta\theta_2 + e^s_{x_3}\,\delta x_3$. Therefore, all the configuration efforts are zero except $e^s_{\theta_1} = \tau$.

Lagrange's equations:

The kinetic coenergy of the system is

$$T^* = \frac{1}{2}m_1(\dot{x}_1^2 + \dot{y}_1^2) + \frac{1}{2}I_1\dot{\theta}_1^2 + \frac{1}{2}m_2(\dot{x}_2^2 + \dot{y}_2^2) + \frac{1}{2}I_2\dot{\theta}_2^2 + \frac{1}{2}m_3\dot{x}_3^2,$$

the potential energy is $V = 0$, and the dissipation function is $D = 0$. To simplify the enumeration of the equations of motion let us define the displacement vector q as $q = [x_1, y_1, \theta_1, x_2, y_2, \theta_2, x_3]^T$, and the corresponding flow vector f as $f = [\dot{x}_1, \dot{y}_1, \dot{\theta}_1, \dot{x}_2, \dot{y}_2, \dot{\theta}_2, \dot{x}_3]^T$. Then equation (4.5) gives

$$
\begin{aligned}
m_1\ddot{x}_1 + \lambda_1 &= 0,\\
m_1\ddot{y}_1 + \lambda_2 &= 0,\\
I_1\ddot{\theta}_1 + \lambda_1\frac{l_1}{2}\sin\theta_1 - \lambda_2\frac{l_1}{2}\cos\theta_1 + \lambda_3 l_1\sin\theta_1 &\\
-\lambda_4 l_1\cos\theta_1 - \lambda_5 l_1\sin\theta_1 + \lambda_6 l_1\cos\theta_1 &= \tau,\\
m_2\ddot{x}_2 + \lambda_3 &= 0,\\
m_2\ddot{y}_2 + \lambda_4 &= 0,\\
I_2\ddot{\theta}_2 + \lambda_3\frac{l_2}{2}\sin\theta_2 - \lambda_4\frac{l_2}{2}\cos\theta_2 &\\
-\lambda_5 l_2\sin\theta_2 + \lambda_6 l_2\cos\theta_2 &= 0,\\
m_3\ddot{x}_3 - \lambda_5 &= 0,\\
x_1 - \frac{l_1}{2}\cos\theta_1 &= 0,\\
y_1 - \frac{l_1}{2}\sin\theta_1 &= 0,\\
x_2 - (l_1\cos\theta_1 + \frac{l_2}{2}\cos\theta_2) &= 0,\\
y_2 - (l_1\sin\theta_1 + \frac{l_2}{2}\sin\theta_2) &= 0,\\
l_1\cos\theta_1 + l_2\cos\theta_2 - x_3 &= 0,\\
l_1\sin\theta_1 + l_2\sin\theta_2 &= 0.
\end{aligned}
$$

Given a consistent set of initial conditions these differential-algebraic equations can be solved to determine the trajectory of the slider-crank mechanism.

Example 4.3.

The figure below shows a planar fourbar mechanism. The crank AB has mass m_1 and moment of inertia I_1 about the center of mass. The length of the crank is l_1, and the center of mass is at distance $l_1/2$ from A. The connecting rod BC has mass m_2 and moment of inertia I_2 about the center of mass. The length of the connecting rod is l_2, and the center of mass is a distance $l_2/2$ from B. The follower DC has mass m_3 and moment of inertia I_3 about the center of mass. The length of the follower is l_3 and the center of mass is at distance $l_3/2$ from D. The distance from A to D is l_4, and a torque τ is applied to the crank AB. The equations of motion for this system can be determined as follows.

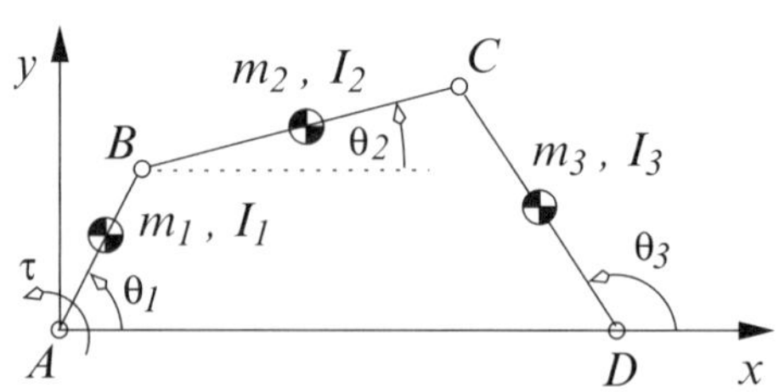

Kinematic analysis:

The angle θ_1 gives the orientation of the crank with respect to the x-axis, Let (x_2, y_2) be the coordinate of the center of mass of the connecting rod BC, and let θ_2 be the angle between the x-axis and the connecting rod. Finally, the angle θ_3 gives the orientation of the follower DC with respect to the x-axis. Thus, the configuration displacement variables for the system are, θ_1, x_2, y_2, θ_3, and θ_3.

A mobility analysis of the system shows that there is $n = 1$ degree of freedom. Typically, the angle θ_1 is selected as the independent displacement. Here, however, we have chosen $N = 5$ configuration displacements to describe the system. Therefore, we must determine $m_1 = N - n = 4$ constraint equations that relate these configuration displacements.

Two of the constraints provide the coordinates of the center of mass of the connecting rod, i.e.,

$$\phi_1 = x_2 - (l_1 \cos\theta_1 + \frac{l_2}{2} \cos\theta_2) = 0,$$
$$\phi_2 = y_2 - (l_1 \sin\theta_1 + \frac{l_2}{2} \sin\theta_2) = 0.$$

The vector loop closure equation $\bar{AB} + \bar{BC} = \bar{AD} + \bar{DC}$ yields the other two constraints, i.e.,

$$\phi_3 = l_1 \cos\theta_1 + l_2 \cos\theta_2 - l_3 \cos\theta_3 - l_4 = 0,$$
$$\phi_4 = l_1 \sin\theta_1 + l_2 \sin\theta_2 - l_3 \sin\theta_3 = 0.$$

From the constraints ϕ_1 and ϕ_2 it can be seen that the variables x_2 and y_2 can be easily eliminated from the problem description. However, the determination of θ_2 and θ_3 from ϕ_3 and ϕ_4 is non-trivial. Here, we choose to keep all the configuration coordinates in the system description.

Applied effort analysis:

The virtual work done by the applied torque, τ, is equal to the work done by the applied efforts, i.e., $\delta\mathcal{W} = \tau\,\delta\theta_1 = e^s_{\theta_1}\,\delta\theta_1$. Therefore, all the configuration efforts are zero except $e^s_{\theta_1} = \tau$.

Lagrange's equations:

The kinetic coenergy of the system is

$$T^* = \frac{1}{2}(I_1 + m_1 l_1^2/4)\dot{\theta}_1^2 + \frac{1}{2}m_2(\dot{x}_2^2 + \dot{y}_2^2) + \frac{1}{2}I_2\dot{\theta}_2^2 + \frac{1}{2}(I_3 + m_3 l_3^2/4)\dot{\theta}_3^2,$$

the potential energy is $V = 0$, and the dissipation function is $D = 0$. To apply equation (4.5) directly, define the displacement vector q as $q =$

$[\theta_1, x_2, y_2, \theta_2, \theta_3]^T$, and the flow vector f as $f = [\dot{\theta}_1, \dot{x}_2, \dot{y}_2, \dot{\theta}_2, \dot{\theta}_3]^T$. Then equation (4.5) gives

$$
\begin{aligned}
(I_1 + m_1 l_1^2/4)\ddot{\theta}_1 + \lambda_1 l_1 \sin\theta_1 - \lambda_2 l_1 \cos\theta_1 - \lambda_3 l_1 \sin\theta_1 + \lambda_4 l_1 \cos\theta_1 &= \tau, \\
m_2 \ddot{x}_2 + \lambda_1 &= 0, \\
m_2 \ddot{y}_2 + \lambda_2 &= 0, \\
I_2 \ddot{\theta}_2 + \lambda_1 \frac{l_2}{2} \sin\theta_2 - \lambda_2 \frac{l_2}{2} \cos\theta_2 - \lambda_3 l_2 \sin\theta_2 + \lambda_4 l_2 \cos\theta_2 &= 0, \\
(I_3 + m_3 l_3^2/4)\ddot{\theta}_3 + \lambda_3 l_3 \sin\theta_3 - \lambda_4 l_3 \cos\theta_3 &= 0, \\
x_2 - (l_1 \cos\theta_1 + \frac{l_2}{2} \cos\theta_2) &= 0, \\
y_2 - (l_1 \sin\theta_1 + \frac{l_2}{2} \sin\theta_2) &= 0, \\
l_1 \cos\theta_1 + l_2 \cos\theta_2 - l_3 \cos\theta_3 - l_4 &= 0, \\
l_1 \sin\theta_1 + l_2 \sin\theta_2 - l_3 \sin\theta_3 &= 0.
\end{aligned}
$$

These 9 differential-algebraic equations can be solved to determine the trajectory of the system.

4.3 Lagrange's Equation with Flow Constraints

We now turn our attention to dynamic systems that include flow constraints in the Pfaffian form

$$\psi_j(q, f, t) = \sum_{i=1}^{N} a_{ji}(q, t)\, f_i + a_j(q, t) = 0, \quad j = 1, 2, \cdots, m_2, \qquad (a)$$

where $q=[q_1, q_2, \cdots, q_N]^T$ are the displacements, and $f=[f_1, f_2, \cdots, f_N]^T$ are the flows. Also, the coefficients a_{ji} and a_j are functions of the displacements, q, and the time, t. It is assumed that the m_2 constraints are independent. In particular, let a_{ji} be the (j, i)-th element of the $m_2 \times N$ matrix A, i.e., $A=[a_{ji}]$. Then, it is assumed the matrix A has rank m_2 along the solution trajectory. It is further assumed that the configuration variables are arranged such that the first m_2 columns of A are linearly independent. (Note that this rearrangement of variables is only required to simplify the derivation presented in this section. In the application of the theory we are not required to perform this manipulation of the system variables.)

If the constraints are holonomic then equation (a) can be integrated to find a displacement constraint of the form (4.2), and the results of Section 4.2 can be applied. If the constraints are nonholonomic, or we choose not to integrate the flow constraints to obtain displacement constraints then, the analysis of the system proceeds as follows.

As in the previous section (and Section 3.2) it can be shown that the variational form of the first law of thermodynamics leads to

$$\sum_{i=1}^{N}\left[\frac{d}{dt}\frac{\partial T^*}{\partial f_i}-\frac{\partial T^*}{\partial q_i}+\frac{\partial D}{\partial f_i}+\frac{\partial V}{\partial q_i}-e_i^s\right]\delta q_i=0, \tag{b}$$

where T^* is the kinetic coenergy, V is the potential energy, D is the dissipation function, and e_i^s is the applied effort associated with the i-th configuration displacement. Due to the flow constraints, (a), the virtual displacements δq_i are related via the m_2 independent equations

$$\sum_{i=1}^{N} a_{ji}(q,t)\,\delta q_i=0, \quad j=1,2,\cdots,m_2. \tag{c}$$

Equation (c) is obtained by considering a variation of the flow constraints (a), and can be written in matrix form as

$$A\,\delta q=[A_D \quad A_I]\begin{bmatrix}\delta q_D\\ \delta q_I\end{bmatrix}=A_D\,\delta q_D+A_I\,\delta q_I=0,$$

where A_D contains the first m_2 columns of A, $q_D=[q_1,q_2,\cdots,q_{m_2}]^T$, and $q_I=[q_{m_2+1},q_{m_2+2},\cdots,q_N]^T$. We have arranged the variables so that A_D is nonsingular hence,

$$\delta q_D=-A_D^{-1}A_I\,\delta q_I.$$

This last equation indicates the virtual displacements δq_D are dependent on the virtual displacements δq_I.

To account for the constraints (c) in equation (b) we will employ the method of Lagrange multipliers. That is, multiply the j-th equation in (c) by μ_j and add the result to equation (b) to get

$$\sum_{i=1}^{N}\left[\frac{d}{dt}\frac{\partial T^*}{\partial f_i}-\frac{\partial T^*}{\partial q_i}+\frac{\partial D}{\partial f_i}+\frac{\partial V}{\partial q_i}-e_i^s+\sum_{j=1}^{m_2}\mu_j a_{ji}\right]\delta q_i=0. \tag{d}$$

Here, μ_j are functions of the time, and are called the Lagrange multipliers. We will select these m_2 multipliers so that

$$\frac{d}{dt}\frac{\partial T^*}{\partial f_i}-\frac{\partial T^*}{\partial q_i}+\frac{\partial D}{\partial f_i}+\frac{\partial V}{\partial q_i}-e_i^s+\sum_{j=1}^{m_2}\mu_j a_{ji}=0, \quad i=1,2,\cdots,m_2. \tag{e}$$

Thus, we have used the Lagrange multipliers to eliminate the first m_2 terms of the summation in equation (d). As a result equation (d) becomes

$$\sum_{i=1+m_2}^{N}\left[\frac{d}{dt}\frac{\partial T^*}{\partial f_i}-\frac{\partial T^*}{\partial q_i}+\frac{\partial D}{\partial f_i}+\frac{\partial V}{\partial q_i}-e_i^s+\sum_{j=1}^{m_2}\mu_j a_{ji}\right]\delta q_i=0. \qquad (f)$$

By construction the $N-m_2$ virtual displacements, δq_i, $i=m_2+1,\cdots,N$, can all be chosen independently. Since the virtual displacements in (f) can be varied arbitrarily, this equation will be satisfied only if

$$\frac{d}{dt}\frac{\partial T^*}{\partial f_i}-\frac{\partial T^*}{\partial q_i}+\frac{\partial D}{\partial f_i}+\frac{\partial V}{\partial q_i}-e_i^s+\sum_{j=1}^{m_2}\mu_j a_{ji}=0, \quad i=m_2+1,m_2+2,\cdots,N. \qquad (g)$$

Combining equations (e), (g) and (a) we see that the trajectory of the constrained dynamic system satisfies

$$\frac{d}{dt}\frac{\partial T^*}{\partial f_i}-\frac{\partial T^*}{\partial q_i}+\frac{\partial D}{\partial f_i}+\frac{\partial V}{\partial q_i}+\sum_{j=1}^{m_2}\mu_j a_{ji}-e_i^s=0, \quad i=1,2,\cdots,N,$$

$$\sum_{i=1}^{N}a_{ji}(q,t)\,f_i+a_j(q,t)=0, \quad j=1,2,\cdots,m_2. \qquad (4.6)$$

Given a set of initial conditions, that are consistent with the constraints, these $N+m_2$ differential-algebraic equations can be integrated to determine the N displacements, q, and the m_2 Lagrange multiplier, μ.

Note that holonomic flow constraints can be treated directly using equation (4.6), or they can be integrated to obtain displacement constraints which are then used in equation (4.5). On the other hand, the derivatives of the displacement constraints (4.2) are in the Pfaffian form, in fact, $a_{ji}=\partial\phi_j/\partial q_i$ and $a_j=\partial\phi_j/\partial t$. Therefore, equation (4.6) can be used to analyze systems with displacement constraints. However, converting the displacement constraints to flow constraints must be used with some care, since there is no assurance that the original displacement constraints will be satisfied exactly. This issue is considered in more detail in Chapter 5.

4.3.1 Application to flow constrained systems

This section presents some examples that illustrate the application of equation (4.6) to the modeling of dynamic systems that have flows constraints which can be written in the Pfaffian form. The first example has integrable flow constraints, and the other examples all have nonholonomic constraints.

Example 4.4.
The network shown in the figure below can be analyzed using equation (4.6) as follows.

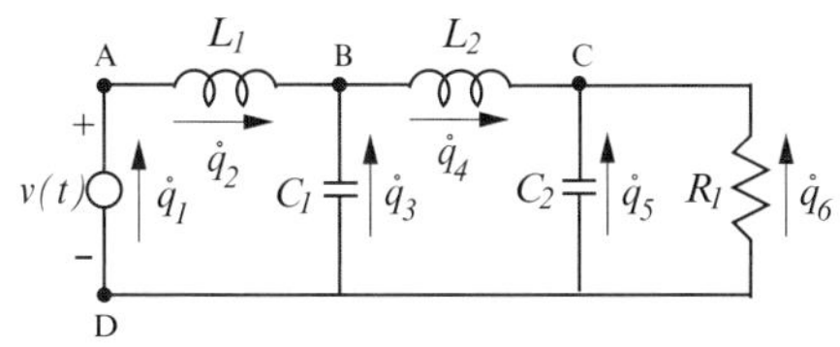

Kinematic analysis:

Six configuration flows are assigned to the model, i.e., $\dot{q}_1, \dot{q}_2, \cdots, \dot{q}_6$. These variables are not all independent, in fact there are $n = B - N + 1 = 6 - 4 + 1 = 3$ independent flow variables. (Recall that, B is the number of branches, and N is the number of nodes in the network.) This implies that there are three independent constraint equations that relate the 6 flow variables. These constraints can be found by applying Kirchhoff's current law to the nodes A, B and C of thew network. Doing so gives the 3 flow constraints

$$\begin{aligned}\psi_1 &= \dot{q}_1 - \dot{q}_2 = 0,\\ \psi_2 &= \dot{q}_2 + \dot{q}_3 - \dot{q}_4 = 0,\\ \psi_3 &= \dot{q}_4 + \dot{q}_5 + \dot{q}_6 = 0.\end{aligned}$$

Clearly these constraints are integrable and we can use them to (i) obtain displacement constraints of the form (4.2), and/or (ii) eliminate three of the flow variables for the problem description. Here however, we will retain all the variables in the problem description.

Applied effort analysis:

The virtual work done by the applied voltage $v(t)$ is $\delta \mathcal{W} = v(t)\, \delta q_1$. Thus, the applied efforts $e^s_{q_i}$ are all zero except $e^s_{q_1} = v(t)$.

Lagrange's equations:

The kinetic coenergy for the system is

$$T^* = \frac{1}{2}(L_1 \dot{q}_2^2 + L_2 \dot{q}_4^2),$$

the potential energy is

$$V = \frac{q_3^2}{2C_1} + \frac{q_5^2}{2C_2},$$

and the dissipation function is

$$D = \frac{1}{2} R_1 \dot{q}_6^2.$$

In terms of the Pfaffian form, (4.1), the constraints ψ_1, ψ_2 and ψ_3 have the coefficients

$$A = [a_{ji}] = \begin{bmatrix} 1 & -1 & 0 & 0 & 0 & 0 \\ 0 & 1 & 1 & -1 & 0 & 0 \\ 0 & 0 & 0 & 1 & 1 & 1 \end{bmatrix},$$

and $a_j = 0$, $j = 1, 2, \cdots, 6$. Hence, Lagrange's equations of motion, (4.6), are

$$\begin{aligned} \mu_1 &= v(t), \\ L_1 \ddot{q}_2 - \mu_1 + \mu_2 &= 0, \\ \frac{q_3}{C_1} + \mu_2 &= 0, \\ L_2 \ddot{q}_4 - \mu_2 + \mu_3 &= 0, \\ \frac{q_5}{C_2} + \mu_3 &= 0, \\ R_1 \dot{q}_6 + \mu_3 &= 0, \\ \dot{q}_1 - \dot{q}_2 &= 0, \\ \dot{q}_2 + \dot{q}_3 - \dot{q}_4 &= 0, \\ \dot{q}_4 + \dot{q}_5 + \dot{q}_6 &= 0. \end{aligned}$$

To solve these differential-algebraic equations we require a set of initial conditions that are consistent with the constraints ψ_1, ψ_2 and ψ_3. These equations of motion are relatively simple and it is not difficult to reduce these to three ordinary differential equations involving three independent flow variables (see Problem 6).

Example 4.5.

The figure below shows a fixed rectangular coordinate system x-y-z. The x-axis make an angle α with respect to the horizontal. A thin disk with radius r rolls without slipping on the inclined x-y plane. The origin of the x_1-y_1-z_1 rectangular coordinate system is located at the geometric center of the disk. Also, the x_1-axis and z_1-axis are in the radial direction of the disk. The disk has a mass m, a moment of inertia I_1 about the y_1-axis, and a moment of inertia I_2 about the x_1-axis and the z_1-axis. The acceleration due to gravity, g, acts downward as shown, i.e., perpendicular to the horizontal and at angle α with respect to the z-axis.

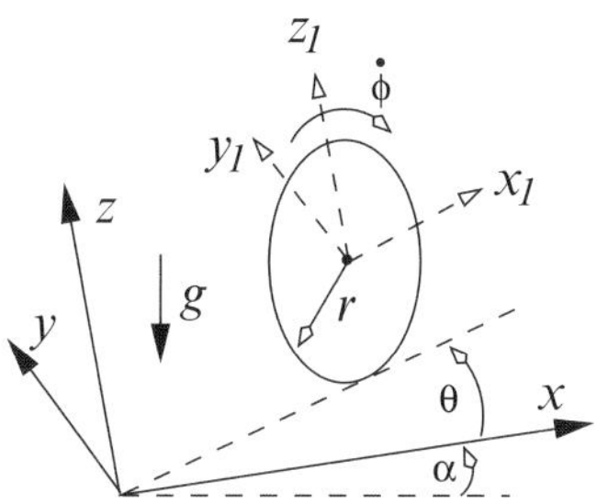

Kinematic analysis:

The coordinate of the center of mass of the disk, with respect to the x-y-z coordinate system, is defined by x, y. (The z coordinate of the center of mass is r at all times.) The orientation of the disk is defined by the angles θ and ϕ. Since the disk is rolling without slipping the velocity of the center of mass must satisfy the nonholonomic constraints

$$\begin{aligned} \psi_1 &= \dot{x} - r\dot{\phi}\cos\theta = 0, \\ \psi_2 &= \dot{y} - r\dot{\phi}\sin\theta = 0. \end{aligned} \tag{a}$$

These two equations can not be integrated to eliminate the x and y displacements from the system description. Hence, all four displacements (x, y, θ and ϕ) are retained in the model.

Applied effort analysis:

The weight of the disk is given by the vector $\bar{w} = mg(-\sin\alpha\,\hat{\imath} - \cos\alpha\,\hat{k})$, where $\hat{\imath}$ is the unit vector along the x-axis, and $\hat{k}$ is the unit vector along the z-axis. Since the force along the z-axis does no virtual work, the virtual work done by the weight is $\delta\mathcal{W} = -mg\sin\alpha\,\delta x$. Therefore, the applied efforts are all zero except $e_x^s = -mg\sin\alpha$.

Lagrange's equations:

The kinetic coenergy for the system is

$$T^* = \frac{1}{2}m(\dot{x}^2 + \dot{y}^2) + \frac{1}{2}I_1\dot{\phi}^2 + \frac{1}{2}I_2\dot{\theta}^2 = \frac{1}{2}(I_1 + mr^2)\dot{\phi}^2 + \frac{1}{2}I_2\dot{\theta}^2.$$

The potential energy is $V = 0$, and the dissipation function is $D = 0$. To use equation (4.6) let $q = [x, y, \phi, \theta]^T$. Also, using the notation of equation (4.1) to describe the constraints (a) implies that

$$\begin{aligned} &a_{11} = 1, \quad a_{12} = 0, \quad a_{13} = -r\cos\theta, \quad a_{14} = 0, \quad a_1 = 0, \\ &a_{21} = 0, \quad a_{22} = 1, \quad a_{23} = -r\sin\theta, \quad a_{24} = 0, \quad a_2 = 0. \end{aligned}$$

Putting these in equation (4.6) gives

$$\begin{aligned}
\mu_1 - mg\sin\alpha &= 0, \\
\mu_2 &= 0, \\
(I_1 + mr^2)\ddot{\phi} - \mu_1 r\cos\theta - \mu_2 r\sin\theta &= 0, \\
I_2\ddot{\theta} &= 0, \\
\dot{x} - r\dot{\phi}\cos\theta &= 0, \\
\dot{y} - r\dot{\phi}\sin\theta &= 0.
\end{aligned}$$

These differential-algebraic equations can be reduced to the differential equations

$$\begin{aligned}
(I_1 + mr^2)\ddot{\phi} + mgr\sin\alpha\cos(\theta_0 + \omega t) &= 0, \\
\dot{x} - r\dot{\phi}\cos(\theta_0 + \omega t) &= 0, \\
\dot{y} - r\dot{\phi}\sin(\theta_0 + \omega t) &= 0.
\end{aligned}$$

To obtain this result we have used the fact that $\theta = \theta_0 + \omega t$, where θ_0 and ω are constants, and t is the time. These differential equations can easily be solved given initial conditions for x, y and ϕ.

Example 4.6.

The figure below shows a two-wheeled vehicle that moves in the x-y plane. The wheels at A and B roll without slipping in the plane. Both wheels have the same diameter and are connected by an axle through their centers. The vehicle has mass m, and moment of inertia I about the center of mass in the z direction.

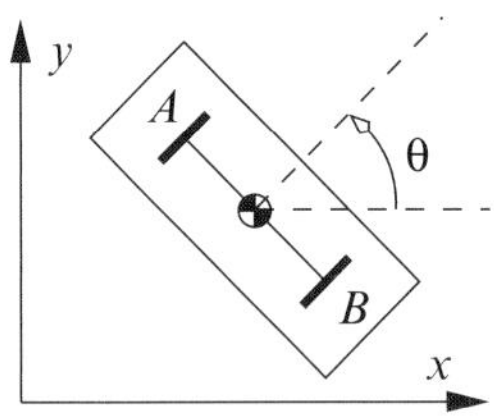

Kinematic analysis:

The coordinate of the center of mass of the vehicle is x, y. The angle $\theta + \frac{\pi}{2}$ gives the orientation the axle with respect to the x-axis. Since the wheels roll without slipping the x and y velocities of the center of mass must satisfy the nonholonomic constraint

$$\psi_1 = \dot{y} - \dot{x}\tan\theta = \dot{y}\cos\theta - \dot{x}\sin\theta = 0. \qquad (a)$$

This constraint indicates the the velocity of the vehicle is normal to the axle at all times.

Applied effort analysis:

There are no applied efforts for this problem.

Lagrange's equations:

The kinetic coenergy of the systems is

$$T^* = \frac{1}{2}m(\dot{x}^2 + \dot{y}^2) + \frac{1}{2}I\dot{\theta}^2,$$

the potential energy is $V = 0$, and the dissipation function is $D = 0$. To apply equation (4.6) define the displacement vector $q = [x, y, \theta]^T$. Also, note that the coefficients of the flow constraint (a) in the Pfaffian form (4.1) are

$$a_{11} = -\sin\theta, \quad a_{12} = \cos\theta, \quad a_{13} = 0, \quad a_1 = 0.$$

Using these terms the equations of motion can be written as

$$\begin{aligned} m\ddot{x} - \mu_1 \sin\theta &= 0, \\ m\ddot{y} + \mu_1 \cos\theta &= 0, \\ I\ddot{\theta} &= 0, \\ \dot{y}\cos\theta - \dot{x}\sin\theta &= 0. \end{aligned}$$

Given a set of initial conditions these 4 differential-algebraic equations can be solved to determine x, y, θ and μ_1.

Example 4.7.

The figure below shows a three-wheeled vehicle that moves in the x-y plane. The wheels at A, B and C roll without slipping, and the vehicle has mass m, and moment of inertia I about the center of mass in the z direction.

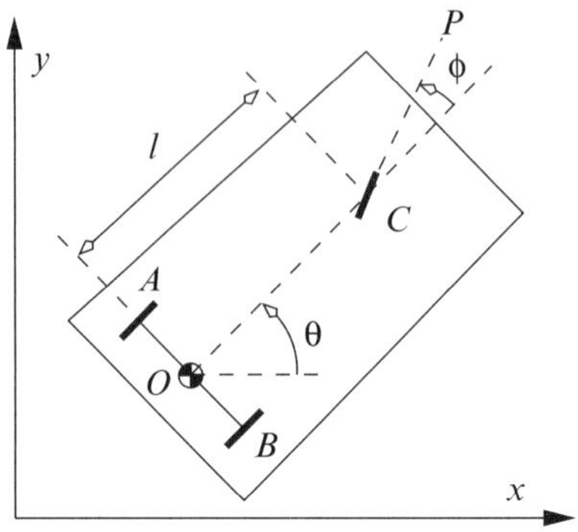

Kinematic analysis:

The coordinates of the center of mass of the vehicle are x, y, and the orientation of the vehicle with respect to the x-axis is given by the angle θ. The angle ϕ measures the orientation of the steering wheel at C relative to the body of the vehicle. Since the wheels A and B roll without slipping the velocity of the center of mass satisfies the nonholonomic constraint

$$\psi_1 = \dot{y} - \dot{x}\tan\theta = \dot{y}\cos\theta - \dot{x}\sin\theta = 0. \qquad (a)$$

This constraint indicates that the velocity of the center of mass is directed along the line OC at all times. Also, since the wheel at C rolls without slipping the components of the velocity of the point C satisfy the nonholonomic constraint

$$\frac{v_{C_y}}{v_{C_x}} = \tan(\theta + \phi), \qquad (b)$$

where v_{C_x} is the x component of the velocity at C, and v_{C_y} is the y component of the velocity at C. From the figure above it can be seen that

$$\begin{aligned} v_{C_x} &= \frac{d}{dt}(x + l\cos\theta) = \dot{x} - l\dot{\theta}\sin\theta, \\ v_{C_y} &= \frac{d}{dt}(y + l\sin\theta) = \dot{y} + l\dot{\theta}\cos\theta. \end{aligned}$$

Using these in equation (b) gives the constraint

$$\psi_2 = \dot{y}\cos(\theta + \phi) - \dot{x}\sin(\theta + \phi) + l\dot{\theta}\cos\phi = 0. \qquad (c)$$

Applied effort analysis:

There are no applied efforts for this problem.

Lagrange's equations:

The kinetic coenergy of the system is

$$T^* = \frac{1}{2}m(\dot{x}^2 + \dot{y}^2) + \frac{1}{2}I\dot{\theta}^2,$$

the potential energy is $V = 0$, and the dissipation function is $D = 0$. If we define the displacement vector q as $q = [x, y, \theta]^T$ then, the coefficients of the flow constrains (a) and (c) can be given in the Pfaffian form (4.1) as

$$\begin{array}{llll} a_{11} = -\sin\theta, & a_{12} = \cos\theta, & a_{13} = 0, & a_1 = 0, \\ a_{21} = -\sin(\theta+\phi), & a_{22} = \cos(\theta+\phi), & a_{23} = l\cos\phi, & a_2 = 0, \end{array}$$

Note that the steering angle, ϕ, is treated as a 'control' input to the system. With these definitions the equations of motion (4.6) can be written as

$$\begin{aligned} m\ddot{x} - \mu_1 \sin\theta - \mu_2 \sin(\theta+\phi) &= 0, \\ m\ddot{y} + \mu_1 \cos\theta + \mu_2 \cos(\theta+\phi) &= 0, \\ I\ddot{\theta} + \mu_2 l \cos\phi &= 0, \\ \dot{y}\cos\theta - \dot{x}\sin\theta &= 0, \\ \dot{y}\cos(\theta+\phi) - \dot{x}\sin(\theta+\phi) + l\dot{\theta}\cos\phi &= 0. \end{aligned}$$

These 5 differential-algebraic equations can be integrated to find the trajectory of the system given initial conditions, that are consistent with the constraints, and a steering angle ϕ.

4.4 Lagrange's Equation with Effort Constraints and Dynamic Constraints

In this text we use dynamic constraint equations and effort constraint equations to account for regulated effort/flow sources in the systems model. Some system elements that can be modeled as regulated sources include, DC motors, diodes, transistors, operational amplifiers and Coulomb friction. Also, dynamic control system elements can easily be accommodated using this approach.

The dynamic constraints are in the form of differential equations as described by equation (4.3), and the effort constraints are algebraic equations as described by equation (4.4). If the system contains regulated sources that are described using these constraint equations then we can simply append the dynamic constraints and effort constraints to the Lagrange's equation (4.5), or (4.6). Note however, that the efforts in the dynamic constraints, and the efforts in the effort constraints must be accounted for in the virtual work expression.

The examples below illustrate how dynamic constraints and effort constraints can be included in the systems model.

Example 4.8.

A simple model of a full-wave rectifier is shown in the figure below. The network shows two diodes connected to voltage sources $v_1(t)$ and $v_2(t)$. These voltages are the output from a center tapped transformer, and are such that $v_2(t)$ is 180° out of phase with $v_1(t)$. The circuit also includes a capacitor, C, and a load resistor R.

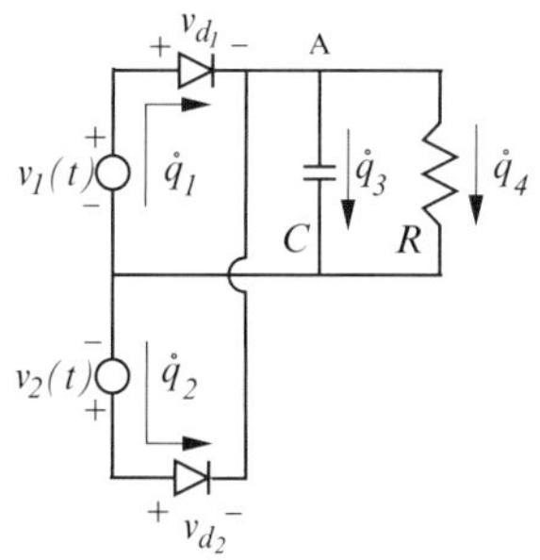

Kinematic analysis:

We have assigned flow coordinates $\dot{q}_1$, $\dot{q}_2$, $\dot{q}_3$ and $\dot{q}_4$ to the model. These flows are not all independent since, at node A, Kirchhoff's current law requires

$$\psi_1 = \dot{q}_1 + \dot{q}_2 - (\dot{q}_3 + \dot{q}_4) = 0.$$

Applied effort analysis:

The voltage sources behave according to the equations $v_1(t) = V \sin \omega t$ and $v_2(t) = V \sin(\omega t + \pi)$ where V and ω are constants. The diodes in the network satisfy the effort constraints

$$\Gamma_1 = \dot{q}_1 - I_s(\exp^{\alpha v_{d1}} - 1) = 0,$$
$$\Gamma_2 = \dot{q}_2 - I_s(\exp^{\alpha v_{d2}} - 1) = 0,$$

where $I_s > 0$ and $\alpha > 0$ are constant and depend on the particular diode employed. In these equations v_{d_1} is the voltage across the diode at the top of the network, and v_{d_2} is the voltage across the diode at the bottom of the network.

The virtual work of all the applied voltages is thus,

$$\delta\mathcal{W} = (v_1 - v_{d_1})\,\delta q_1 + (v_2 - v_{d_1})\,\delta q_2 = e_1^s\,\delta q_1 + e_2^s\,\delta q_2.$$

Lagrange's equations:

The kinetic coenergy for the system is $T^* = 0$, the potential energy is $V = q_3^2/(2C)$, and the dissipation function is $D = R\dot{q}_4^2/2$. Using these terms it can be seen that the equations of motion for the system are

$$\begin{aligned}
\mu_1 &= v_1 - v_{d_1},\\
\mu_1 &= v_2 - v_{d_2},\\
\frac{q_3}{C} - \mu_1 &= 0,\\
R\dot{q}_4 - \mu_1 &= 0,\\
\dot{q}_1 + \dot{q}_2 - (\dot{q}_3 + \dot{q}_4) &= 0,\\
\dot{q}_1 - I_s(\exp^{\alpha v_{d_1}} - 1) &= 0,\\
\dot{q}_2 - I_s(\exp^{\alpha v_{d_2}} - 1) &= 0.
\end{aligned}$$

The first 5 equations are due to (4.6), and the last 2 equations are the effort constraints Γ_1 and Γ_2. These 7 differential-algebraic equations can be solved to determine the variables q_1, q_2, q_3, q_4, μ_1, v_{d_1}, and v_{d_2}.

Example 4.9.

An electrical network with an NPN bipolar junction transistor is shown in the figure (a) below. We would like to find the equations that describe the behavior of this system.

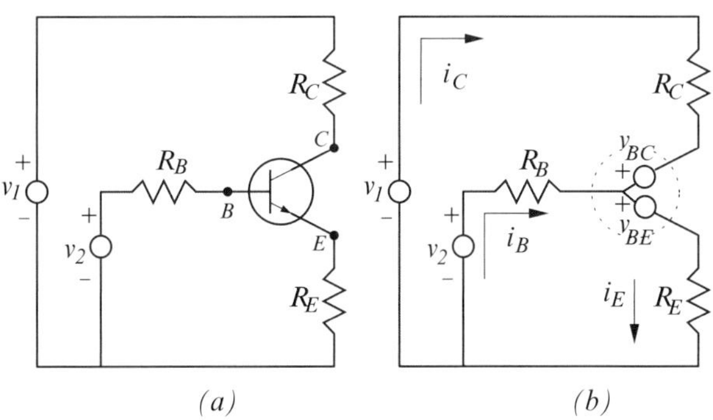

Kinematic analysis:

As described in Chapter 1 we will treat the transistor as voltage regulated current source. In particular, the currents at the base (B), collector (C) and emitter (E) are determined by the voltages at these nodes. An equivalent circuit model for the transistor is shown in the figure (b). In this equivalent network model v_{BE} and v_{BC} denote the base-emitter and base-collector voltages, respectively. Also, we have assigned the flow variables (i.e., the currents) $i_B = \dot{q}_B$, $i_C = \dot{q}_C$ and $i_E = \dot{q}_E$ to the branches of the network. We note that this circuit has two independent loops hence, only two of the assigned currents can be treated as independent variables. By summing the flows into the transistor we get the flow constraint equations

$$\psi = i_B + i_C - i_E = 0.$$

Applied effort analysis:

Using the Ebers-Moll model for the transistor the voltages v_{BE}, v_{BC}, and currents i_E, i_C must satisfy the effort constraint equations

$$\begin{bmatrix} \Gamma_1 \\ \Gamma_2 \end{bmatrix} = \begin{bmatrix} i_E \\ i_C \end{bmatrix} - \begin{bmatrix} I_{ES} & -I_S \\ I_S & -I_{CS} \end{bmatrix} \begin{bmatrix} e^{\alpha v_{BE}} - 1 \\ e^{\alpha v_{BC}} - 1 \end{bmatrix} = \begin{bmatrix} 0 \\ 0 \end{bmatrix}$$

where the nonnegative parameters I_{ES}, I_{CS}, I_S, and α are properties of the transistor.

The virtual work done by the efforts in the equivalent circuit is

$$\delta \mathcal{W} = (v_1 + v_{BC})\, \delta q_C + v_2\, \delta q_B - v_{BE}\, \delta q_E.$$

Lagrange's equations:

The kinetic coenergy, potential energy and dissipation function for the circuit are, $T^* = 0$, $V = 0$, and

$$D = \frac{1}{2}(R_B \dot{q}_B^2 + R_C \dot{q}_C^2 + R_E \dot{q}_E^2),$$

respectively. Hence, Lagrange's equations of motion (3.9) yields

$$\begin{aligned} R_C \dot{q}_C + \mu &= v_1 + v_{BC}, \\ R_B \dot{q}_B + \mu &= v_2, \\ R_E \dot{q}_E - \mu &= -v_{BE}. \end{aligned}$$

Given a set of consistent initial conditions, these three differential equations along with the effort constraints $\Gamma_1 = 0$, and $\Gamma_2 = 0$, and the flow constraint $\psi = 0$, can be solved to determine the six variables $q_B(t)$, $q_C(t)$, $q_E(t)$, $v_{BE}(t)$, $v_{BC}(t)$, and $\mu(t)$.

Example 4.10.

This example develops the equations of motion for a DC motor system with proportional plus integral (PI) position control. The motor model includes the inductance of the coil L, and the resistance of the coil, R. The rotor has moment of inertia I, and is subject to torsional damping which is modeled as a linear torsional damper with damping coefficient b. In this system the angular displacement of the motor is required to follow a desired trajectory $\theta_d(t)$. To accomplish this the input voltage for the motor is determined from a PI feedback control law.

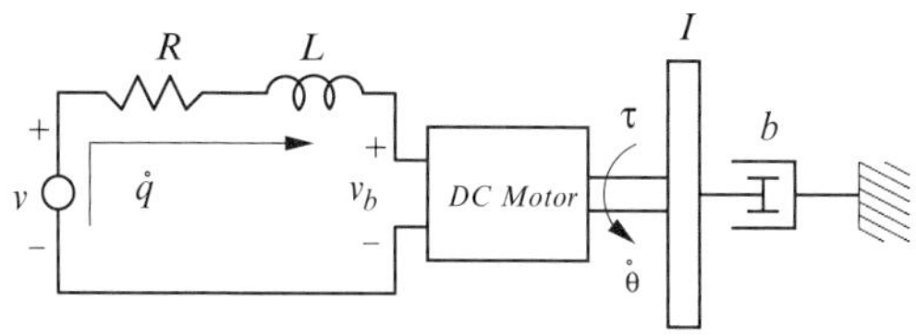

Kinematic analysis:

The generalized flow variables for the model are the motor current, $\dot{q}$, and the shaft speed, $\dot{\theta}$.

Applied effort analysis:

The efforts associated with this model are (i) the back emf voltage, v_b, (ii) the motor torque, τ, and (iii) the input voltage, $v(t)$. The back emf and the motor torque are determined by the effort constraint equations

$$\begin{aligned} \Gamma_1 &= v_b - K_b\dot{\theta} = 0, \\ \Gamma_2 &= \tau - K_t\dot{q} = 0, \end{aligned}$$

where K_b is the back emf constant, and K_t is the torque constant for the motor.

The voltage input for the motor is determined from the control law

$$v(t) = K_p(\theta - \theta_d) + K_i \int (\theta - \theta_d)\, dt. \qquad (a)$$

In this equation K_p is the proportional controller gain, and K_i is the integral controller gain. The first term in equation (a) produces an input voltage that is proportional to the error between the motor angle, θ, and the desired angle θ_d. The second term in equation (a) produces a voltage that is proportional to the integral of the error between the motor angle and the desired angle. If we introduce the dynamic variable s so that

$$\frac{ds}{dt} = \theta - \theta_d,$$

then equation (a) can be rewritten as the constraints

$$\begin{aligned} \dot{s} &= \Sigma_1 = (\theta - \theta_d), \\ \Gamma_3 &= v - K_p(\theta - \theta_d) - K_i s = 0, \end{aligned}$$

where Σ_1 is a dynamic constraint, and Γ_3 is an effort constraint.

The virtual work done by the efforts is

$$\delta\mathcal{W} = v\,\delta q - v_b\,\delta q + \tau\,\delta\theta = (v - v_b)\,\delta q + \tau\,\delta\theta.$$

Therefore, the generalized efforts are $e_q^s = v - v_b$ and $e_\theta^s = \tau$.

Lagrange's equations:

The kinetic coenergy for the system is

$$T^* = \frac{1}{2}(L\dot{q}^2 + I\dot{\theta}^2),$$

the potential energy is $V = 0$, and the dissipation function is

$$D = \frac{1}{2}(R\dot{q}^2 + b\dot{\theta}^2).$$

Using these we can obtain the equations of motion for the system as

$$\begin{aligned} L\ddot{q} + R\dot{q} &= v - v_b, \\ I\ddot{\theta} + b\dot{\theta} &= \tau, \\ \dot{s} - (\theta - \theta_d) &= 0, \\ v_b - K_b\dot{\theta} &= 0, \\ \tau - K_t\dot{q} &= 0, \\ v - K_p(\theta - \theta_d) - K_i s &= 0. \end{aligned}$$

In these equations we have chosen to retain the dynamic constraint and the effort constraints. The first two equations are due to (3.9), the third equation is the dynamic constraint, the fourth equation is the effort constraint due to the back emf voltage, the fifth equation is the effort constraint due to the motor torque, and the sixth equation is the effort constraint that defines the voltage input due to the PID control.

4.5 The Lagrangian Differential-Algebraic Equations

The results of the previous sections can be combined to yield a systematic approach for modeling multidiscipline dynamics systems. In particular, consider a dynamic system described with N configuration displacement variables, $q = [q_1, q_2, \cdots, q_N]^T$, and corresponding flow variables, $f = [f_1, f_2, \cdots, f_N]^T$. In addition, the system is required to satisfy the displacement and flow constraints

$$\phi_j(q,t) = 0, \quad j = 1, 2, \cdots, m_1, \tag{a}$$

$$\psi_j(q,f,t) = 0, \quad j = 1, 2, \cdots, m_2, \tag{b}$$

where $N > (m_1 + m_2)$, and the flow constraints, ψ_j, are in the Pfaffian form. The system also includes effort constraints

$$\Gamma_j(q,f,s,e,t), \quad j = 1, 2, \cdots, m_3, \tag{c}$$

where $e = [e_1, e_2, \cdots, e_{m_3}]^T$ are the regulated efforts, and dynamic constraints

$$\dot{s}_j = \Sigma_j(q,f,s,e,t), \quad j = 1, 2, \cdots, m_4, \tag{d}$$

where $s = [s_1, s_2, \cdots, s_{m_4}]^T$ are the dynamic variables.

Then we can use the results of Sections 4.2 and 4.3 to show that equations of motion for the system becomes

$$\begin{aligned}
0 &= \frac{d}{dt}\frac{\partial T^*}{\partial f_i} - \frac{\partial T^*}{\partial q_i} + \frac{\partial D}{\partial f_i} + \frac{\partial V}{\partial q_i} + \sum_{j=1}^{m_1} \lambda_j \frac{\partial \phi_j}{\partial q_i} + \sum_{j=1}^{m_2} \mu_j \frac{\partial \psi_j}{\partial f_i} - e_i^s, \\
&\qquad i = 1, 2, \cdots, N, \\
0 &= \phi_i(q,t), \quad i = 1, 2, \cdots, m_1, \\
0 &= \psi_i(q,f,t), \quad i = 1, 2, \cdots, m_2, \\
0 &= \dot{s}_i - \Sigma_i(q,f,s,e,t), \quad i = 1, 2, \cdots, m_4, \\
0 &= \Gamma_i(q,f,s,e,t), \quad i = 1, 2, \cdots, m_3.
\end{aligned} \tag{4.7}$$

In these equations $T^*(q,f,t)$ is the kinetic coenergy of the system, $V(q)$ is the potential energy of the system, and $D(f)$ is the dissipation function for the system. The virtual work done by the system efforts is encompassed in configuration efforts, e_i^s, such that $\delta\mathcal{W} = \sum_{i=1}^{N} e_i^s \, \delta q_i$. The configuration efforts, e_i^s include the regulated efforts, e_i, from the dynamic constraints and effort constraints. Note however, that e_i^s *does not* include the efforts due to capacitors and dampers that are represented in the potential energy $V(q)$, and the dissipation function, $D(f)$. The variables λ_j, $j = 1, 2, \cdots, m_1$, are the Lagrange multipliers associated with the displacement constraints, and the variables μ_j, $j = 1, 2, \cdots, m_2$, are the Lagrange multipliers associated with the flow constraints.

Equation (4.7) is called the *Lagrangian Differential-Algebraic Equations* (LDAEs). These equations can be stated compactly as

$$\begin{aligned}
\dot{q} - f &= 0, \\
M\dot{f} + \phi_q^T \lambda + \psi_f^T \mu + \Upsilon &= 0, \\
\dot{s} - \Sigma &= 0, \\
\phi &= 0, \\
\psi &= 0, \\
\Gamma &= 0,
\end{aligned} \tag{4.8}$$

where

$$M = \frac{\partial^2 T^*}{\partial f^2}, \quad \phi = [\phi_1, \phi_2, \cdots, \phi_{m_1}]^T, \quad \phi_q = \frac{\partial \phi}{\partial q}, \quad \lambda = [\lambda_1, \lambda_2, \cdots, \lambda_{m_1}]^T,$$

$$\psi = [\psi_1, \psi_2, \cdots, \psi_{m_2}]^T, \quad \psi_f = \frac{\partial \psi}{\partial f}, \quad \mu = [\mu_1, \mu_2, \cdots, \mu_{m_2}]^T,$$

$$\Upsilon = \left[\frac{\partial}{\partial q} \left(\frac{\partial T^*}{\partial f} \right) \right] f + \frac{\partial}{\partial t} \left(\frac{\partial T^*}{\partial f} \right) - \frac{\partial T^*}{\partial q} + \frac{\partial V}{\partial q} + \frac{\partial D}{\partial f} - e^s,$$

$$e^s = [e_1^s, e_2^s, \cdots, e_N^s]^T, \quad \Gamma = [\Gamma_1, \Gamma_2, \cdots, \Gamma_{m_3}]^T, \quad \Sigma = [\Sigma_1, \Sigma_2, \cdots, \Sigma_{m_4}]^T.$$

This system consists of $2N + m_1 + m_2 + m_3 + m_4$ differential-algebraic equations in the variables q, f, λ, μ, s and e. The first equation in this sequence is a simple statement that the flow variables are the time derivative of the displacement variables. In the second equation the matrix M is called the inertia matrix. It should be clear that M represents the coefficients of terms that are quadratics of f in the kinetic coenergy, T^*. The matrix ϕ_q is the Jacobian of the displacement constraints with respect to the displacement variables, q. The matrix ψ_f is the Jacobian of the flow constraints with respect to the flow variables, f. The third, fourth, fifth and sixth equations are the displacement constraints, flow constraints, dynamic constraints, and effort constraints, respectively.

It is interesting to note that equations of motion for all of the physical dynamic systems considered in this text can be put in the form of equation (4.8). Moreover, equations (4.8) are well suited for automatic generation via computer programs. Such a program would be required to compute the partial derivatives of T^*, V, D, ϕ, and ψ, needed to construct M, ϕ_q, ψ_f and Υ. This approach to systems modeling is particularly attractive for physical systems that involve a large number of variables, or systems that contain nonlinear constraints. For 'simple' dynamic systems however, it may be advantageous to eliminate the algebraic equations from the LDAEs. This is the subject of the next section.

4.5.1 The Underlying ODEs

In this section we show how a special class of LDAEs can be reduced to ordinary differential equations (ODEs). Here we consider an *autonomous* LDAEs of the form

$$\begin{aligned} \dot{q} - f &= 0, \\ M(q,f)\dot{f} + \phi_q(q)^T\lambda + \Upsilon(q,f) &= 0, \\ \phi(q) &= 0. \end{aligned} \qquad (a)$$

This system of differential-algebraic equations is called autonomous because the time variable does not appear explicitly in these equations. Note also that there are no flow constraints, dynamic constraints or effort constraints in this system. Moreover, we assume that the matrix $M(q,f)$ is nonsingular, and the displacement constraints, $\phi(q)$, are linearly independent along the solution to the LDAEs. This system of $2N + m_2$ differential algebraic equations can be reduced to a system of $2N$ differential equations by eliminating the Lagrange multipliers.

To do so we differentiate the displacement constraint with respect to time twice to get

$$\begin{aligned} \frac{d\phi(q,t)}{dt} &= \phi_q f = 0, \\ \frac{d^2\phi(q,t)}{dt^2} &= \phi_q \dot{f} + \left(\phi_q f\right)_q f = 0. \end{aligned} \qquad (b)$$

Since M is nonsingular the second equation in (a) gives

$$\dot{f} = -M^{-1}(\phi_q^T\lambda + \Upsilon).$$

Using this in (b) we find that the Lagrange multipliers must satisfy

$$\lambda = \left[\phi_q M^{-1}\phi_q^T\right]^{-1}\left[\left(\phi_q f\right)_q f - \phi_q M^{-1}\Upsilon\right] \qquad (c)$$

The underlying ODEs for the system is then obtained by putting (c) in the second equation of (a), and ignoring the constraint $\phi = 0$, i.e.,

$$\begin{aligned} \dot{q} &= f, \\ \dot{f} &= -M^{-1}\phi_q(q)^T\left[\phi_q M^{-1}\phi_q^T\right]^{-1}\left[\left(\phi_q f\right)_q f - \phi_q M^{-1}\Upsilon\right] \\ &\quad - M^{-1}\Upsilon. \end{aligned}$$

Example 4.11.

Consider the simple pendulum shown below. The one degree of freedom system is modeled using displacement coordinates q_1 and q_2 (the location of the mass m). The corresponding flows are f_1 and f_2. The displacements are related by the constraint equation $\phi = q_1^2 + q_2^2 - l^2 = 0$.

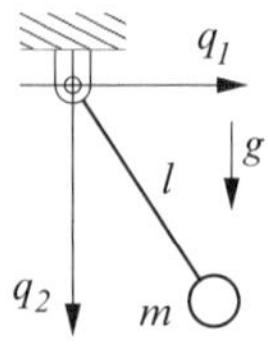

In this case the kinetic coenergy, potential energy and dissipation function are, $T^* = m(f_1^2 + f_2^2)/2$, $V = 0$, and $D = 0$, respectively. The virtual work done by the weight is $\delta W = mg\delta q_2$.

The Lagrangian DAEs for the system are thus

$$\begin{aligned} \dot{q}_1 &= f_1, \\ \dot{q}_2 &= f_2, \\ m\dot{f}_1 + 2q_1\lambda &= 0, \\ m\dot{f}_2 + 2q_2\lambda &= mg, \\ q_1^2 + q_2^2 - l^2 &= 0. \end{aligned}$$

Taking two time derivatives of the displacement constraint equation gives

$$\begin{aligned} \dot{\phi} &= 2q_1 f_1 + 2q_2 f_2 = 0, \\ \ddot{\phi} &= 2f_1^2 + 2f_2^2 + 2q_1\dot{f}_1 + 2q_2\dot{f}_2 = 0. \end{aligned}$$

However, from the Lagrangian DAE we know that

$$\dot{f}_1 = -2q_1\lambda/m, \qquad \dot{f}_2 = -2q_2\lambda/m + g.$$

Using these in $\ddot{\phi} = 0$ shows that the Lagrange multiplier must satisfy

$$\lambda = \frac{m}{2l^2}(f_1^2 + f_2^2 + q_2 g).$$

Hence, the underlying ODEs for this model is

$$\begin{aligned} \dot{q}_1 &= f_1, \\ \dot{q}_2 &= f_2, \\ \dot{f}_1 + \frac{q_1}{l^2}(f_1^2 + f_2^2 + q_2 g) &= 0, \\ \dot{f}_2 + \frac{q_2}{l^2}(f_1^2 + f_2^2 + q_2 g) &= g. \end{aligned}$$

Note that the underlying ODEs do not satisfy the displacement constraint $\phi = q_1^2 + q_2^2 - l^2 = 0$, explicitly. As a result the numerical solution of the underlying ODEs may 'drift' away from the displacement constraint. This issue will be discussed further in Chapter 5.

References

1. B. C. Fabien and R. A. Layton, "Modeling and simulation of physical systems III: An approach for modeling dynamic constraints," *IASTED International Conference, Applied Modeling and Simulation*, 260–263, 1997.
2. H. Goldstein, *Classical Mechanics*, 2nd ed., Addison-Wesley, 1980.
3. D. T. Greenwood, *Principles of Dynamics*, 2nd ed., Prentice-Hall, 1988.
4. E. J. Haug, *Intermediate Dynamics*. Prentice-Hall, 1992.
5. R. W. Jensen and L. R. McNamee, *Handbook of Circuit Analysis Languages and Techniques*, Prentice-Hall, 1976.
6. R. A. Layton and B. C. Fabien, "Modeling and simulation of physical systems I: An introduction to Lagrangian DAEs," *4th IASTED International Conference in Robotics and Manufacturing*, 197–200, 1996.
7. S. W. McCuskey, *Introduction to Advanced Dynamics*, Addison-Wesley, 1959.
8. L. Meirovitch, *Methods of Analytical Dynamics*, McGraw-Hill, 1970.
9. D. A. Wells, *Theory and Problems of Lagrangian Dynamics*, Schaum's Outline Series, McGraw-Hill, 1967.
10. M. R. Spiegel, *Theory and Problems of Theoretical Mechanics*, Schaum's Outline Series, McGraw-Hill, 1967.

Problems

1. Use x_1, y_1, x_2 and y_2 as the only configuration displacements to derive Lagrange's equation of motion for the double-pendulum shown in Section 4.1.
2. Derive Lagrange's equation of motion for the electrical circuit in Section 4.1 using q_0, q_1, q_2 and q_3 as the only configuration displacements.
3. Repeat Example 4.1 using the angle between the rod AB and the x-axis as the generalized coordinate.
4. Derive Lagrange's equation of motion for the slider-crank mechanism in Example 4.2 using θ_1, θ_2 and x_3 as the only configuration displacements.
5. Repeat Example 4.3 using θ_1, θ_2 and θ_3 as the only configuration displacements.
6. In Example 4.4 select three for the flow variables as generalized coordinates. Then reduce the differential-algebraic equation of motion to three ordinary differential equations involving the generalized coordinates.

7. Derive Lagrange's equations of motion for the system in Example 3.12 (Chapter 3) using x and θ as the configuration displacements.
8. The model of a planar R-R robot is shown in the figure here. The link AB can rotate about the revolute joint at A in the x-y plane, and the link BC can rotate about the revolute joint at B in the x-y plane. The link AB has length l_1, mass m_1, and moment of inertia I_1 about its center of mass, which is in the geometric center of the link. The link BC has length l_2, mass m_2, and moment of inertia I_2 about its center of mass, which is in the geometric center of the link. The torques τ_1 and τ_2 are applied to AB and BC, respectively. The coordinate of the center of mass of link AB is x_1, y_1, and the coordinate of the center of mass of link BC is x_2, y_2. Derive Lagrange's equation of motion for the system using x_1, y_1, α, x_2, y_2, and β as the system coordinates. (Neglect gravity.)

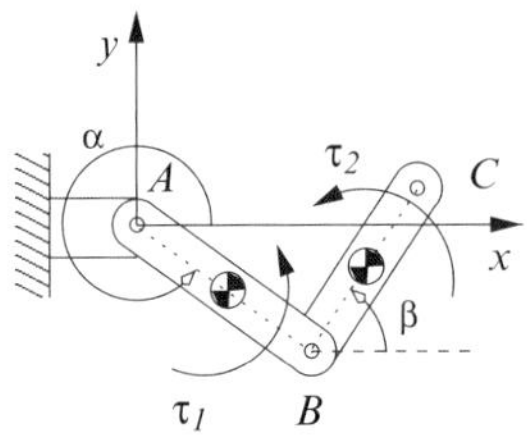

9. Derive Lagrange's equation of motion for the systems shown below. The uniform rod AB has mass m_1 and moment of inertia I_1. The slider at D can be treated as a point with mass m_2. Neglect the weight of the rod DC. A torque τ is applied to the rod AB and gravity acts in the direction shown.

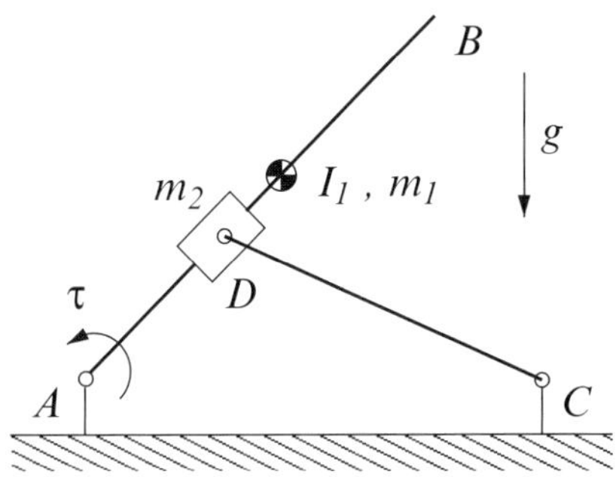

10. A torque τ applied to the link AB drives the mechanism shown below. The center of mass for link AB is located at A, and the link has mass m_1 and moment of inertia I_1. The link BD has mass m_2 and moment of inertia I_2 located at the geometric center of the link. Neglect the mass/inertia of the slider at C. Derive Lagrange's equation of motion for this system.

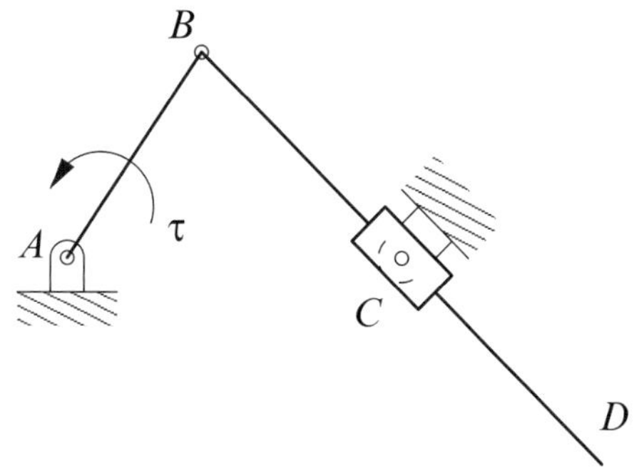

11. Derive Lagrange's equation of motion for the mechanism shown below. The link AB is subject to the input torque τ. Assume that the links AB, ED CF and FG are uniform, i.e., the center of mass is located at the geometric center of the link. The center of mass for the link BCD is located at the midpoint of the line BD.

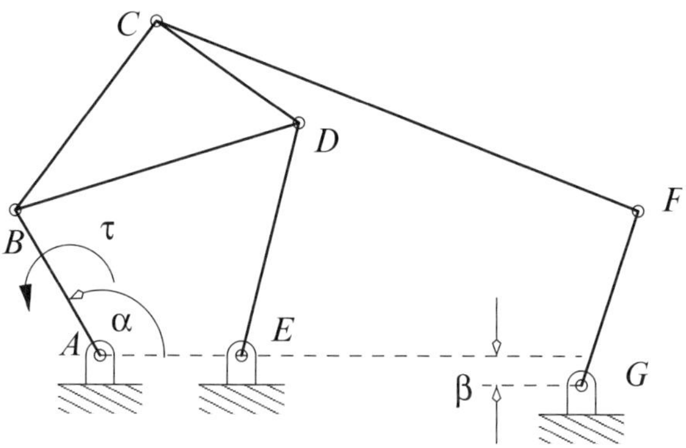

12. The gears at A and B have radius r_1 and r_2, respectively. The link AB is subject to an input torque τ. Derive Lagrange's equation of motion for this system. Assume that the links AB, CD and DE are uniform. Also, the gear at B has its mass center at the point B.

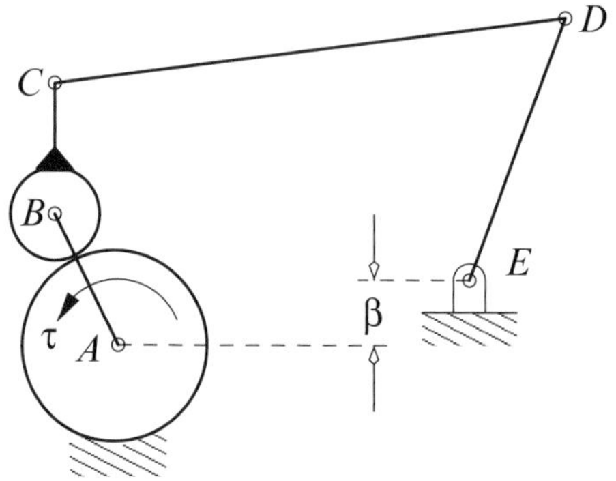

13. Derive Lagrange's equation of motion for the electrical networks in Chapter 2, Problem 14. Do not eliminate the 'excess' flows, or efforts.
14. Consider the Lagrangian DAEs

$$\begin{aligned} \dot{q} - f &= 0, \\ M(q, f, t)\dot{f} + \phi_q(q, t)^T \lambda + \Upsilon(q, f, t) &= 0, \\ \phi(q, t) &= 0. \end{aligned}$$

Find the underlying ODEs for this system by eliminating the Lagrange multiplier λ.
15. Consider the Lagrangian DAEs

$$\begin{aligned} \dot{q} - f &= 0, \\ M(q, f, t)\dot{f} + \phi_q(q, t)^T \lambda + \psi_f(q, f, t)^T \mu + \Upsilon(q, f, t) &= 0, \\ \phi(q, t) &= 0, \\ \psi(q, f, t) &= 0. \end{aligned}$$

Find the underlying ODEs for this system by eliminating the Lagrange multipliers λ and μ.

Chapter 5
Numerical Solution of ODEs and DAEs

It has been shown in the previous chapters that the equations of motion for dynamic systems can be written as systems of ordinary differential equations (ODEs), or differential-algebraic equations (DAEs). In this chapter we present some numerical methods for the solution of these ODEs and DAEs, with emphasis on the single step Runge-Kutta methods.

In Section 5.1 we present some basic results from the numerical solution of ordinary differential equations. In particular, we introduce some first-order solution techniques as well as the concepts of stability, stiffness, error estimation and step size control. Section 5.2 presents the higher-order Taylor series and Runge-Kutta methods for the solution of ordinary differential equations. Here we consider both explicit and implicit Runge-Kutta methods. Section 5.3 considers the numerical solution of differential-algebraic equations. Here, we give specific details for an implicit Runge-Kutta method that works well, when applied to both ODEs and DAEs. Finally, in Section 5.4 we outline a multistep method based on the backward differentiation formula (BDF).

5.1 First-order Methods for ODEs

This section presents some basic numerical methods for the solution of ordinary differential equations. To do so we introduce some important definitions and concepts relevant to the numerical solution of ODEs and DAEs. Here we are concerned with the numerical solution of the ordinary differential equations

$$\dot{y} = f(y,t), \quad t_i \le t \le t_f \tag{5.1}$$

$$y(t_i) = y_i, \tag{5.2}$$

where, t is the time, $y(t) \in \mathcal{R}^n$, is a vector of n time dependent state variables, $f(y,t) \in \mathcal{R}^n$, are the system equations, t_i is the initial time, t_f is the final

B. Fabien, *Analytical System Dynamics: Modeling and Simulation*,
DOI 10.1007/978-0-387-85605-6_5,

time. Equation (5.2) indicates that at the initial time, the state vector must be equal to y_i. We would like to find an approximate solution to ordinary differential equations (5.1) in the interval $t_i \leq t \leq t_f$.

The problem (5.1)-(5.2) is called an initial value problem, because the solution to the differential equations (5.1) is specified at the initial time. (This type of problem is distinct from two-point (or multi-point) boundary value problem where the solution is specified at two (or more) points in the interval $t_i \leq t \leq t_f$). Throughout this text it is assumed that $f(y,t)$ is continuously differentiable, with respect to its arguments, along the solution to the initial value problem.

5.1.1 The explicit Euler method

Let $t^{(0)} = t_i$ and $t^{(1)} = t^{(0)} + h$, where h is some small increment in the time. In the algorithms developed below we call h the step size. In this section we also let $t^{(k)} = t^{(k-1)} + h$ for $k = 0, 1, 2, \cdots$. Thus, $t^{(k)}$ represents a set of discrete time points that are equally spaced. Later we will relax this condition and allow nonuniform spacing of the time points $t^{(k)}$.

Now let $y(t^{(k)})$ be the exact solution to the initial value problem at time $t^{(k)}$. Then a Taylor series expansion of $y(t^{(k+1)})$ about $t^{(k)}$ yields

$$\begin{aligned} y(t^{(k+1)}) &= y(t^{(k)}) + \frac{dy(t^{(k)})}{dt}h + \frac{d^2y(t^{(k)})}{dt^2}\frac{h^2}{2} + \cdots \\ &= y(t^{(k)}) + f(y(t^{(k)}), t^{(k)})h + \frac{df(y(t^{(k)}), t^{(k)})}{dt}\frac{h^2}{2} + \cdots . \end{aligned} \tag{5.3}$$

Neglecting terms of order h^2 and higher we get an approximate solution to the initial value problem at $t^{(k+1)}$ as the discretization formula

$$y^{(k+1)} = y^{(k)} + f(y^{(k)}, t^{(k)})h. \tag{5.4}$$

Here, $y^{(k)}$ represents the numerical solution at time $t^{(k)}$, i.e., $y^{(k)} \approx y(t^{(k)})$.

Another approach to deriving (5.4) is to approximate the derivative $\dot{y}(t^{(k)})$ using

$$\dot{y}(t^{(k)}) = \frac{1}{h}(y^{(k+1)} - y^{(k)}). \tag{5.5}$$

Putting (5.5) into (5.1) with $t = t^{(k)}$ and $y(t) = y^{(k)}$ gives (5.4). The discrete equation (5.5) is called the *forward Euler formula*, and (5.4) is called the *explicit Euler method.*

Algorithm 5.1.1 uses the explicit Euler method to find the approximate solution of the initial value problem (5.1)-(5.2). This algorithm computes the approximate solution at N discrete time points $t^{(k)}$, $k = 1, 2, \cdots, N$. Starting with the initial condition $y^{(0)} = y(t_i)$ the method marches forward using the prior solution, $y^{(k)}$, to compute the solution at the next time step, i.e., $y^{(k+1)}$. The explicit Euler method is called a *single step* method because it only uses the solution at time $t^{(k)}$ to compute the solution at time $t^{(k+1)}$.

Algorithm 5.1.1 Explicit Euler Method

Input: An integer $N > 0$, $h = (t_f - t_i)/N$, $t^{(0)} = t_i$, $y^{(0)} = y(t_i) = y_i$.
Output: $y^{(k)}$, $k = 1, 2, \cdots, N$.

```
1: for k = 0, 1, ..., N - 1 do
2:     y^(k+1) = y^(k) + h f(y^(k), t^(k))
3:     t^(k+1) = t^(k) + h
4: end for
```

Example 5.1.

In this example we will apply the explicit Euler method to the scalar ordinary differential equation

$$\dot{y} = -y,$$

with initial condition $y(0) = 2$. The exact (analytical) solution to this initial value problem is $y(t) = 2e^{-t}$. Applying the explicit Euler method to this ordinary differential equation gives the iteration

$$\begin{aligned} y^{(k+1)} &= y^{(k)} + hf(y^{(k)}, t^{(k)}) \\ &= y^{(k)} - hy^{(k)} \\ &= (1 - h)y^{(k)}. \end{aligned}$$

Let $E^{(k)} = |y(t^{(k)}) - y^{(k)}|$, i.e., $E^{(k)}$ is the error between the analytical solution and the numerical solution at time $t^{(k)}$. The table below shows the errors produced by Algorithm 5.1.1 at different times, using step size $h = 0.1$ and step size $h = 0.01$.

t	$h = 0.1$ $E^{(k)}$	$h = 0.01$ $E^{(k)}$
0.1	0.009	0.0009
0.2	0.017	0.0016
0.3	0.023	0.0022
0.4	0.028	0.0026
0.5	0.032	0.003
⋮	⋮	⋮

These results show that (i) a smaller step size produces a smaller error, and (ii) in the interval $0 \le t \le 0.5$, the error increases as the time increases.

From the results of this example we may be tempted to make h very small so that the error $E^{(k)}$ becomes insignificant. This is inadvisable for two reasons. First, as the step size becomes very small we will require a large number of steps to complete the integration from t_i to t_f. Thus, the method becomes less efficient as h gets smaller. Second, there is a limit to how small we can make the step size. In computers, real numbers are represented by a finite number of digits and not all real numbers can be represented exactly. An unfortunate consequence of this inexact representation is that two distinct real numbers may have the same representation in the computer. Specifically, all digital computers have a parameter called the 'machine precision', or the 'machine epsilon', $\epsilon_M > 0$, which is defined as the largest number for which $1+\epsilon_M = 1$. Thus, 1 and $1+\epsilon_M$ have the same representation in the computer. If we make our step size $h \approx \epsilon_M$ then, $t^{(k+1)} = t^{(k)} + h = t^{(k)}$, and it becomes meaningless to attempt an integration from step k to $k+1$ in this case.

In the following sections we will address the issue of selecting a step size, h that provides a sufficiently accurate solution and yields an efficient numerical procedure.

Stability

Consider the application of the explicit Euler method to the scalar ODE $\dot{y} = \lambda y$, where $\lambda < 0$ and $y(0) = 1$. Then (5.5) gives

$$y^{(k+1)} = y^{(k)} + hf(y^{(k)}, t^{(k)}) = y^{(k)} + h\lambda y^{(k)} = (1 + h\lambda)y^{(k)}. \qquad (a)$$

For this problem we know that the exact solution is $y(t) = e^{\lambda t}$, and that

$$\lim_{t\to\infty} y(t) = 0, \qquad (b)$$

because $\lambda < 0$. However, the numerical solution produced by the explicit Euler method does not share the property (b) for all step sizes $h > 0$. For example, suppose $\lambda = -1$ and $h = 3$ then (a) gives $y^{(k+1)} = -2y^{(k)}$. So for $k = 0, 1, 2, \cdots$ we get

$$\begin{aligned} y^{(1)} &= -2 \\ y^{(2)} &= -2y^{(1)} = 4 \\ y^{(3)} &= -2y^{(2)} = -8 \\ y^{(4)} &= -2y^{(3)} = 16 \\ &\vdots \end{aligned}$$

Thus, in the limit as $k \to \infty$, (i.e., $t \to \infty$) we get $y^{(k)} \to \infty$. As a result the numerical method is unstable, even though the ODE itself has a stable solution.

A close examination of (a) shows that the explicit Euler method is stable only when $|1+h\lambda| \leq 1$. Thus, we must select $h \leq -2/\lambda$ for the explicit Euler method to be stable for this example problem.

We can establish an important property of numerical integration formulas by considering the scalar ordinary differential equation $\dot{y} = \lambda y$, where λ is a complex variable. Specifically, the application of the explicit Euler method to this problem gives

$$y^{(k+1)} = (1 + h\lambda)y^{(k)} = (1 + z)y^{(k)} = R(z)y^{(k)} \qquad (c)$$

Here, $z = h\lambda$ is a complex variable, and $R(z) = 1 + z$ is called the stability function for the numerical integration formula. The discretization equation (c) will be stable provided that $|R(z)| \leq 1$. The region in the complex plane for which $|R(z)| \leq 1$ is called the region of absolute stability. Figure 5.1 shows the region of absolute stability for the explicit Euler method. This is the shaded region defined by a unit circle centered at $z = -1$. Thus, all values of h for which z is not in the shaded region leads to unstable iterations.

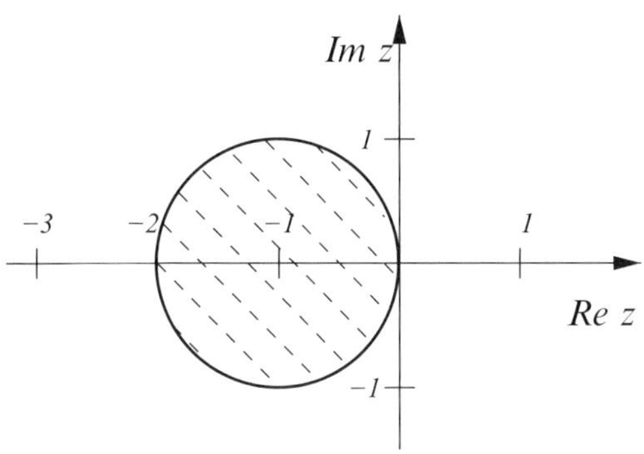

Fig. 5.1 Explicit Euler method region of absolute stability

Stiff differential equations

According to Hairer and Wanner (1996) stiff differential equations are systems for which explicit methods do not work. For stiff systems explicit methods are very inefficient because they require very small step sizes to maintain stability of the method.

As an example consider the differential equation $\dot{y} = \lambda(y - t^2) + 2t$, with $\lambda < 0$ and initial condition $y(0) = 1$. The analytical solution to this problem is $y(t) = e^{\lambda t} + t^2$.

Now, applying the explicit Euler method to this differential equation yields

$$y^{(k+1)} = y^{(k)} + hf(y^{(k)}, t^{(k)})$$

$$= (1 + \lambda h)y^{(k)} - \lambda h(t^{(k)})^2 + 2ht^{(k)}.$$

On the other hand, writing the analytical solution $y(t^{(k+1)})$ as a Taylor series about $t^{(k)}$ gives

$$\begin{aligned} y(t^{(k+1)}) &= y(t^{(k)}) + h(\lambda e^{\lambda t^{(k)}} + 2t^{(k)}) + O(h^2) \\ &= y(t^{(k)}) + \lambda h(y(t^{(k)}) - (t^{(k)})^2) + 2ht^{(k)} + O(h^2) \\ &= (1 + \lambda h)y(t^{(k)}) - \lambda h(t^{(k)})^2 + 2ht^{(k)} + O(h^2). \end{aligned}$$

Neglecting the term $O(h^2)$ we see that the error between the analytical solution and the numerical solution is

$$\begin{aligned} E^{(k+1)} &= y(t^{(k+1)}) - y^{(k+1)} \\ &= (1 + \lambda h)E^{(k)}, \end{aligned}$$

where $E^{(k)} = y(t^{(k)}) - y^{(k)}$. Thus, the error will eventually vanish provided that $|1 + \lambda h| < 1$, i.e., we must select the step size h so that $h < -2/\lambda$.

From the analytical solution we can see that if $\lambda \ll -1$ then the first term in the solution decays rapidly and has very little influence on the steady state response. However, the analysis above shows that if $\lambda \ll -1$ then the explicit Euler method must take small steps in order to maintain stability. Thus, although the transient term $e^{\lambda t}$ does not dominate the analytical solution Differential equations that have this property are called stiff.

Stiff differential equations typically have solutions that include rapidly decaying transient terms. Explicit methods are inefficient when applied to these systems because, the stability of the method requires very small step sizes to accurately represent the transient terms even though these terms have little influence on the steady state solution.

5.1.2 The implicit Euler method

The implicit Euler method provides improved stability properties when compared to the explicit Euler method. Recall that in the explicit Euler method the ODE (5.1) is approximated using

$$\frac{1}{h}(y^{(k+1)} - y^{(k)}) = f(y^{(k)}, t^{(k)}). \tag{a}$$

In the implicit Euler method the ODE is approximated as

$$\frac{1}{h}(y^{(k+1)} - y^{(k)}) = f(y^{(k+1)}, t^{(k+1)}). \tag{b}$$

In this case the right hand side of (5.1) is evaluated at $t^{(k+1)}$, whereas in the explicit method the right hand side of the ODE is evaluated at $t^{(k)}$.

Equation (b) yields the discretization formula

$$y^{(k+1)} = y^{(k)} + hf(y^{(k+1)}, t^{(k+1)}). \tag{5.6}$$

Given $y^{(k)}$ we can use (5.6) to determine $y^{(k+1)}$. Note however that $y^{(k+1)}$ appears on both sides of this equation. Thus, the approximation (b) leads to an implicit equation to determine $y^{(k+1)}$.

Discretization formula (5.6) can be written as

$$\Psi(y^{(k+1)}) = y^{(k+1)} - y^{(k)} - hf(y^{(k+1)}, t^{(k+1)}) = 0. \tag{5.7}$$

This system of equations must be solved to determine $y^{(k+1)}$. To do so we can employ Newton's method.

Newton's Method

Let $y^{(k+1)}_{(j)}$ be the j-th approximate solution to equations $\Psi(y^{(k+1)}) = 0$. Then Newton's method proceeds according to the following algorithm.

Input: Given $y^{(k+1)}_{(0)}$ and a small number $\epsilon_N > 0$.
Output: $y^{(k+1)}$ that solves $\Psi(y^{(k+1)}) = 0$.
1: **for** $j = 0, 1, \cdots$ **do**
2: **if** $\|\Psi(y^{(k+1)}_{(j)})\| \leq \epsilon_N$ **stop**, $y^{(k+1)} = y^{(k+1)}_{(j)}$
3: Solve $D\Psi \Delta y = -\Psi(y^{(k+1)}_{(j)})$
4: Set $y^{(k+1)}_{(j+1)} = y^{(k+1)}_{(j)} + \Delta y$
5: **end for**

Here, $y^{(k+1)}_{(0)}$ is an initial estimate of the solution to the equations (5.7), and ϵ_N is a small number that represents the convergence tolerance. The algorithm terminates in step 2 if $\Psi(y^{(k+1)}_{(j)})$ is sufficiently close to zero, as defined by the tolerance ϵ_N. Otherwise, the algorithm computes a correction Δy for the estimate $y^{(k+1)}_{(j)}$ by solving the linear equation

$$D\Psi \Delta y = -\Psi(y^{(k+1)}_{(j)}),$$

where

$$D\Psi = \frac{\partial}{\partial y^{(k+1)}_{(j)}} \Psi(y^{(k+1)}_{(j)}) = I - h\frac{d}{dy} f(y^{(k+1)}_{(j)}, t^{(k)})$$

is an n by n matrix, with I being the identity matrix, and df/dy being the Jacobian of the function $f(y, t)$.

It can be shown that if $y^{(k+1)}_{(0)}$ is sufficiently close to the solution $y^{(k+1)}$ then Newton's method converges quadratically (Dennis and Schnabel (1983)).

That is, at the j-th iteration of the Newton's method we have

$$\|y_{(j+1)}^{(k+1)} - y^{(k+1)}\| \leq \Omega_0 \|y_{(j)}^{(k+1)} - y^{(k+1)}\|^2$$

for some constant Ω_0.

In practice we sometimes use the 'simplified' Newton's method, where the Jacobian $D\Psi$ is evaluated once at $j = 0$. Thus, $D\Psi$ is fixed for iterations $j = 0, 1, \cdots$. In this case it can be shown that for $y_{(0)}^{(k+1)}$ sufficiently close to the solution $y^{(k+1)}$ the simplified Newton's iteration converges at a rate that is at least linear. That is, at the j-th iteration of the simplified Newton's method we have

$$\|y_{(j+1)}^{(k+1)} - y^{(k+1)}\| \leq \Omega_1 \|y_{(j)}^{(k+1)} - y^{(k+1)}\|,$$

for some constant $0 < \Omega_1 < 1$.

The implicit Euler algorithm

Using the Newton's method described above an implicit Euler method is given by Algorithm 5.1.2. It can be seen that the main computational effort

---

Algorithm 5.1.2 Implicit Euler Method

Input: An integer $N > 0$, $h = (t_f - t_i)/N$, $t^{(0)} = t_i$, $y^{(0)} = y(t_i) = y_i$.
Output: $y^{(k)}$, $k = 1, 2, \cdots, N$.
1: **for** $k = 0, 1, \cdots, N - 1$ **do**
2: $\quad t^{(k+1)} = t^{(k)} + h$
3: $\quad$ Solve $\Psi(y^{(k+1)}) = 0$ via Newton's method to obtain $y^{(k+1)}$
4: **end for**

---

in this algorithm is the solution of the equations $\Psi(y^{(k+1)}) = 0$ at each step. In step 3 we can use $y^{(k)}$ is the initial estimate of the solution to the implicit equations, i.e., we set $y_{(0)}^{(k+1)} = y^{(k)}$. Note that the implicit Euler algorithm is also a single step method, because we only use the result at the most recent step ($y^{(k)}$) to compute the result at the next step ($y^{(k+1)}$).

Comparing algorithms 5.1.1 and 5.1.2 it should be clear that the explicit Euler method is easier to implement than implicit Euler method. However, we can show that the implicit Euler method has better stability properties than the explicit Euler method. In addition, the implicit Euler method is more efficient than the explicit Euler method when applied to stiff differential equations.

To show these results first consider the application of the implicit Euler method to the ODE $\dot{y} = \lambda y$, where $\lambda < 0$ and $y(0) = 1$. Then (5.6) gives

$$y^{(k+1)} = y^{(k)} + hf(y^{(k+1)}, t^{(k+1)})$$

$$\begin{aligned} y^{(k+1)} &= y^{(k)} + h\lambda y^{(k+1)} \\ y^{(k+1)} - h\lambda y^{(k+1)} &= y^{(k)} \\ y^{(k+1)} &= y^{(k)}/(1 - h\lambda). \end{aligned}$$

Since $\lambda < 0$ we see that $1/(1-h\lambda) < 1$ for all $h > 0$. As a result $\lim_{k\to\infty} y^{(k)} = 0$, i.e., the implicit Euler method is stable for all $h > 0$. Recall that, for this problem, the explicit Euler method is stable only if $h \leq -2/\lambda$.

Next, consider the stiff differential equation $\dot{y} = \lambda(y - t^2) + 2t$, with initial condition $y(0) = 1$, and $\lambda \ll -1$. Recall that the analytical solution to this problem is $y(t) = e^{\lambda t} + t^2$. Writing $y(t^{(k)})$ as a Taylor series about $t^{(k+1)}$ gives

$$y(t^{(k+1)}) = y(t^{(k)}) + \lambda h(y(t^{(k+1)}) - (t^{(k+1)})^2) + 2ht^{(k+1)} + O(h^2). \qquad (c)$$

If we apply the implicit Euler method to this ODE we obtain

$$y^{(k+1)} = y^{(k)} + \lambda h(y^{(k+1)} - (t^{(k+1)})^2) + 2ht^{(k+1)}. \qquad (d)$$

The error between the analytical solution and the numerical solution is $E^{(k)} = y(t^{(k)}) - y^{(k)}$. Therefore, using (c) and (d) we get

$$E^{(k+1)} = E^{(k)} + \lambda h E^{(k+1)} + O(h^2).$$

Neglecting the term $O(h^2)$ gives

$$E^{(k+1)} = E^{(k)}/(1 - \lambda h).$$

Since $\lambda \ll -1$ we see that $1/(1 - \lambda h) < 1$ for all $h > 0$. Thus, $\lim_{k\to\infty} E^{(k)} = 0$, i.e., the implicit Euler method is stable for all positive step sizes. This should be compared to the explicit Euler method which is only stable when $h \leq -2/\lambda$, and thus requires very small step sizes when $\lambda \ll -1$. Since the implicit Euler method is unconditionally stable we can use arbitrarily large step sizes and still maintain stability of the integration method.

Finally, we can determine the region of absolute stability of the implicit Euler method by considering the scalar ODE $\dot{y} = \lambda y$, where λ is a complex variable. In this case the implicit Euler method gives

$$\begin{aligned} y^{(k+1)} &= y^{(k)} + h\lambda y^{(k+1)} \\ &= \frac{1}{1 - h\lambda} y^{(k)} \\ &= \frac{1}{1 - z} y^{(k)} \\ &= R(z) y^{(k)}, \end{aligned}$$

where $z = h\lambda$ and $R(z) = 1/(1 - z)$. Now recall that the region of absolute stability is defined as the region of the complex plane where $|R(z)| \leq 1$. Figure

5.2 shows the region of absolute stability of the implicit Euler method. This is shaded region defined by the entire complex plane, except for the region that includes the unit circle centered at $z = 1$. This should be compared with Fig. 5.1 which shows the region of absolute stability for the explicit Euler method.

If the region of absolute stability includes the entire left hand complex plane then the numerical method is said to be 'A-stable'. Thus, for A-stable methods $|R(z)| \leq 1$ for all z where $\text{Real}(z) \leq 0$. From Fig. 5.2 it is clear that the implicit Euler method is A-stable.

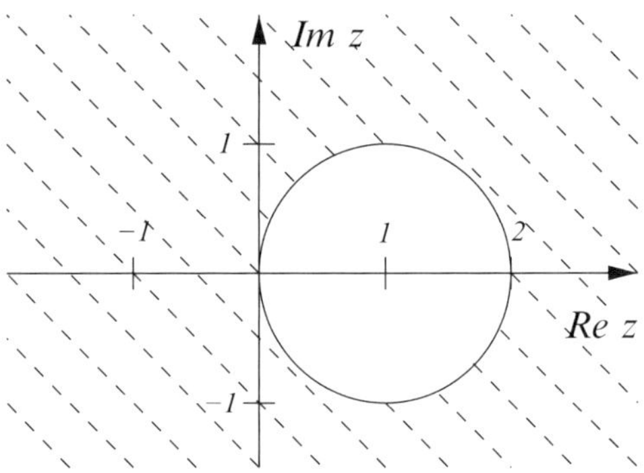

Fig. 5.2 Implicit Euler method region of absolute stability

5.1.3 Integration errors

The usefulness of a numerical method for integrating differential equations depends not only on its stability properties but also on its accuracy. In this section we will outline a basic procedure for evaluating the accuracy of numerical integration methods. Although the discussion will focus on the explicit and implicit Euler methods the techniques used here can be applied to other methods.

In the previous sections we have seen that the explicit and implicit Euler methods were developed by approximating the derivative term in the differential equations. Due to this approximation there will be a discrepancy between the exact and numerical solutions of the differential equation. Here we will use three error measures to evaluate the performance of the numerical integration technique. Specifically, we will consider the *local discretization error*, the *local error*, and the *global error*.

Local discretization error

The local discretization error is also called the local truncation error, and is defined as the error that results when the exact solution is applied to the discretization formula. The explicit Euler formula (5.4) can be written as

$$\Phi(y^{(k+1)}) = y^{(k+1)} - y^{(k)} - hf(y^{(k)}, t^{(k)}) = 0.$$

Applying the exact solution $y(t^{(k)})$ to this formula gives

$$\Phi(y(t^{(k+1)})) = y(t^{(k+1)}) - y(t^{(k)}) - hf(y(t^{(k)}), t^{(k)}) = \delta^{(k+1)},$$

where $\delta^{(k+1)}$ is defined as the local discretization error at time $t^{(k+1)}$. To quantify $\delta^{(k+1)}$ we use the facts that

$$y(t^{(k)}) = y(t^{(k+1)}) - h\frac{d}{dt}y(t^{(k+1)}) + \frac{h^2}{2}\frac{d^2}{dt^2}y(t^{(k+1)}) + O(h^3). \qquad (a)$$

and

$$\begin{aligned} f(y(t^{(k)}), t^{(k)}) &= \frac{d}{dt}y(t^{(k)}) \\ &= \frac{d}{dt}y(t^{(k+1)}) - h\frac{d^2}{dt^2}y(t^{(k+1)}) + \frac{h^2}{2}\frac{d^3}{dt^3}y(t^{(k+1)}) + O(h^3). \end{aligned}$$

Putting these into $\Phi(y(t^{(k+1)}))$ gives

$$\delta^{(k+1)} = \frac{h^2}{2}\frac{d^2}{dt^2}y(t^{(k+1)}) + O(h^3).$$

Thus, the leading term in the local discretization error of the explicit Euler method is of order h^2.

A similar result can be derived for the implicit Euler method. In particular, using the exact solution $y(t^{(k+1)})$ in the implicit Euler formula (5.7) gives the discretization error

$$\Psi(y(t^{(k+1)})) = y(t^{(k+1)}) - y(t^{(k)}) - hf(y(t^{(k+1)}), t^{(k+1)}) = \delta^{(k+1)}.$$

Putting (a) into $\Psi(y(t^{(k+1)}))$ gives the local discretization error

$$\delta^{(k+1)} = -\frac{h^2}{2}\frac{d^2}{dt^2}y(t^{(k+1)}) + O(h^3).$$

The order of an integration method is defined in terms of the local discretization error. If the leading term in the local discretization error is of order h^{p+1} then, the integration method is said to be of order p. Thus, the explicit and implicit Euler methods are of order 1, i.e., the are first-order methods.

Local error

The local error at time $t^{(k+1)}$ is defined as

$$\eta^{(k+1)} = y(t^{(k+1)}) - y^{(k+1)}, \tag{5.8}$$

where it is assumed that the numerical solution coincides with the exact solution at $t^{(k)}$, i.e., $y(t^{(k)}) = y^{(k)}$. Hence, $\eta^{(k+1)}$ is a measure of the error made in the single step from $t^{(k)}$ to $t^{(k+1)}$.

In the case of the explicit Euler method the local error can be determined by writing $\Phi(y(t^{(k+1)}))$ in a Taylor series about the point $y^{(k+1)}$. Doing so gives

$$\begin{aligned}\Phi(y(t^{(k+1)})) &= \Phi(y^{(k+1)}) + D\Phi(y^{(k+1)})[y(t^{(k+1)}) - y^{(k+1)}] \\ &\quad + \text{higher order terms} \\ &= \delta^{(k+1)}.\end{aligned}$$

But, $\Phi(y^{(k+1)}) = 0$ because $y^{(k)} = y(t^{(k)})$. Also, $D\Phi(y^{(k+1)}) = d\Phi/dy^{(k+1)} = I$. Hence, if we neglect the higher order terms we get

$$y(t^{(k+1)}) - y^{(k+1)} = \eta^{(k+1)} = \delta^{(k+1)}. \tag{b}$$

Therefore, in the case of the explicit Euler method, the local error is equal to the local discretization error.

For the implicit Euler method the local error can be determined by writing $\Psi(y(t^{(k+1)}))$ in a Taylor series about the point $y^{(k+1)}$. That is,

$$\begin{aligned}\Psi(y(t^{(k+1)})) &= \Psi(y^{(k+1)}) + D\Psi(y^{(k+1)})[y(t^{(k+1)}) - y^{(k+1)}] \\ &\quad + \text{higher order terms} \\ &= \delta^{(k+1)},\end{aligned}$$

where $D\Psi(y^{(k+1)}) = d\Psi/dy^{(k+1)} = I - h\,\frac{d}{dy}f(y^{(k+1)}, t^{(k+1)})$, and $\Psi(y^{(k+1)}) = 0$ because $y^{(k)} = y(t^{(k)})$. By neglecting the higher order terms we can see that the local error is

$$y(t^{(k+1)}) - y^{(k+1)} = \eta^{(k+1)} = \left[I - h\frac{d}{dy}f(y^{(k+1)}, t^{(k+1)})\right]^{-1} \delta^{(k+1)}. \tag{c}$$

If the term $h\,df/dy$ is small then the local error for the explicit and implicit Euler methods will be the same. For stiff differential equations however, the term $h\,df/dy$ can be very large, and as a result the local error must be determined by scaling the discretization error as shown in (c).

Global Error

Consider the ODE (5.1) with initial condition $y(t^{(0)}) = y^{(0)}$ then the global error at time $t^{(k)}$ is defined as

$$E^{(k)} = y(t^{(k)}) - y^{(k)}. \tag{5.9}$$

Using the explicit Euler method in (5.9) we get

$$\begin{aligned} E^{(k+1)} &= y(t^{(k+1)}) - (y^{(k)} + hf(y^{(k)}, t^{(k)})) \\ &= y(t^{(k+1)}) - (y(t^{(k)}) - E^{(k)} + hf(y(t^{(k)}) - E^{(k)}, t^{(k)})) \\ &= E^{(k)} + y(t^{(k+1)}) - y(t^{(k)}) - hf(y(t^{(k)}), t^{(k)}) \\ &\quad + hf_y E^{(k)} + O(\|E^{(k)}\|^2) \\ &= (I + hf_y)E^{(k)} + \delta^{(k+1)} + O(\|E^{(k)}\|^2) \\ &\approx (I + hf_y)E^{(k)} + \delta^{(k+1)}, \end{aligned}$$

where $f_y = \frac{d}{dy} f(y(t^{(k)}), t^{(k)})$. This result shows that the global error at $t^{(k+1)}$ is a function of the global error at $t^{(k)}$ and the local discretization error $\delta^{(k+1)}$. Thus, the global error is a propagation of the previous errors as well as the current the local discretization error.

In practice controlling the global error when integrating initial value problems can be very expensive. This involves solving the problem once to establish an estimate of the global error. If the desired global error is not achieved the problem is solved a second time with reduced step sizes.

Since controlling the global error is too expensive most numerical methods regulate the local error instead. This is done by requiring that the local error be less than some specified solution tolerance at each step. The details of how such schemes are implemented will be discussed below.

5.2 Higher-order Methods for ODEs

The previous section introduced the explicit and implicit Euler methods which are both first-order, i.e., the leading term in the local discretization error is $O(h^2)$. These methods tend to be inefficient when the interval $[t_i, t_f]$ is large and the desired solution tolerance is small. Since, in this case small step sizes, h, will be required to maintain a small local error. It is therefore desirable to have integration methods that allow us to take large steps while maintaining small errors. This can be accomplished if the local discretization error of the method is of a high order. In this section we consider some single step higher-order numerical integration methods.

5.2.1 The Taylor series method

To develop this method let t be some time where the solution to the ODE is known. Let h be some small fixed increment in the time. Then, expanding the exact solution $y(t+h)$ in a Taylor series about the time t, gives

$$y(t+h) = y(t) + h\frac{d}{dt}y(t) + \frac{h^2}{2}\frac{d^2}{dt^2}y(t) + \frac{h^3}{6}\frac{d^3}{dt^3}y(t) + \cdots + \frac{h^p}{p!}\frac{d^p}{dt^p}y(t) + R(\xi), \tag{5.10}$$

where $R(\xi) = \frac{h^{p+1}}{(p+1)!}\frac{d^{p+1}}{dt^{p+1}}y(\xi)$ is called the remainder, with ξ being some point between t and $t+h$, and $p > 0$ being some integer. Here we assume that $y(t)$ and $f(y(t), t))$ are smooth enough to assure that the representation (5.10) is valid.

Now let $y^{(k)}$ approximate the exact solution at time t, and let $y^{(k+1)}$ approximate the exact solution at time $t+h$. Then by neglecting the remainder in (5.10), and using the fact that $f(y(t), t) = dy(t)/dt$, we can obtain the explicit discretization formula

$$y^{(k+1)} = y^{(k)} + hT_p(y^{(k)}, t^{(k)}), \tag{5.11}$$

where

$$T_p(y^{(k)}, t^{(k)}) = f(y^{(k)}, t^{(k)}) + \frac{h}{2}\frac{d}{dt}f(y^{(k)}, t^{(k)}) + \cdots + \frac{h^{p-1}}{p!}\frac{d^{p-1}}{dt^{p-1}}f(y^{(k)}, t^{(k)}). \tag{5.12}$$

The evaluation of $T_p(y^{(k)}, t^{(k)})$ requires the total derivatives of $f(y^{(k)}, t^{(k)})$. For example in the case of a scalar ODE

$$T_3 = f + \frac{h}{2}\frac{df}{dt} + \frac{h^2}{3!}\frac{d^2 f}{dt^2},$$

where

$$\begin{aligned}\frac{df}{dt} &= \frac{\partial f}{\partial y}f + \frac{\partial f}{\partial t}\\ \frac{d^2 f}{dt^2} &= \frac{\partial^2 f}{\partial y^2}f^2 + \left(\frac{\partial f}{\partial y}\right)^2 f + 2\frac{\partial^2 f}{\partial t \partial y}f + \frac{\partial f}{\partial y}\frac{\partial f}{\partial t} + \frac{\partial^2 f}{\partial t^2}.\end{aligned}$$

Here, it is understood that all functions are evaluated at the point $(y^{(k)}, t^{(k)})$.

From (5.10) it can be seen that the local discretization error for the formula (5.12) is

$$\begin{aligned}\delta^{(k+1)} &= \frac{h^{p+1}}{(p+1)!}\frac{d^{p+1}}{dt^{p+1}}y(\xi)\\ &= \frac{h^{p+1}}{(p+1)!}\frac{d^p}{dt^p}f(y(\xi), \xi),\end{aligned}$$

where $t^{(k)} < \xi < t^{(k+1)}$. Thus the formula (5.11) defines a p-th order explicit method.

The formula (5.12) can be used to construct a Taylor series method for the numerical integration of ordinary differential equations. Such a procedure is given in Algorithm 5.2.1. Note that the explicit Euler method (Algorithm 5.1.1) is a special case of the Taylor series method. In particular, when $p = 1$ the formula (5.12) defines the explicit Euler method.

Algorithm 5.2.1 Taylor Series Method

Input: An integer $p > 0$, and integer $N > 0$, $h = (t_f - t_i)/N$, $t^{(0)} = t_i$, $y^{(0)} = y(t_i) = y_i$.
Output: $y^{(k)}$, $k = 1, 2, \cdots, N$.

```
for k = 0, 1, ..., N − 1 do
    y^(k+1) = y^(k) + h T_p(y^(k), t^(k))
    t^(k+1) = t^(k) + h
end for
```

5.2.2 *Explicit Runge-Kutta methods*

The implementation of higher order Taylor series methods require the total derivatives of $f(y,t)$ which can be be tedious to compute. The explicit Runge-Kutta methods attempt to duplicate the higher order Taylor series methods without computing the higher order derivatives of $f(y,t)$.

To illustrate how these Runge-Kutta methods can be constructed we consider the development of a method that has a local discretization error of order h^3. For the sake of simplicity we restrict our attention to scalar ODEs.

An example of an explicit Runge-Kutta method is

$$y^{(k+1)} = y^{(k)} + h\left[b_1 k_1 + b_2 k_2\right], \tag{a}$$

where

$$\begin{aligned} k_1 &= f(y^{(k)}, t^{(k)}) \\ k_2 &= f(y^{(k)} + h a_{21} k_1, t^{(k)} + c_2 h). \end{aligned}$$

Here, the method coefficients b_1, b_2, a_{21} and c_2 are selected to ensure that the local discretization error of (a) is of order h^3.

The formula (a) defines a '2-stage' Runge-Kutta method, where k_1 and k_2 are called the stage derivatives. Comparing (a) to a Taylor series expansion of $y(t^{(k+1)})$ we see that the terms $b_1 k_1 + b_2 k_2$ must coincide with $f + (h/2)\frac{d}{dt} f$ for this method to have a local discretization error of order h^3. Hence, the combination $b_1 k_1 + b_2 k_2$ is used to approximate the total derivative of f, and this is to be accomplished using only function evaluations.

The local discretization error for the formula (a) is computed as

$$\delta^{(k+1)} = y(t^{(k+1)}) - y(t^{(k)}) - h\left[b_1\bar{k}_1 + b_2\bar{k}_2\right], \tag{b}$$

where

$$\begin{aligned}\bar{k}_1 &= f(y(t^{(k)}), t^{(k)})\\ \bar{k}_2 &= f(y(t^{(k)} + ha_{21}\bar{k}_1), t^{(k)} + c_2h).\end{aligned}$$

Now, write $y(t^{(k+1)})$, $\bar{k}_1$ and $\bar{k}_2$ as Taylor polynomials centered at the point $(y(t^{(k)}), t^{(k)})$ to get

$$\begin{aligned}y(t^{(k+1)}) &= y(t^{(k)}) + hf + \frac{h^2}{2}\frac{df}{dt} + \frac{h^3}{6}\frac{d^2f}{dt^2} + O(h^4)\\ &= y(t^{(k)}) + hf + \frac{h^2}{2}\left[\frac{\partial f}{\partial y}f + \frac{\partial f}{\partial t}\right]\\ &\quad + \frac{h^3}{6}\left[\frac{\partial^2 f}{\partial y^2}f^2 + \left(\frac{\partial f}{\partial y}\right)^2 f + 2\frac{\partial^2 f}{\partial t\partial y}f + \frac{\partial f}{\partial y}\frac{\partial f}{\partial t} + \frac{\partial^2 f}{\partial t^2}\right]\\ &\quad + O(h^4)\\ \bar{k}_1 &= f\\ \bar{k}_2 &= f + hc_2\frac{\partial f}{\partial t} + \frac{1}{2}h^2c_2^2\frac{\partial^2 f}{\partial t^2} + ha_{21}\frac{\partial f}{\partial y}f + \frac{1}{2}h^2a_{21}^2\frac{\partial^2 f}{\partial y^2}f^2\\ &\quad + a_{21}c_2h^2\frac{\partial^2 f}{\partial y\partial t}f + O(h^3).\end{aligned}$$

Using these in (b) yields

$$\begin{aligned}\delta^{(k+1)} &= [(1 - b_1 - b_2)f]\,h + \left[\left(\frac{1}{2} - b_2a_{21}\right)\frac{\partial f}{\partial y}f + \left(\frac{1}{2} - b_2c_2\right)\frac{\partial f}{\partial t}\right]h^2\\ &\quad + \left[\left(\frac{1}{6} - \frac{1}{2}b_2a_{21}^2\right)\frac{\partial^2 f}{\partial y^2}f^2 + \left(\frac{1}{6} - \frac{1}{2}b_2c_2^2\right)\frac{\partial^2 f}{\partial t^2}\right.\\ &\qquad \left. + \left(\frac{1}{3} - b_2a_{21}c_2\right)\frac{\partial^2 f}{\partial y\partial t}f + \frac{1}{6}\left(\frac{\partial f}{\partial y}\right)^2 f + \frac{1}{6}\frac{\partial f}{\partial y}\frac{\partial f}{\partial t}\right]h^3\\ &\quad + O(h^4).\end{aligned}$$

From this result it can be seen that the local discretization error, $\delta^{(k+1)}$, will be of order h^2 if

$$1 - b_1 - b_2 = 0. \tag{c}$$

If (c) holds along with

$$\frac{1}{2} - b_2a_{21} = 0, \tag{d}$$

and

$$\frac{1}{2} - b_2c_2 = 0 \qquad (e)$$

then, the local discretization error will be of order h^3. It is also evident from $\delta^{(k+1)}$ that there is no way for us to select b_1, b_2, a_{21} and c_2 so that the coefficient of h^3 is made to vanish, for all possible functions $f(y(t), t)$. Thus, the highest possible order that can be achieved by this 2-stage Runge-Kutta method is order 2.

Equations (c), (d) and (e) are called the *order conditions* for the Runge-Kutta method. These equations provide us with three equations for the four coefficients b_1, b_2, a_{21} and c_2. Therefore, one of the coefficients can be selected arbitrarily and as a result an infinite number of solutions are possible.

Two well known solutions to these order conditions are;

- The explicit midpoint method

$$b_1 = 0,\ b_2 = 1,\ a_{21} = \frac{1}{2},\ c_2 = \frac{1}{2}.$$

- Huen's method/Explicit trapezoidal method

$$b_1 = \frac{1}{2},\ b_2 = \frac{1}{2},\ a_{21} = 1,\ c_2 = 1.$$

It is easy to verify that both of these methods yield local discretization error of order h^3, and are thus called second-order methods.

The general s-stage explicit Runge-Kutta method has the form

$$\begin{aligned}
y^{(k+1)} &= y^{(k)} + h\sum_{i=1}^{s} b_i k_i,\\
k_i &= f(Y_i, \tau_i),\\
Y_i &= y^{(k)} + h\sum_{j=1}^{i-1} a_{ij}k_j,\\
\tau_i &= t^{(k)} + c_i h.
\end{aligned} \qquad (5.13)$$

The coefficients of the method are given in $A \in \mathcal{R}^{s\times s}$, $b \in \mathcal{R}^s$ and $c \in \mathcal{R}^s$. Here, the (i, j)-th element of the matrix A is denoted as a_{ij}, etc. For explicit Runge-Kutta methods the matrix A is strictly lower triangular, i.e., $a_{ij} = 0$ if $j \geq i$.

The coefficients of a Runge-Kutta method can be written compactly as a *Butcher tableau*,

$$\begin{array}{c|c} c & A \\ \hline & b^T \end{array}.$$

For example, the explicit Euler method, the explicit midpoint method and the explicit trapezoidal method are written as

$$\begin{array}{c|c} 0 & 0 \\ \hline & 1 \end{array}, \quad \begin{array}{c|cc} 0 & 0 & 0 \\ 1/2 & 1/2 & 0 \\ \hline & 0 & 1 \end{array}, \quad \text{and} \quad \begin{array}{c|cc} 0 & 0 & 0 \\ 1 & 1 & 0 \\ \hline & 1/2 & 1/2 \end{array},$$

respectively.

Using the approach outlined above it is possible to derive the order conditions for higher order Runge-Kutta methods. As shown above, the 2-stage Runge-Kutta formula is at best a second-order method. To achieve higher-order methods we must use more stages in the Runge-Kutta formula. However, as the number of stages increases so does the complexity of the analysis required to determine the order conditions. In Problem 5 the reader is asked to derive the order conditions for a 3-stage, third-order explicit Runge-Kutta method. The order conditions for a 4-stage, fourth-order Runge-Kutta method are;

$$\begin{aligned}
\sum_{i=1}^{4} b_i &= 1, \\
\sum_{i=1}^{4} b_i c_i &= \frac{1}{2}, \\
\sum_{i=1}^{4} b_i c_i^2 &= \frac{1}{3}, \\
\sum_{i=1}^{4}\sum_{j=1}^{4} b_i a_{ij} c_j &= \frac{1}{6}, \\
\sum_{i=1}^{4} b_i c_i^3 &= \frac{1}{4}, \\
\sum_{i=1}^{4}\sum_{j=1}^{4} b_i c_i a_{ij} c_j &= \frac{1}{8}, \\
\sum_{i=1}^{4}\sum_{j=1}^{4} b_i a_{ij} c_j^2 &= \frac{1}{12}, \\
\sum_{i=1}^{4}\sum_{j=1}^{4}\sum_{k=1}^{4} b_i a_{ij} a_{jk} c_k &= \frac{1}{24}.
\end{aligned} \tag{5.14}$$

These 8 nonlinear equations must be solved to determine the coefficients of the fourth-order method.

Perhaps the most famous Runge-Kutta method is the 'classical' fourth-order formula given by the Butcher tableau

0	0	0	0	0
1/2	1/2	0	0	0
1/2	0	1/2	0	0
1	0	0	1	0
	1/6	1/3	1/3	1/6

This 4-stage method is often called 'the Runge-Kutta method' but, as we have seen there are many other Runge-Kutta methods. In fact, there are several popular explicit fourth-order Runge-Kutta methods.

The development of explicit Runge-Kutta methods of order $p \geq 5$ is very complex. One of the difficulties is that the number of order conditions that must be solved becomes very large. For example, a method of order $p = 5$ has 17 order conditions, a method of order $p = 6$ has 37 order conditions, and a method of order $p = 7$ has 85 order conditions. Moreover, to realize a method of order $p = 5$ requires at least 6 stages, a method of order $p = 6$ requires at least 7 stages, and a method of order 7 requires at least 9 stages. The interested reader should consult Butcher (1987), and Hairer, Norsett and Wanner (1993) for additional details.

Error estimation and step size adaptation

An important consideration for the efficient integration of differential equations is the ability of adjust the step size so that the local error satisfies a desired tolerance at each step. To see how such schemes are possible consider the numerical integration of the scalar ODE $\dot{y} = f(y(t), t)$, with initial condition $y(t_i) = y_i$. Here we will use two Runge-Kutta methods, the first method has order p, and the second method has order $p-1$. From the analysis in the previous sections we know that at the k-th step the p-th order method produces a local error

$$y(t^{(k+1)}) - y^{(k+1)} = Ch^{p+1} + O(h^{p+2}), \tag{a}$$

where $y^{(k+1)}$ is the result of the Runge-Kutta formula, $y(t^{(k+1)})$ is the exact solution to the ODE, and C is a constant that depends on the Runge-Kutta formula and the function $f(y(t), t)$. On the other hand, the method of order $p-1$ produces a local error

$$y(t^{(k+1)}) - \hat{y}^{(k+1)} = \hat{C}h^{p} + O(h^{p+1}), \tag{b}$$

where $\hat{y}^{(k+1)}$ is the result of the $(p-1)$-th order Runge-Kutta formula, and $\hat{C}$ is a constant that depends on the Runge-Kutta formula and the function $f(y(t), t)$. We assume that both methods have the same initial condition for the k-th step, i.e., $y^{(k)} = \hat{y}^{(k)} = y(t^{(k)})$.

From (a) we see that $y(t^{(k+1)}) = y^{(k+1)} + Ch^{p+1} + O(h^{p+2})$. Using this result in (b) we obtain a computable estimate of the local error, i.e.,

$$\eta^{(k+1)} = y(t^{(k+1)}) - \hat{y}^{(k+1)} = y^{(k+1)} - \hat{y}^{(k+1)} = \hat{C}h^p + O(h^{p+1}). \qquad (c)$$

At each step we would like $|\eta^{(k+1)}|$ to be less than or equal to some specified tolerance ϵ. If $|\eta^{(k+1)}| \leq \epsilon$ then the integration from $t^{(k)}$ to $t^{(k+1)}$ is considered to be successful, and we accept $y^{(k+1)}$ as the numerical solution to the ODE at $t^{(k+1)}$. However, if $|\eta^{(k+1)}| > \epsilon$ then we need to repeat the integration with a smaller step size so that the local error tolerance is not violated.

The computation of a new step size is accomplished using the local error estimate given by (c). Suppose $|\eta^{(k+1)}| > \epsilon$ then

$$|\eta^{(k+1)}| = |\hat{y}^{(k+1)} - y^{(k+1)}| \approx |\hat{C}h^p| > \epsilon.$$

Now, let $\bar{h}$ be the step size that satisfies the condition $|\bar{\eta}^{(k+1)}| = |\hat{C}\bar{h}^p| = \epsilon$. Then, dividing $|\bar{\eta}^{(k+1)}|$ by $|\eta^{(k+1)}|$ gives

$$\left(\frac{\bar{h}}{h}\right)^p = \frac{\epsilon}{|\eta^{(k+1)}|},$$

which shows that $\bar{h}$ and h are related via

$$\bar{h} = h\left(\frac{\epsilon}{|\eta^{(k+1)}|}\right)^{\frac{1}{p}}. \qquad (d)$$

Thus, if $|\eta^{(k+1)}| > \epsilon$ then (d) computes a new step size $\bar{h} < h$ which is used to repeat the integration of the ODE at time $t^{(k)}$. If on the other hand $|\eta^{(k+1)}| \leq \epsilon$ then (d) computes a new step size $\bar{h} \geq h$ which can be used to advance the solution from time $t^{(k+1)}$, i.e., $t^{(k+2)} = t^{(k+1)} + \bar{h}$.

In practice the new step size $\bar{h}$ is restricted to ensure that it does not vary to greatly from the old step size h. This can be done using a formula of the type

$$\bar{h} = h\min(\mathtt{fac}_1, \max(\mathtt{fac}_0, \beta(|\eta^{(k+1)}|/\epsilon)^{-1/p})), \qquad (e)$$

where $0 < \mathtt{fac}_0 < \mathtt{fac}_1$ and $0 < \beta < 1$. Typical values for these parameters are $\beta = 0.9$, $\mathtt{fac}_0 = 0.2$ and $\mathtt{fac}_1 = 5$, which ensures that $0.2h \leq \bar{h} \leq 5h$.

If $y(t)$ is an n-dimensional vector then the error estimation and step size computation must take account of the possible variability in the scaling of the elements of $y(t)$. In this case the step size computation formula (e) is replaced with

$$\sigma = \left[\sqrt{\frac{1}{n}\sum_{i=1}^{n}\left(\frac{|\eta_i^{(k+1)}|}{\mathtt{atol}_i + \mathtt{rtol}_i|y_i^{(k+1)}|}\right)^2}\right]^{-1/p}$$

$$\bar{h} = h\min(\mathtt{fac}_1, \max(\mathtt{fac}_0, \beta\sigma)). \qquad (5.15)$$

Here, η_i^{k+1} is the i-th element of the vector $\eta^{k+1} = \hat{y}^{(k+1)} - y^{(k+1)}$, $\mathtt{atol} \in \mathcal{R}^n$ is the absolute error tolerance, and $\mathtt{rtol} \in \mathcal{R}^n$ is the relative error tolerance.

Note that the elements of `atol` and `rtol` are nonnegative. In this formulation, for the local error to be acceptable it is sufficient that $|\eta_i^{k+1}| \leq \texttt{atol}_i + \texttt{rtol}_i |y_i^{(k+1)}|$. If $\texttt{rtol}_i = 0$ then we require that $|\eta_i^{k+1}|$ not exceed the absolute error measure $\texttt{atol}_i$. This is called absolute error control. If $\texttt{atol}_i = 0$ then the local error will be acceptable if $(|\eta_i^{k+1}|/|y_i^{(k+1)}|) \leq \texttt{rtol}_i$. This is called relative error control.

Embedded Runge-Kutta Formulas

The local error estimation and step size control scheme described in the previous section requires Runge-Kutta methods of order p and $p-1$. A practical and efficient approach to implementing such a strategy is to use so called 'embedded' Runge-Kutta methods. In this approach $y^{(k+1)}$ is computed by (5.13), and $\hat{y}^{(k+1)}$ is computed from the formula

$$\hat{y}^{(k+1)} = y^{(k)} + h \sum_{i=1}^{s} \hat{b}_i k_i, \tag{5.16}$$

where the coefficients $\hat{b}_i$, $i = 1, 2, \cdots, s$ are selected to ensure that the local error $y(t^{(k+1)}) - \hat{y}^{(k+1)} = O(h^p)$. The efficiency of this method is derived from the fact that both $y^{(k+1)}$ and $\hat{y}^{(k+1)}$ are computed using the same stage derivatives k_i, $i = 1, 2, \cdots, s$.

The Butcher tableau for these embedded methods are of the form

$$\begin{array}{c|c} c & A \\ \hline & b^T \\ & \hat{b}^T \end{array}.$$

An example for such a method can be obtained by combining the explicit Euler methods with the explicit trapezoidal method to get

$$\begin{array}{c|cc} 0 & 0 & 0 \\ 1 & 1 & 0 \\ \hline & 1/2 & 1/2 \\ & 1 & 0 \end{array}.$$

For this embedded method we compute

$$\begin{aligned} y^{(k+1)} &= y^{(k)} + h\left(\frac{1}{2}k_1 + \frac{1}{2}k_2\right) \\ \hat{y}^{(k+1)} &= y^{(k)} + hk_1 \\ k_1 &= f(y^{(k)}, t^{(k)}) \\ k_2 &= f(y^{(k)} + hk_1, t^{(k)} + h). \end{aligned}$$

The local error for the solution $y^{(k+1)}$ is $O(h^3)$ and the local error for the solution $\hat{y}^{(k+1)}$ is $O(h^2)$.

Over the years researchers have developed many ingenious embedded Runge-Kutta methods. Here we provide the coefficients for one of the many embedded Runge-Kutta formulas that can be found in the literature. The Butcher tableau for this particular method is given by Cash and Karp (1990) as

$$
\begin{array}{c|cccccc}
0 & & & & & & \\
\frac{1}{5} & \frac{1}{5} & & & & & \\
\frac{3}{10} & \frac{3}{40} & \frac{9}{40} & & & & \\
\frac{3}{5} & \frac{3}{10} & -\frac{9}{10} & \frac{6}{5} & & & \\
1 & -\frac{11}{54} & \frac{5}{2} & -\frac{70}{27} & \frac{35}{27} & & \\
\frac{7}{8} & \frac{1631}{55296} & \frac{175}{512} & \frac{575}{13824} & \frac{44275}{110592} & \frac{253}{4096} & \\
\hline
 & \frac{37}{378} & 0 & \frac{250}{621} & \frac{125}{594} & 0 & \frac{512}{1771} \\
 & \frac{2825}{27648} & 0 & \frac{18575}{48384} & \frac{13525}{55296} & \frac{277}{14336} & \frac{1}{4}
\end{array}
$$

Note that the terms not included in the tableau are all zero. This method has six stages with $y^{(k+1)} = y^{(k)} + h\sum_{i=1}^{6} b_i k_i$ being a 5-th order formula, i.e., $p = 5$. The embedded formula $\hat{y}^{(k+1)} = y^{(k)} + h\sum_{i=1}^{6} \hat{b}_i k_i$ is a 4-th order formula.

Embedded Runge-Kutta formulas of this type are used to construct very efficient computer programs for the numerical solution of non-stiff differential equations. Algorithm 5.2.2 indicates how such programs can be organized to integrate the system of differential equations (5.1)-(5.2). This algorithm assumes that we are given an s stage explicit, embedded Runge-Kutta method with coefficients A, c, b and $\hat{b}$ as described above. Moreover, we assume that the pair of formulas provide methods of order p and $p-1$.

In addition to the Runge-Kutta coefficients the algorithm uses the following parameters.

- An initial step size, h_0. A good estimate for the initial step size is $h_0 = 0.01\epsilon < |t_f - t_i|$, where ϵ is a measure of the desired tolerance, say $\epsilon = \max(\texttt{atol}_i, \texttt{rtol}_i)$.

Algorithm 5.2.2 Explicit Runge-Kutta Method with Adaptive Step Sizes

Input: An initial time t_i, a final time t_f, the initial condition $y(t_i)$, an initial step size h_0, absolute error tolerance $\mathtt{atol} \in \mathcal{R}^n$, relative error tolerance $\mathtt{rtol} \in \mathcal{R}^n$, step size adjustment parameters $0 < \beta < 1$, $\mathtt{fac}_1 > \mathtt{fac}_0 > 0$, the minimum allowable step size $h_{\min}$, and the maximum allowable number of iterations $\mathtt{MAX\_ITER} > 0$.

Output: $y(t_f)$

$y = y(t_i)$, $t = t_i$, and $h = h_0$
$k_1 = f(y, t)$
if $t + h > t_f$, **then** $h = t_f - t$ **end if**
for $\mathtt{ITER} = 0, 1, \cdots, \mathtt{MAX\_ITER}$ **do**
  for $i = 2, 3, \cdots, s$ **do**
    $Y = y + h \sum_{j=1}^{i-1} a_{ij} k_j$
    $\tau = t + c_i h$
    $k_i = f(Y, \tau)$
  end for
  $\bar{y} = y + h \sum_{i=1}^{s} b_i k_i$
  $\eta = h \sum_{i=1}^{s} (b_i - \hat{b}_i) k_i$
  $\sigma = \sqrt{\frac{1}{n} \sum_{i=1}^{n} \left(\frac{|\eta_i|}{\mathtt{atol}_i + \mathtt{rtol}_i |\bar{y}_i|} \right)^2}$
  $\bar{h} = h \min(\mathtt{fac}_1, \max(\mathtt{fac}_0, \beta \sigma^{-1/p}))$
  if $\sigma \leq 1$ **then**
    if $t = t_f$, **then** $y(t_f) = \bar{y}$, **STOP end if**
    $t = t + h$
    $y = \bar{y}$
    $k_1 = f(y, t)$
    if $t + \bar{h} > t_f$, **then** $\bar{h} = t_f - t$ **end if**
  end if
  $h = \bar{h}$
  if $h < h_{\min}$, **then STOP end if**
end for
STOP, too many iterations

We can develop another strategy for selecting an initial step size by assuming that the local error for a $(p-1)$-th order method can be approximated by

$$\|\eta\| = \hat{C} h_0^p \approx \left[\|h_0 \frac{dy(t_i)}{dt}\| \right]^p = \left[\|h_0 f(y(t_i), t_i)\| \right]^p .$$

If we set $\|\eta\|$ equal to the desired tolerance, ϵ, then we get an estimate of the initial step size as

$$h_0 = \frac{\epsilon^{1/p}}{\|f(y(t_i), t_i)\|} .$$

- Step size adjustment parameters β, $\mathtt{fac}_0$ and $\mathtt{fac}_1$. Typical values for these parameters are $\beta = 0.9$, $\mathtt{fac}_0 = 0.2$ and $\mathtt{fac}_1 = 5$.
- The maximum number of iterations allowed $\mathtt{MAX\_ITER} > 0$, and the minimum allowable step size $h_{\min}$. These parameters are used as safeguards to ensure that the algorithm does not perform too many step. If the algo-

rithm requires a large number of steps then this may be an indication that the differential equations are stiff, and the explicit Runge-Kutta method is not an appropriate method to use to solve such differential equations.

We can see that the algorithm will terminate if (i) $t = t_f$, i.e., we have reached the final time, (ii) $h < h_{\min}$, i.e., the step size is too small, or, (iii) `ITER` $>$ `MAX_ITER`, i.e., too many steps have been performed.

Algorithm 5.2.2 is in fact quite simple to implement in most computer programming languages. For this reason the explicit Runge-Kutta method is a favorite among engineers and scientist. We note however that these methods are not efficient when applied to stiff differential equations. In fact it is easy to demonstrate that these explicit Runge-Kutta methods are not A-stable (See Problems 3 and 4). Thus these methods are inefficient when applied to very stiff systems. For stiff differential equations and differential-algebraic equations we must employ implicit methods.

5.2.3 Implicit Runge-Kutta methods

The s-stage implicit Runge-Kutta method for the solution of the ODE (5.1)-(5.2) has the form

$$\begin{aligned} y^{(k+1)} &= y^{(k)} + h\sum_{i=1}^{s} b_i k_i, \\ k_i &= f(Y_i, \tau_i), \\ Y_i &= y^{(k)} + h\sum_{j=1}^{s} a_{ij} k_j, \\ \tau_i &= t^{(k)} + c_i h. \end{aligned} \tag{5.17}$$

Here, $k_i \in \mathcal{R}^n$ is called a stage derivative, and $Y_i \in \mathcal{R}^n$ is called a stage value, for $i = 1, 2, \cdots, s$. Particular methods are defined by the coefficients $A \in \mathcal{R}^{s\times s}$, $b \in \mathcal{R}^s$, and $c \in \mathcal{R}^s$. In implicit methods the matrix A may have non-zero terms on or above the diagonal. Recall the explicit Runge-Kutta methods are defined by matrices A that are strictly lower triangular. If we compare (5.13) and (5.17) we will see that the stage value Y_i are defined implicitly in (5.17).

The coefficients of the implicit Runge-Kutta method can be determined using the same procedure used to find the coefficients of explicit methods. Specifically, we construct the discretization error for the formula (5.17), and determine the conditions necessary for the method to have order p. We then select the coefficients A, b and c so that these order conditions are satisfied.

To facilitate the construction of implicit Runge-Kutta methods researchers often use the simplified order conditions developed by Butcher (1964).

Namely,

$$\sum_{i=1}^{s} b_i c_i^{q-1} = \frac{1}{q}, \quad q = 1, 2, \cdots, p; \tag{5.18}$$

$$\sum_{j=1}^{s} a_{ij} c_j^{q-1} = \frac{c_i^q}{q}, \quad i = 1, 2, \cdots, s, \ q = 1, 2, \cdots, \eta; \tag{5.19}$$

$$\sum_{i=1}^{s} b_i c_i^{q-1} a_{ij} = \frac{b_j}{q}(1 - c_j^q), \quad j = 1, 2, \cdots, s, \ q = 1, 2, \cdots, \zeta. \tag{5.20}$$

(These should be compared with (5.14)). Butcher (1964) shows that if the coefficients A, b and c are selected so that (5.18), (5.19) and (5.20) are satisfied with $p \leq \eta + \zeta + 1$ and $p \leq 2\eta + 2$, the the method is of order p.

Table 5.2.3 shows the Butcher tableau for some frequently used implicit Runge-Kutta methods. The table presents the method coefficients along with the number of stages, s, and the order of the method, p. The Radau IIA method of order $p = 1$ is the implicit Euler method, and the Lobatto IIIA method of order $p = 2$ is the implicit trapezoidal method. It can be shown that all of the methods in Table 5.2.3 are A-stable. (Note that not all implicit methods are A-stable.)

The coefficient matrices A for the Radau IIA methods are invertible, while the A matrices for the Lobatto IIIA methods are not invertible. Also, in both the Radau IIA and Lobatto IIIA methods we have $b_i = a_{si}$, for $i = 1, 2, \cdots, s$. This property is called 'stiff accuracy' and is important for the solution of differential-algebraic equations (Hairer and Wanner (1996)).

In addition, the Radau IIA methods have a property called L-stability. A discretization formula is said to be L-stable if it is A-stable and $\lim_{z \to -\infty} R(z) = 0$, where $R(z)$ is the stability function for the method, and z is a complex variable. L-stable methods are able to quickly damp out the transient response of very stiff components in the solution of the differential equation (see Problem 12).

Example 5.2.
Consider the implicit trapezoidal method

$$\begin{array}{c|cc} 0 & 0 & 0 \\ 1 & \frac{1}{2} & \frac{1}{2} \\ \hline & \frac{1}{2} & \frac{1}{2} \end{array}\,.$$

The discretization formula for this method is

$$y^{(k+1)} = y^{(k)} + hb_1 k_1 + hb_2 k_2$$

Radau IIA methods

$$\begin{array}{c|c} 1 & 1 \\ \hline & 1 \end{array}$$

$s = 1,\ p = 1$

$$\begin{array}{c|cc} \frac{1}{3} & \frac{5}{12} & -\frac{1}{12} \\ 1 & \frac{3}{4} & \frac{1}{4} \\ \hline & \frac{3}{4} & \frac{1}{4} \end{array}$$

$s = 2,\ p = 3$

$$\begin{array}{c|ccc} \frac{4-\sqrt{6}}{10} & \frac{88-7\sqrt{6}}{360} & \frac{296-169\sqrt{6}}{1800} & \frac{-2+3\sqrt{6}}{225} \\ \frac{4+\sqrt{6}}{10} & \frac{296+169\sqrt{6}}{1800} & \frac{88+7\sqrt{6}}{360} & \frac{-2-3\sqrt{6}}{225} \\ 1 & \frac{16-\sqrt{6}}{36} & \frac{16+\sqrt{6}}{36} & \frac{1}{9} \\ \hline & \frac{16-\sqrt{6}}{36} & \frac{16+\sqrt{6}}{36} & \frac{1}{9} \end{array}$$

$s = 3,\ p = 5$

Lobatto IIIA methods

$$\begin{array}{c|cc} 0 & 0 & 0 \\ 1 & \frac{1}{2} & \frac{1}{2} \\ \hline & \frac{1}{2} & \frac{1}{2} \end{array}$$

$s = 2,\ p = 2$

$$\begin{array}{c|ccc} 0 & 0 & 0 & 0 \\ \frac{1}{2} & \frac{5}{24} & \frac{1}{3} & -\frac{1}{24} \\ 1 & \frac{1}{6} & \frac{2}{3} & \frac{1}{6} \\ \hline & \frac{1}{6} & \frac{2}{3} & \frac{1}{6} \end{array}$$

$s = 3,\ p = 4$

$$\begin{array}{c|cccc} 0 & 0 & 0 & 0 & 0 \\ \frac{5-\sqrt{5}}{10} & \frac{11+\sqrt{5}}{120} & \frac{25-\sqrt{5}}{120} & \frac{25-13\sqrt{5}}{120} & \frac{-1+\sqrt{5}}{120} \\ \frac{5+\sqrt{5}}{10} & \frac{11-\sqrt{5}}{120} & \frac{25+13\sqrt{5}}{120} & \frac{25+\sqrt{5}}{120} & \frac{-1-\sqrt{5}}{120} \\ 1 & \frac{1}{12} & \frac{5}{12} & \frac{5}{12} & \frac{1}{12} \\ \hline & \frac{1}{12} & \frac{5}{12} & \frac{5}{12} & \frac{1}{12} \end{array}$$

$s = 4,\ p = 6$

Table 5.1 Implicit Runge-Kutta methods

$$
\begin{aligned}
&= y^{(k)} + \frac{h}{2}k_1 + \frac{h}{2}k_2 \\
k_1 &= f(y^{(k)} + ha_{11}k_1 + ha_{12}k_2, t^{(k)} + c_1 h) \\
&= f(y^{(k)}, t^{(k)}) \\
k_2 &= f(y^{(k)} + ha_{21}k_1 + ha_{22}k_2, t^{(k)} + c_2 h) \\
&= f(y^{(k)} + \frac{h}{2}k_1 + \frac{h}{2}k_2, t^{(k+1)})
\end{aligned}
$$

Applying this method to the scalar ODE $\dot{y} = \lambda y$ gives

$$
\begin{aligned}
k_1 &= \lambda y^{(k)} \\
k_2 &= \lambda \left(1 + \frac{h\lambda}{2}\right) y^{(k)} + \frac{h\lambda}{2} k_2 \\
&= \lambda \left[1 - \frac{h\lambda}{2}\right]^{-1} \left(1 + \frac{h\lambda}{2}\right) y^{(k)}
\end{aligned}
$$

$$
\begin{aligned}
y^{(k+1)} &= y^{(k)} + \frac{h\lambda}{2} y^{(k)} + \frac{h\lambda}{2} \frac{(1 + h\lambda/2)}{(1 - h\lambda/2)} y^{(k)} \\
&= \frac{1 + h\lambda/2}{1 - h\lambda/2} y^{(k)} = \frac{1 + z/2}{1 - z/2} y^{(k)},
\end{aligned}
$$

where $z = h\lambda$. Therefore, the stability function is

$$
R(z) = \frac{1 + z/2}{1 - z/2}.
$$

Thus, $\lim_{z \to -\infty} |R(z)| = 1$, and we can conclude that this method is not L-stable. (Also, see Problem 12).

5.2.4 An implicit Runge-Kutta Algorithm for ODEs

This section presents an algorithm that implements an implicit Runge-Kutta method for ordinary differential equations. For the most part we will follow the techniques described in Hairer and Wanner (1996). To develop this implicit Runge-Kutta algorithm we will use (5.17), and define the increment in the stage value as

$$
Z_i = Y_i - y^{(k)}, \tag{5.21}
$$

where $Z_i \in \mathcal{R}^n$, $i = 1, 2, \cdots, s$. Using this we see that the s-stage implicit Runge-Kutta method (5.17) must satisfy

$$\phi(Z) = \begin{bmatrix} Z_1 - h\sum_{j=1}^{s} a_{1j} f(Z_j + y^{(k)}, \tau_j) \\ Z_2 - h\sum_{j=1}^{s} a_{2j} f(Z_j + y^{(k)}, \tau_j) \\ \vdots \\ Z_s - h\sum_{j=1}^{s} a_{sj} f(Z_j + y^{(k)}, \tau_j) \end{bmatrix} = Z - h(A \otimes I)F(Z) = 0, \quad (5.22)$$

where $Z = [Z_1^T,\ Z_2^T,\ \cdots,\ Z_s^T]^T \in \mathcal{R}^{ns}$, $\phi(Z) \in \mathcal{R}^{ns}$, $F(Z) = [f(Z_1 + y^{(k)}, \tau_1)^T,\ f(Z_2 + y^{(k)}, \tau_2)^T, \cdots, f(Z_s + y^{(k)}, \tau_s)^T]^T$, and $\otimes$ denotes the Kronecker or tensor product. Moreover, if A is invertible then

$$y^{(k+1)} = y^{(k)} + \sum_{i=1}^{s} d_i Z_i, \qquad (5.23)$$

where $[d_1,\ d_2,\ \cdots,\ d_s] = [b_1,\ b_2,\ \cdots,\ b_s]A^{-1}$.

The implicit equations in (5.22) are solved using the simplified Newton's method. Specifically, given an initial estimate $Z^{(0)}$ we perform the iterations

for $q = 0, 1, \cdots$ **do**

  Solve the linear system

$$D\phi\, \Delta Z^{(q)} = -\phi(Z^{(q)}) \qquad (5.24)$$

  to find the correction $\Delta Z^{(q)}$.

  Set $Z^{(q+1)} = Z^{(q)} + \Delta Z^{(q)}$

  if $\|\Delta Z^{(q)}\|$, or $\|\phi(Z^{(q+1)})\|$ is sufficiently small **then STOP end if**

end for

In the simplified Newton's method the Jacobian approximation, $D\phi$, is given by

$$D\phi = \begin{bmatrix} I - ha_{11}J & -ha_{12}J & \cdots & -ha_{1s}J \\ -ha_{21}J & I - ha_{22}J & \cdots & -ha_{2s}J \\ \vdots & & \ddots & \vdots \\ -ha_{s1}J & -ha_{12}J & \cdots & I - ha_{ss}J \end{bmatrix} = I - hA \otimes J, \qquad (5.25)$$

where

$$J = \frac{\partial}{\partial y} f(y^{(k)}, t^{(k)}).$$

Note that the simplified Newton's method provides considerable computational saving because we need only evaluate the Jacobian approximation once for all iterations $q = 0, 1, \cdots$. (See Problem 14.) In the practical implementation of this simplified Newton's method we must establish a criteria for terminating the iterations, and we must also develop an efficient method for solving the linear system (5.24). These issues are discussed next.

Simplified Newton's method termination criteria. If we assume that the simplified Newton's method converges linearly then there is a constant $0 < \Theta < 1$ such that

$$\|\Delta Z^{(q+1)}\| \leq \Theta \|\Delta Z^{(q)}\|. \qquad (a)$$

Moreover, if Z^* is the solution to the system (5.22), then

$$Z^{(q+1)} - Z^* = Z^{(q+1)} - Z^{(q+2)} + Z^{(q+2)} - Z^{(q+3)} + \cdots. \qquad (b)$$

Taking the norm of both sides of (b), using the triangle inequality, and employing (a) gives

$$\begin{aligned} \|Z^{(q+1)} - Z^*\| &\leq \Theta(1 + \Theta + \Theta^2 + \cdots)\|\Delta Z^{(q)}\| \\ &\leq \frac{\Theta}{1-\Theta}\|\Delta Z^{(q)}\|, \end{aligned}$$

where we have used the fact that $\sum_{i=0}^{\infty} \Theta^i = 1/(1-\Theta)$. Thus, we terminate the simplified Newton's method if

$$\frac{\Theta}{1-\Theta}\|\Delta Z^{(q)}\| \leq \texttt{ctol}, \qquad (5.26)$$

where $\texttt{ctol} > 0$ is some convergence tolerance, since this would indicate that $Z^{(q+1)}$ is sufficiently close to the solution Z^*. In practice we estimate the rate of convergence, Θ, as

$$\Theta = \frac{\Delta Z^{(q+1)}}{\Delta Z^{(q)}}, \quad q > 0.$$

For $q = 0$ we use $\Theta = 1/2$. If for some iteration $q > 0$ we obtain $\Theta \geq 1$, then the simplified Newton's iteration is terminated, and the method will be considered to have failed.

It is also important to limit the number of Newton's iterations performed. Hence, we only carry out the iterations for $q = 0, 1, \cdots, \texttt{q\_MAX}$, where q_MAX is the maximum number of simplified Newton's iterations allowed. If (5.26) does not hold when $q = \texttt{q\_MAX}$, the Newton's method will be considered to have failed.

Solution of the linear system. The q-th iteration of the simplified Newton's method requires the solution of the linear system (5.24). Since the Jacobian approximation remains unchanged for all iterations we can solve this linear system with one LU factorization of the matrix $D\phi = I - hA \otimes J$. A direct LU factorization of $D\phi$ requires on the order of $\frac{2}{3}(ns)^3$ operations.

The cost of solving the linear system (5.24) can be reduced by transforming A into a Jordan canonical form. By doing so we can reduce (5.24) to a diagonal or block diagonal system of equations of smaller dimension. For example, suppose there is an invertible matrix T such that $\Gamma = T^{-1}AT$ is in the

Jordan canonical form. That is, Γ is a matrix with one by one, or two by two block diagonal entries. Moreover, all the elements of Γ are real.

Using the transformation matrix T we can rewrite (5.24) as

$$(I - h\Gamma \otimes J)\Delta \bar{Z} = -\bar{\phi},$$

where $\Delta Z = (T \otimes I)\Delta \bar{Z}$, and $\bar{\phi} = (T^{-1} \otimes I)\phi$. In which case the block diagonal elements of $I - h\Gamma \otimes J$ has dimension n by n or $2n$ by $2n$. Hence, if we use real arithmetic, the LU factorization of $D\phi$ will require on the order of $\frac{2}{3}n^3$ operations for the n by n blocks, and on the order of $\frac{16}{3}n^3$ operations for the $2n$ by $2n$ blocks. For $s > 2$ the factorization described above will lead to significant computational savings.

Example 5.3.
Consider the Radau IIA method with $s = 3$. For this method the matrix A is nonsingular, and it has a real eigenvalue, 0.27489, and a complex conjugate pair of eigenvalues, $0.16256 \pm i0.18495$. Using the transformation matrix

$$T = \begin{bmatrix} 0.09123 & -0.12846 & 0.02731 \\ 0.24172 & 0.18564 & -0.34825 \\ 0.96605 & 0.90940 & 0.00000 \end{bmatrix},$$

it can be shown that

$$\Gamma = T^{-1}AT = \begin{bmatrix} 0.27489 & 0 & 0 \\ 0 & 0.16256 & -0.18495 \\ 0 & 0.18495 & 0.16256 \end{bmatrix}.$$

Therefore, the matrix $I - h\Gamma \otimes J$ has the form

$$\begin{bmatrix} I - \gamma_1 hJ & 0 & 0 \\ 0 & I - \gamma_2 hJ & \gamma_3 hJ \\ 0 & -\gamma_3 hJ & I - \gamma_2 hJ \end{bmatrix},$$

where $\gamma_1 = 0.27489$, $\gamma_2 = 0.16256$ and $\gamma_3 = 0.18495$. Hence, in this case we will require on the order of $6n^3$ operations to factor $I - h\Gamma \otimes J$, whereas the matrix $I - hA \otimes J$ requires on the order of $18n^3$ operations for an LU factorization.

Additional computational savings can be realized by writing the lower $2n$ by $2n$ partition of $I - h\Gamma \otimes J$ as the complex system

$$(I - (\gamma_2 + i\gamma_3)hJ)(\Delta \bar{Z}_2 + i\Delta \bar{Z}_3) = -(\bar{\phi}_2 + i\bar{\phi}_3).$$

Here, $\Delta \bar{Z}_2 \in \mathcal{R}^n$ contains elements $n+1$ to $2n$ of $\Delta \bar{Z}$, $\Delta \bar{Z}_3 \in \mathcal{R}^n$ contains elements $2n+1$ to $3n$ of $\Delta \bar{Z}$, etc. Using this approach the factorization of

$(I - hA \otimes J)$ requires $\frac{2}{3}n^3$ operations on the real matrix $(I - \gamma_1 hJ)$ and $\frac{2}{3}n^3$ operations on the complex variable matrix $(I - (\gamma_2 + i\gamma_3)hJ)$.

Local error estimate. As noted above, the efficient implementation of numerical integration methods requires an estimate of the local error so that the step size can be adjusted to satisfy the desired error tolerance. The implicit Runge-Kutta methods given in Table 5.2.3 do not include embedded formulas to estimate the local error. However, we can derive error estimation formulas using the approach described by Hairer and Wanner (1993), and de Swart and Söderlind (1997). We will illustrate this technique using the Radau IIA, $s = 3$ method.

First we note that we can compute a numerical solution to the ODE at time $t^{(k+1)}$ via (5.17). Doing so gives the approximate solution $y^{(k+1)}$. Here we compute a second approximate solution, $\hat{y}^{(k+1)}$, using

$$\hat{y}^{(k+1)} = y^{(k)} + h\hat{b}_0 f(y^{(k)}, t^{(k)}) + h \sum_{i=1}^{s} \hat{b}_i k_i + h\hat{\gamma} f(\hat{y}^{(k+1)}, t^{(k+1)}). \tag{5.27}$$

Note that (5.27) is an implicit equation for the solution estimate $\hat{y}^{(k+1)}$. In this formula the stage derivatives k_i, $i = 1, 2, \cdots, s$ are defined in (5.17), and are the same values used to compute $y^{(k+1)}$. The coefficients in (5.27), i.e., $\hat{\gamma}$, and $\hat{b}_i$, $i = 0, 1, \cdots, s$ are selected so that the discretization error is of order $s + 1$. To meet this criteria the coefficients must satisfy the order conditions

$$\hat{C}\hat{b} = \left[1 - \hat{b}_0, \frac{1}{2}, \cdots, \frac{1}{s}\right]^T - \hat{\gamma}\mathbf{1}, \tag{5.28}$$

where $\hat{C} \in \mathcal{R}^{s \times s}$, the i, j-th element of the matrix $\hat{C}$ is given by $\hat{C}_{ij} = c_j^{i-1}$, $j = 1, 2, \cdots, s$, $\hat{b} = [\hat{b}_1, \hat{b}_2, \cdots, \hat{b}_s]^T$, and $\mathbf{1} = [1, 1, \cdots, 1]^T \in \mathcal{R}^s$. The coefficients c_j, $j = 1, 2, \cdots, s$ are associated with a specific implicit Runge-Kutta method. In the case of the Radau IIA, $s = 3$ method we have $c_1 = \frac{16-\sqrt{6}}{36}$, $c_2 = \frac{16+\sqrt{6}}{36}$ and $c_3 = \frac{1}{9}$.

In (5.27) the variables $\hat{\gamma}$ and $\hat{b}_0$ are two free parameters. In which case it is convenient to select $\hat{\gamma}$ as a real eigenvalue of the matrix A. Thus, for the Radau IIA, $s = 3$ method we use $\hat{\gamma} = \gamma_1 = 0.27489$. In our implementation of the implicit Runge-Kutta method we also use $\hat{b}_0 = 0.02$. The remaining coefficients, $\hat{b}$, are determined by solving the linear system (5.28). (Note that de Swart and Söderlind (1997) describe the method used to select $\hat{b}_0$.)

To implement the formula (5.27) it is sufficient to perform one iteration of a simplified Newton's method on the system

$$\hat{\phi}(\hat{y}^{(k+1)}) = \hat{y}^{(k+1)} - y^{(k)} - h\hat{b}_0 f(y^{(k)}, t^{(k)})$$

$$-h\sum_{i=1}^{s}\hat{b}_i k_i - h\hat{\gamma}f(\hat{y}^{(k+1)}, t^{(k+1)}) = 0.$$

Taking $y^{(k+1)}$ as the estimate of the solution, the simplified Newton's method gives

$$\begin{aligned}\hat{y}^{(k+1)} &= y^{(k+1)} + h\left[I - h\hat{\gamma}J\right]^{-1}\left\{\sum_{i=1}^{s}(\hat{b}_i - b_i)k_i\right.\\ &\qquad \left. + \hat{b}_0 f(y^{(k)}, t^{(k)}) + \hat{\gamma}f(y^{(k+1)}, t^{(k+1)})\right\}\\ &= y^{(k+1)} + \left[I - h\hat{\gamma}J\right]^{-1}\left\{\sum_{i=1}^{s} e_i Z_i + h\hat{b}_0 f(y^{(k)}, t^{(k)})\right.\\ &\qquad \left. + h\hat{\gamma}f(y^{(k+1)}, t^{(k+1)})\right\}, \end{aligned} \tag{5.29}$$

where $[e_1,\ e_2,\ \cdots,\ e_s] = [(\hat{b}_1 - b_1),\ (\hat{b}_2 - b_2),\ \cdots, (\hat{b}_s - b_s)]A^{-1}$. Also, we use the approximation $\partial f(\hat{y}^{(k+1)}, t^{(k+1)})/\partial y \approx J = \partial f(y^{(k)}, t^{(k)})/\partial y$. Thus, the local error estimate is

$$\begin{aligned}\eta^{(k+1)} &= y^{(k+1)} - \hat{y}^{(k+1)}\\ &= -\left[I - h\hat{\gamma}J\right]^{-1}\left\{\sum_{i=1}^{s} e_i Z_i + h\hat{b}_0 f(y^{(k)}, t^{(k)})\right.\\ &\qquad \left. + h\hat{\gamma}f(y^{(k+1)}, t^{(k+1)})\right\}. \end{aligned} \tag{5.30}$$

In the case of the Radau IIA, $s = 3$ method $\eta^{(k+1)} = O(h^{s+1}) = O(h^4)$.

Using (5.30) the step size can be adjusted as

$$\begin{aligned}\sigma &= \left[\sqrt{\frac{1}{n}\sum_{i=1}^{n}\left(\frac{|\eta_i^{(k+1)}|}{\mathtt{atol}_i + \mathtt{rtol}_i|y_i^{(k+1)}|}\right)^2}\,\right]^{-1/4}\\ \bar{h} &= h\min(\mathtt{fac}_1, \max(\mathtt{fac}_0, \beta\sigma)). \end{aligned} \tag{5.31}$$

where $\mathtt{atol} \in \mathcal{R}^n$ is the absolute error tolerance, $\mathtt{rtol} \in \mathcal{R}^n$ is the relative error tolerance, and the step size adjustment parameters $0 < \beta < 1$, $\mathtt{fac}_1 > \mathtt{fac}_0 > 0$. As discussed in the previous section, if the local error estimate satisfies the desired tolerance, the new step size at $t^{(k+1)}$ is $\bar{h} \geq h$. Otherwise, the integration from $t^{(k)}$ is repeated with a new step size $\bar{h} < h$.

With these ingredients an implicit Runge-Kutta method for the solution of the ODE (5.1)-(5.2) can be implemented as outlined in Algorithm 5.2.3.

Starting at $t = t_i$ the algorithm executes the loop defined by steps 2 through 20 until $t = t_f$. In step 3 the algorithm solves the equations (5.22) via the simplified Newton's method. For these Newton iterations we use $Z^{(0)} = 0$ however, other initial estimates are possible (see Section 5.3.5 below, and Olsson and Söderlind, (1998)). If the simplified Newton's iterations fail to converge the step size is reduced by a factor 0.5 and the Newton's iterations are repeated. If the Newton's iterations converge then the local error estimate is computed along with a new step size $\bar{h}$. If the local error is sufficiently small, i.e., $\sigma \leq 1$, then the solution advances to time $t + h$ with $y(t + h) = \bar{y}$. Otherwise, if $\sigma > 1$ the equations (5.22) are solved again at time t, with a new step size $\bar{h} < h$.

Algorithm 5.2.3 Implicit Runge-Kutta Method

Input: An initial time t_i, a final time t_f, the initial condition $y(t_i)$, an initial step size h_0, absolute error tolerance $\texttt{atol} \in \mathcal{R}^n$, relative error tolerance $\texttt{rtol} \in \mathcal{R}^n$, step size adjustment parameters $0 < \beta < 1$, $\texttt{fac}_1 > \texttt{fac}_0 > 0$, the minimum allowable step size $h_{\min}$, and the maximum allowable number of iterations $\texttt{MAX\_ITER} > 0$.

Output: $y(t_f)$

1: $y = y(t_i)$, $t = t_i$, and $h = h_0$
2: **for** $\texttt{ITER} = 0, 1, \cdots, \texttt{MAX\_ITER}$ **do**
3:   **if** the simplified Newton's method, (5.24), converges **then**
4:     $\bar{y} = y + \sum_{i=1}^{s} d_i Z_i$
5:     $\eta = -\left[I - h\hat{\gamma}J\right]^{-1}\left\{\sum_{i=1}^{s} e_i Z_i + h\hat{b}_0 f(y,t) + h\hat{\gamma} f(\bar{y}, t+h)\right\}$
6:     $\sigma = \left[\sqrt{\frac{1}{n}\sum_{i=1}^{n}\left(\frac{|\eta_i|}{\texttt{atol}_i + \texttt{rtol}_i|\bar{y}_i|}\right)^2}\right]^{-1/4}$
7:     $\bar{h} = h\min(\texttt{fac}_1, \max(\texttt{fac}_0, \beta\sigma))$
8:     **if** $\sigma \leq 1$ **then**
9:       **if** $t = t_f$, **then** $y(t_f) = \bar{y}$, **STOP end if**
10:       $t = t + h$
11:       $y = \bar{y}$
12:       **if** $t + \bar{h} > t_f$, **then** $\bar{h} = t_f - t$ **end if**
13:     **else**
14:       $h = \bar{h}$
15:     **end if**
16:   **else**
17:     $h = h/2$
18:   **end if**
19:   **if** $h < h_{\min}$, **then STOP end if**
20: **end for**
21: **STOP**, too many iterations

5.3 Numerical Solution of DAEs

We now consider the extension of the methods described above to systems of differential-algebraic equations (DAEs) of the form

$$\Phi(y, \dot{y}, t) = 0, \tag{5.32}$$

where t is the time, $y(t) \in \mathcal{R}^n$ is the state vector, and $\dot{y} = dy/dt$ is the state derivative. It is assumed that $\Phi(y, \dot{y}, t) \in \mathcal{R}^n$ is continuously differentiable with respect to all of its arguments. The system (5.32) is sometimes called implicit differential equations (IDEs).

We note that the equation (5.32) includes ordinary differential equations (5.1), since we can write

$$\Phi(y, \dot{y}, t) = \dot{y} - f(y, t) = 0.$$

The description (5.32) also includes the Lagrangian DAEs (4.8). To show this we simply take $y = [q^T, f^T, s^T, e^T, \lambda^T, \mu^T]^T$. Thus, (5.32) can include a combination of differential equations and algebraic equations.

An important structural property of the system (5.32) is the *differentiation index*. To define this quantity consider the system of equations

$$\begin{aligned}
\Phi(y, \dot{y}, t) &= 0 \\
\frac{d}{dt}\Phi(y, \dot{y}, t) &= 0 \\
\frac{d^2}{dt^2}\Phi(y, \dot{y}, t) &= 0 \\
&\vdots \\
\frac{d^p}{dt^p}\Phi(y, \dot{y}, t) &= 0.
\end{aligned} \tag{5.33}$$

Then the differentiation index is the smallest integer $p \geq 0$ for which the system (5.33) can be used to determine an explicit system of ordinary differential equations of the form $\dot{y} = \bar{f}(y, t)$. In the following example we consider the differentiation index for some typical dynamic systems described using the Lagrangian DAEs.

Example 5.4.

(i) Consider the mass-spring-damper system shown here.

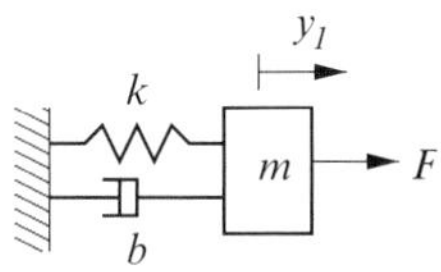

Using y_1 as the generalized displacement, and y_2 as the generalized flow (velocity), the differential equations of motion for this system can be written as the implicit differential equations

$$\begin{aligned}\Phi_1 &= \dot{y}_1 - y_2 = 0 \\ \Phi_2 &= \dot{y}_2 - (k/m)y_1 - (b/m)y_2 - F/m = 0.\end{aligned}$$

Therefore, $\Phi(y, \dot{y}, t) = [\Phi_1, \Phi_2]^T$, where $y = [y_1, y_2]^T$. We can see that $\dot{y} = [\dot{y}_1, \dot{y}_2]^T$ can be determined explicitly from the equations above. Hence, no derivatives of Φ are required, and system has differentiation index $p = 0$. In fact it is easy to see that a system of ordinary differential equations has differentiation index 0.

(ii) Consider the dynamic system described by a damper and an applied force, as shown below.

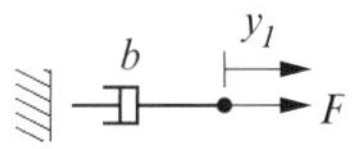

Using y_1 as the generalized displacement, and y_2 as the generalized flow, the differential equations of motion for this system can be written as the implicit differential equations

$$\begin{aligned}\Phi_1 &= \dot{y}_1 - y_2 = 0, \\ \Phi_2 &= by_2 + F = 0.\end{aligned}$$

We note that $\dot{y}_2$ does not appear explicitly in this system of equations. Taking one time derivative of the system however, gives the equation

$$b\dot{y}_2 + \dot{F} = 0,$$

which can be used to determine $\dot{y}_2$. Hence, this system has differentiation index 1.

(iii) Consider the dynamic system described by a spring and an applied force, as shown below.

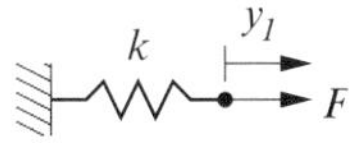

Using y_1 as the generalized displacement, and y_2 as the generalized flow, the differential equations of motion for this system can be written as the implicit differential equations

$$\Phi_1 = \dot{y}_1 - y_2 = 0,$$
$$\Phi_2 = ky_1 + F = 0.$$

We note that $\dot{y}_2$ does not appear explicitly in this system of equations. Taking two time derivatives of the system however, gives the equation

$$k\dot{y}_2 + \ddot{F} = 0,$$

which can be used to determine $\dot{y}_2$. Hence, this system has differentiation index 2.

(iv)Consider the simple pendulum shown below. In this model we will use y_1 and y_2 as the displacements, and $y_3 = \dot{y}_1$, and $y_4 = \dot{y}_2$ as the corresponding flows.

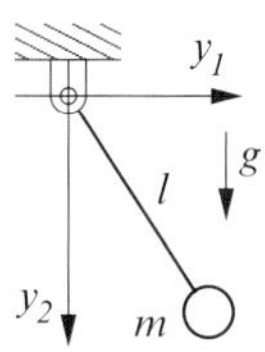

The Lagrangian DAEs for this system can be written as

$$\begin{aligned}
\Phi_1 &= \dot{y}_1 - y_3 = 0,\\
\Phi_2 &= \dot{y}_2 - y_4 = 0,\\
\Phi_3 &= m\dot{y}_3 + 2y_1y_5 = 0,\\
\Phi_4 &= m\dot{y}_4 + 2y_2y_5 - mg = 0,\\
\Phi_5 &= y_1^2 + y_2^2 - l^2 = 0,
\end{aligned}$$

where y_5 is the Lagrange multiplier associated with the displacement constraint $y_1^2 + y_2^2 - l^2 = 0$.
In these differential-algebraic equations $\dot{y}_5$ does not appear explicitly. However, three time derivatives of the equation $\Phi_5 = 0$ produces

$$\dot{y}_5 = \frac{m}{2l^2}(2y_3\dot{y}_3 + 2y_4\dot{y}_4 - \dot{y}_2 g).$$

Therefore, this system has differentiation index 3.

5.3.1 Hessenberg form for the DAE

In many cases it is possible to separate the variables in y so that we can write (5.32) explicitly as coupled differential equations and algebraic equations. For example, it can be seen the Lagrangian DAEs have such a structure. Some important structural arrangements of the DAEs are known as the *Hessenberg forms*, which are defined below. For the sake of simplicity we give the autonomous form of these systems. The non-autonomous form can be deduced by appending the differential equation $\dot{\tau} = 1$, where τ represents the time variable.

- Hessenberg Index-1 Form.
  The DAEs are said to be in Hessenberg index-1 form if they can be written as

$$\begin{aligned} \dot{x} &= f(x, z), \\ 0 &= g(x, z), \end{aligned} \tag{5.34}$$

  where $x(t) \in \mathcal{R}^{n_x}$, $z(t) \in \mathcal{R}^{n_z}$, and the Jacobian, $g_z = \partial g/\partial z$, is nonsingular along the solution to the DAEs. In this formulation x is called the differential variable and z is called the algebraic variable.
  These index-1 DAEs can be reduced to a system of ordinary differential equations by taking one time derivative of the algebraic equation $g(x, z) = 0$. In fact, $\frac{d}{dt}g(x, z) = g_x\dot{x} + g_z\dot{z} = 0$, where $g_x = \partial g/\partial x$. Since g_z is nonsingular we have $\dot{z} = -g_z^{-1}g_x\dot{x} = -g_z^{-1}g_xf$. Hence, only one time derivative of the algebraic equation is required to determine $\dot{z}$ in terms of x and z.
  Clearly we can write (5.34) in the form of (5.32) by using $y = [x^T, z^T]^T$. Also note that the DAEs in Example 5.4-(ii) can be put in Hessenberg index-1 form if we select $x = y_1$ and $z = y_2$.
- Hessenberg Index-2 Form.
  The DAEs are said to be in Hessenberg index-2 form if they can be written as

$$\begin{aligned} \dot{x} &= f(x, z), \\ 0 &= g(x), \end{aligned} \tag{5.35}$$

  where $x(t) \in \mathcal{R}^{n_x}$, $z(t) \in \mathcal{R}^{n_z}$, and the matrix g_xf_z is nonsingular along the solution to the DAEs, where $f_z = \partial f/\partial z$.
  Two time derivatives of the algebraic equation $g(x) = 0$ gives $\dot{z} = -(g_xf_z)^{-1}(g_xf)_xf$, where $(g_xf)_x = \partial(g_xf)/\partial x$. Note that we can write (5.35) in the form of (5.32) by using $y = [x^T, z^T]^T$, and the DAEs in Example 5.4-(iii) can be put in Hessenberg index-2 form if we select $x = y_1$ and $z = y_2$.
- Hessenberg Index-3 Form.
  The DAEs are said to be in Hessenberg index-3 form if they can be written as

$$\begin{aligned} \dot{x} &= f(x, z), \\ \dot{z} &= h(x, z, u), \end{aligned}$$

$$0 = g(x), \tag{5.36}$$

where $x(t) \in \mathcal{R}^{n_x}$, $z(t) \in \mathcal{R}^{n_z}$, $u(t) \in \mathcal{R}^{n_u}$ and the matrix $g_x f_z h_u$ is nonsingular along the solution to the DAEs, where $h_u = \partial h / \partial u$. Taking three time derivatives of the algebraic equation $g(x)$ will allow us to reduce the DAEs to a system of ordinary differential equations (see Problem 15). We can write (5.36) in the form of (5.32) by using $y = [x^T, z^T, u^T]^T$, and the DAEs in Example 5.4-(iv) can be put in Hessenberg index-3 form if we select $x = [y_1, y_2]^T$, $z = [y_3, y_4]^T$ and $u = y_5$.

5.3.2 Implicit Runge-Kutta methods for DAEs

Implicit Runge-Kutta methods can be applied directly to the differential equations (5.32), and the Hessenberg form DAEs (5.34), (5.35) and (5.36). In the case of the differential equations (5.32) the k-th step of an s-stage implicit Runge-Kutta method has the form

$$\begin{aligned}
y^{(k+1)} &= y^{(k)} + h \sum_{i=1}^{s} b_i Y_i', \\
0 &= \Phi(Y_i, Y_i', \tau_i), \ i = 1, 2, \cdots, s, \\
Y_i &= y^{(k)} + h \sum_{j=1}^{s} a_{ij} Y_j', \\
\tau_i &= t^{(k)} + c_i h,
\end{aligned} \tag{5.37}$$

where $Y_i' \in \mathcal{R}^n$ is the stage derivative, and $Y_i \in \mathcal{R}^n$ is the stage value. Therefore, each step of the method requires that we solve the equations $\Phi(Y_i, Y_i', \tau_i) = 0$, $i = 1, 2, \cdots, s$, for $Y_i \in \mathcal{R}^n$ and $Y_i' \in \mathcal{R}^n$.

In the case of the Hessenberg index-1 DAE (5.34) the k-th step of an s-stage implicit Runge-Kutta method can be written as

$$\begin{aligned}
x^{(k+1)} &= x^{(k)} + h \sum_{i=1}^{s} b_i X_i', \\
z^{(k+1)} &= z^{(k)} + h \sum_{i=1}^{s} b_i Z_i', \\
X_i' &= f(X_i, Z_i), \ i = 1, 2, \cdots, s, \\
0 &= g(X_i, Z_i), \ i = 1, 2, \cdots, s, \\
X_i &= x^{(k)} + h \sum_{j=1}^{s} a_{ij} X_j',
\end{aligned}$$

$$Z_i = z^{(k)} + h \sum_{j=1}^{s} a_{ij} Z'_j, \tag{5.38}$$

where $X'_i \in \mathcal{R}^{n_x}$ and $X_i \in \mathcal{R}^{n_x}$ are the stage derivative and stage value for the x variables, respectively. Similarly, $Z'_i \in \mathcal{R}^{n_z}$ and $Z_i \in \mathcal{R}^{n_z}$ are the stage derivative and stage value for the z variables, respectively.

We can develop similar formulas for the Hessenberg index-2 form and the Hessenberg index-3 form. (See Problem 16.)

Hairer, Lubich and Roche (1989), and Hairer and Wanner (1996) have investigated the properties of implicit Runge-Kutta methods applied to DAEs in the Hessenberg form. An important set of results they develop show the relationship between the local error and the structure of the DAEs for a given set of Runge-Kutta coefficients. Here, we summarize the results they obtain for the local error behavior of the Radau IIA methods.

- Hessenberg Index-1 Form.
  Suppose an s-stage Radau IIA Runge-Kutta method is applied to the system (5.34) with *consistent initial conditions* $(x^{(0)}, z^{(0)})$, then the local error in the x and z variables satisfy

$$x^{(1)} - x(t^{(1)}) = O(h^{2s-1}),$$
$$z^{(1)} - z(t^{(1)}) = O(h^{2s-1}),$$

  where $t^{(1)} = t^{(0)} + h$, and $x(t^{(1)}),\ z(t^{(1)})$ is the exact solution to the DAEs at time $t^{(1)}$. The initial conditions are consistent if $g(x^{(0)}, z^{(0)}) = 0$. It is interesting to note that the local error is the same as that obtained when the s-stage Radau IIA method is applied to ordinary differential equations.
- Hessenberg Index-2 Form.
  Suppose an s-stage Radau IIA Runge-Kutta method is applied to the system (5.35) with consistent initial conditions $(x^{(0)}, z^{(0)})$, then the local error in the x and z variables satisfy

$$x^{(1)} - x(t^{(1)}) = O(h^{2s-1}),$$
$$z^{(1)} - z(t^{(1)}) = O(h^{s}),$$

  where $t^{(1)} = t^{(0)} + h$, and $x(t^{(1)}),\ z(t^{(1)})$ is the exact solution to the DAEs at time $t^{(1)}$. In this case the initial conditions are consistent if $g(x^{(0)}) = 0$ and $g_x(x^{(0)}) f(x^{(0)}, z^{(0)}) = 0$. Note here the reduction in the order of the local error of the algebraic variable, z, when compared to Hessenberg index-1 systems.
- Hessenberg Index-3 Form.
  Suppose an s-stage Radau IIA Runge-Kutta method is applied to the system (5.36) with consistent initial conditions $(x^{(0)}, z^{(0)}, u^{(0)})$, then the local error in the x, z and u variables satisfy

$$x^{(1)} - x(t^{(1)}) = O(h^{2s-2}),$$

$$z^{(1)} - z(t^{(1)}) = O(h^s),$$
$$u^{(1)} - u(t^{(1)}) = O(h^{s-1}),$$

where $t^{(1)} = t^{(0)} + h$, and $x(t^{(1)})$, $z(t^{(1)})$, $u(t^{(1)})$ is the exact solution to the DAEs at time $t^{(1)}$. In this case the initial conditions are consistent if $g = 0$, $g_x f = 0$ and $(g_x f)_x f + g_x f_x f + g_x f_z h = 0$, when evaluated at the initial condition $x^{(0)}, z^{(0)}, u^{(0)}$. Again we note that the order of the local error, for the algebraic variables, is significantly reduced here, when compared to the order attained for index-1 systems.

5.3.3 Index reduction

From the results in the previous section we observe that, for the algebraic variables (z and u), the order of the local error decreases as the differentiation index increases. Therefore, we can assert that the computational expense for solving differential-algebraic equations increases with increasing differentiation index. That is, the higher the differentiation index, the greater the computational effort required to achieve a desired solution accuracy for the algebraic variables since, smaller step sizes will be required.

Due to this *order reduction* in the local error it is desirable to reformulate a system of differential-algebraic equations with high differentiation index into a problem with lower differentiation index. One approach to reducing the differentiation index of DAEs in Hessenberg form is to differentiate the algebraic equations. Each such differentiation will reduce the index of the original problem by one, as illustrated in the following example.

Example 5.5.
Consider the index-3 DAEs

$$\dot{x} = f(x, z),$$
$$\dot{z} = h(x, z, u),$$
$$0 = g(x).$$

If we replace the algebraic equation $g(x) = 0$ with $\frac{d}{dt}g(x) = 0$ we obtain the system

$$\dot{x} = f(x, z),$$
$$\dot{z} = h(x, z, u),$$
$$0 = g_x(x)f(x, z),$$

which are index-2 DAEs.

If we replace the equation $g(x) = 0$ with $\frac{d^2}{dt^2}g(x) = 0$, in the original index-3 DAEs, we obtain the system

$$\begin{aligned}
\dot{x} &= f(x, z),\\
\dot{z} &= h(x, z, u),\\
0 &= (g_x(x)f(x, z))_x f(x, z)\\
&\quad + g_x(x)f_x(x, z)f(x, z)\\
&\quad + g_x(x)f_z(x, z)h(x, z, u),
\end{aligned}$$

which are index-1 DAEs.
A third time derivative of the algebraic equation $g(x) = 0$ will lead to system of first-order differential equations in x, z and u.

Unfortunately, the index reduction technique described in Example 5.5 does not always lead to satisfactory numerical results. The main reason for the poor performance of these reduced index DAEs is that they do not explicitly satisfy the constraint $g(x) = 0$. As a result, the numerical solution tends to drift away from this constraint as the time variable increases (see Problem 9).

Nevertheless, there is an index reduction technique for index-3 DAEs that works well in practice. This method can be described as follows.

Consider the index-3 LDAEs

$$\begin{aligned}
\dot{q} - f &= 0,\\
M(q, f)\dot{f} + \phi_q(q)^T\lambda + \Upsilon(q, f) &= 0,\\
\phi(q) &= 0.
\end{aligned} \tag{5.39}$$

We assume that the displacement constraints are linearly independent, and the inertia matrix, M, is nonsingular at all times. Taking two time derivatives of the displacement constraint $\phi = 0$ we can reduce this LDAEs to a system of ordinary differential equations

$$\dot{x} = F(x),$$

where $x = [q^T,\ f^T]^T$ and

$$F(x) = \begin{bmatrix} f \\ -M^{-1}\phi_q(q)^T\left[\phi_q M^{-1}\phi_q^T\right]^{-1}\left[(\phi_q f)_q f - \phi_q M^{-1}\Upsilon\right] - M^{-1}\Upsilon \end{bmatrix}.$$

To ensure that the solution to this ODEs does not drift from the constraints $\phi = 0$ and $\dot{\phi} = \phi_q f = 0$ we solve the system

$$\begin{aligned}
\dot{x} &= F(x),\\
0 &= H(x),
\end{aligned} \tag{5.40}$$

where

$$H(x) = \begin{bmatrix} \phi \\ \phi_q f \end{bmatrix}.$$

This is consider to be an ODE, $\dot{x} = F(x)$, with an *invariant* $H(x) = 0$.

Next we note that the system (5.40) is equivalent to the index-2 DAEs

$$\begin{aligned} \dot{x} &= F(x) - V(x)\nu, \\ 0 &= H(x), \end{aligned} \tag{5.41}$$

where $V(x)$ is a bounded matrix such that $H_x V$ is nonsingular for all time. Here,

$$H_x = \begin{bmatrix} \phi_q & 0 \\ (\phi_q f)_q & \phi_q \end{bmatrix}.$$

In this case, if $x(t)$ is a solution to (5.40) then $x(t)$ and $\nu(t) = 0$ is a solution to (5.41).

If we select

$$V(x) = \begin{bmatrix} \phi_q^T & 0 \\ 0 & \phi_q^T \end{bmatrix}$$

then (5.41) can be written as

$$\begin{aligned} \dot{q} - f + \phi_q^T \nu &= 0, \\ M\dot{f} + \phi_q^T \bar{\lambda} + \Upsilon &= 0, \\ \phi &= 0, \\ \phi_q f &= 0. \end{aligned} \tag{5.42}$$

Therefore, the index-3 DAEs (5.39) are reduced to index-2 DAEs by adding the multipliers, ν, and the constraint $\phi_q f = 0$, to the original system. Note that the Lagrange multiplier, λ, in (5.39) is different from the multiplier $\bar{\lambda}$ in (5.42). The system (5.42) is known as the Gear, Gupta and Leimkuhler (GGL) stabilized index-2 DAEs. (See Gear, Gupta and Leimkuhler, (1985)).

Example 5.6.

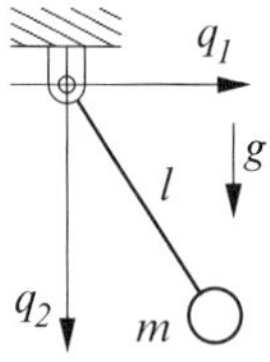

The index-3 Lagrangian differential-algebraic equations for the simple pendulum shown here are

$$\begin{aligned} \dot{q}_1 - f_1 &= 0, \\ \dot{q}_2 - f_2 &= 0, \\ m\dot{f}_1 + 2q_1\lambda &= 0, \\ m\dot{f}_2 + 2q_2\lambda - mg &= 0, \\ q_1^2 + q_2^2 - l^2 &= 0, \end{aligned}$$

where (q_1, q_2) are the displacements, and (f_1, f_2) are the corresponding flows. The displacement constraint is $\phi = q_1^2 + q_2^2 - l^2 = 0$.

This system can be written in GGL index-2 stabilized form as

$$\begin{aligned} \dot{q}_1 - f_1 + 2q_1\nu &= 0, \\ \dot{q}_2 - f_2 + 2q_2\nu &= 0, \\ m\dot{f}_1 + 2q_1\bar{\lambda} &= 0, \\ m\dot{f}_2 + 2q_2\bar{\lambda} - mg &= 0, \\ q_1^2 + q_2^2 - l^2 &= 0, \\ 2q_1f_1 + 2q_2f_2 &= 0, \end{aligned}$$

where ν is an additional multiplier, and the last equation represent the adjoined flow constraint, $\dot{\phi} = 0$. We also emphasize that λ and $\bar{\lambda}$ represent different Lagrange multipliers.

5.3.4 Consistent initial conditions

Successful numerical integration of the differential algebraic equations requires that we begin at consistent initial conditions as discussed in section 5.3.2. Here, we consider the problem of finding consistent initial conditions for the Lagrangian DAEs that includes displacement constraints, flow constraints, effort constraints and dynamic constraints. As shown in Section 4.5 such differential-algebraic equations can be written as

$$\begin{aligned} \dot{q} - f &= 0, \\ M\dot{f} + \phi_q^T\lambda + \psi_f^T\mu + \Upsilon &= 0, \\ \dot{s} - \Sigma &= 0, \\ \phi &= 0, \\ \psi &= 0, \\ \Gamma &= 0, \end{aligned} \tag{5.43}$$

where

$$M = \frac{\partial^2 T^*}{\partial f^2}, \quad \phi = [\phi_1, \phi_2, \cdots, \phi_{m_1}]^T, \quad \phi_q = \frac{\partial \phi}{\partial q}, \quad \lambda = [\lambda_1, \lambda_2, \cdots, \lambda_{m_1}]^T,$$

$$\psi = [\psi_1, \psi_2, \cdots, \psi_{m_2}]^T, \quad \psi_f = \frac{\partial \psi}{\partial f}, \quad \mu = [\mu_1, \mu_2, \cdots, \mu_{m_2}]^T,$$

$$\Upsilon = \left[\frac{\partial}{\partial q}\left(\frac{\partial T^*}{\partial f}\right)\right] f + \frac{\partial}{\partial t}\left(\frac{\partial T^*}{\partial f}\right) - \frac{\partial T^*}{\partial q} + \frac{\partial V}{\partial q} + \frac{\partial D}{\partial f} - e^s,$$

$$e^s = [e_1^s, e_2^s, \cdots, e_N^s]^T, \quad \Gamma = [\Gamma_1, \Gamma_2, \cdots, \Gamma_{m_3}]^T, \quad \Sigma = [\Sigma_1, \Sigma_2, \cdots, \Sigma_{m_4}]^T.$$

In these equations $q(t) \in \mathcal{R}^N$ denotes the displacement variables, $f(t) \in \mathcal{R}^N$ denotes the corresponding flow variables, $e^s \in \mathcal{R}^N$ denotes the known source efforts, $\lambda(t) \in \mathcal{R}^{m_1}$ denote the Lagrange multipliers associated with the displacement constraints $\phi(q) \in \mathcal{R}^{m_1}$, $\mu(t) \in \mathcal{R}^{m_2}$ denote the Lagrange multipliers associated with the flow constraints $\psi(q,f) \in \mathcal{R}^{m_2}$, $e(t) \in \mathcal{R}^{m_3}$ denotes the regulated effort variables, and $s(t) \in \mathcal{R}^{m_4}$ denotes the dynamic variables. The kinetic coenergy is $T^*(q,f,t)$, the potential energy is $V(q)$, and the dissipation function is $D(f)$. The displacement variables must satisfy the constraints $\phi(q) = 0$. The flow variables must satisfy the constraints $\psi(q,f) = 0$. The regulated efforts e must satisfy the constraints $\Gamma(q,f,s,e,t) = 0$. The dynamic constraints satisfy the differential equation $\dot{s} = \Sigma(q,f,s,e,t)$.

This system consists of $2N + m_1 + m_2 + m_3 + m_4$ differential-algebraic equations in the variables q, f, λ, μ, e and s. It is assumed that the displacement, flow and effort constraints are linearly independent. That is, ϕ_q has rank m_1, ψ_f has rank m_2, and Γ_e has rank m_3, along the solution to the LDAEs.

The index-3 Lagrangian DAEs (5.43) can be put in the GGL index-2 form by introducing the multipliers $\nu \in \mathcal{R}^{m_1}$ and the constrains $\phi_q f = 0$ to get

$$\begin{aligned}
\dot{q} - f + \phi_q^T \nu &= 0,\\
M\dot{f} + \phi_q^T \lambda + \psi_f^T \mu + \Upsilon &= 0,\\
\dot{s} - \Sigma &= 0,\\
\phi &= 0,\\
\phi_q f &= 0,\\
\psi &= 0,\\
\Gamma &= 0.
\end{aligned} \tag{5.44}$$

The system (5.44) can be viewed as the implicit differential equation $\Phi(Y, \dot{Y}, t) = 0$, with state $Y = [q^T, f^T, s^T, e^T, \lambda^T, \mu^T, \nu^T]^T$. We are interested in finding initial conditions $Y(t_0), \dot{Y}(t_0)$ such that $\Phi(Y(t_0), \dot{Y}(t_0), t_0) = 0$.

This dynamic system has $n-(m_1+m_2)+m_4$ 'degrees of freedom'. Among the $N = 2n + 2m_2 + m_2 + m_3 + m_4$ state variables we can arbitrarily assign $n-(m_1+m_2)$ displacements, $n-(m_1+m_2)$ flows, and m_4 dynamic variables at the initial time. All the other state variables and the state derivatives must be selected to satisfy (5.44).

A general procedure for finding consistent initial conditions of the Lagrangian DAEs is as follows. Let $\mathcal{I} \subseteq \{1, 2, \cdots, n\}$ denote the indexes of the independent displacement variables (and corresponding flow variables). Hence, the cardinality of $\mathcal{I}$ is $n - (m_1 + m_2)$. Let $Y^{(0)} = [(q^{(0)})^T$, $(f^{(0)})^T$, $(s^{(0)})^T$, $(e^{(0)})^T$, $(\lambda^{(0)})^T$, $(\mu^{(0)})^T$, $(\nu^{(0)})^T]^T$, denote an estimate of the initial state. Also, let $\dot{Y}^{(0)}$ denote an estimate of the initial state derivative. Then the consistent initial state and state derivative are determined by solving the problem

$$\min_{Y,\dot{Y}} \sum_{i=1}^{N}(Y_i - Y_i^{(0)})^2 + (\dot{Y}_i - \dot{Y}_i^{(0)})^2, \tag{5.45}$$

subject to

$$\Phi(Y, \dot{Y}, t_0) = 0, \tag{5.46}$$

$$Y_i - Y_i^{(0)} = 0, \ i \in \mathcal{I}, \tag{5.47}$$

$$Y_{i+n} - Y_{i+n}^{(0)} = 0, \ i \in \mathcal{I}, \tag{5.48}$$

$$Y_{i+2n} - Y_{i+2n}^{(0)} = 0, \ i = 1, 2, \cdots, m_4. \tag{5.49}$$

$$Y_{i+2n+m_1+m_2+m_3+m_4} = 0, \ i = 1, 2, \cdots, m_1. \tag{5.50}$$

The system equations (5.45)-(5.50) defines an equality constrained minimization problem. Here, the cost function (5.45) penalizes the deviation between the initial estimate $(Y^{(0)}, \dot{Y}^{(0)})$ and the actual consistent initial conditions $(Y, \dot{Y})$. The constraint (5.46) indicates that $(Y, \dot{Y})$ must satisfy the implicit differential equations at time $t = t_0$. The constraint (5.47) indicates that the independent displacement variables must satisfy the initial estimate, i.e., $q_i(t_0) = q_i^{(0)}$, $i \in \mathcal{I}$. The constraint (5.48) indicates that the independent flow variables must satisfy the initial estimate, i.e., $f_i(t_0) = f_i^{(0)}$, $i \in \mathcal{I}$. The constraint (5.49) indicates that the dynamic variables must satisfy the initial estimate, i.e., $s_i(t_0) = s_i^{(0)}$, $i = 1, 2, \cdots, m_4$. Finally, the constraint (5.50) indicates that multipliers ν must be zero at the initial time.

If M is nonsingular we can also insist that the Lagrange multipliers λ and μ vanish at the initial time. That is,

$$Y_{i+2n+m_3+m_4} = 0, \ i = 1, 2, \cdots, m_1, \tag{5.51}$$

$$Y_{i+2n+m_1+m_3+m_4} = 0, \ i = 1, 2, \cdots, m_2. \tag{5.52}$$

In which case we adjoin (5.51)-(5.52) to (5.45)-(5.50).

Example 5.7.

Consider the problem of finding consistent initial conditions for the simple pendulum given in Example 5.6. In this case we are required to find initial conditions that satisfy the GGL index-2 formulation

$$\begin{aligned}
\dot{q}_1 - f_1 + 2q_1\nu &= 0, \\
\dot{q}_2 - f_2 + 2q_2\nu &= 0, \\
m\dot{f}_1 + 2q_1\bar{\lambda} &= 0, \\
m\dot{f}_2 + 2q_2\bar{\lambda} - mg &= 0,
\end{aligned}$$

$$q_1^2 + q_2^2 - l^2 = 0,$$
$$2q_1 f_1 + 2q_2 f_2 = 0.$$

This system has 1 degree of freedom, which implies that we can select either q_1 or q_2 arbitrarily. That is there is only one independent displacement variable for this system. We can also arbitrarily assign the flow variable corresponding to the independent displacement variable.

Suppose $m = 2$, $l = 1$, $g = 9.8$, and we assign $q_1(0) = q_1^{(0)} = \sqrt{2}/2$, $f_1(0) = f_1^{(0)} = 0$. Then we can easily formulate the consistent initial condition problem (5.45)-(5.52). However, we must be careful in selecting the estimate for the dependent displacement variable, $q_2^{(0)}$. This is because for any given q_1 the displacement constraint $\phi = q_1^2 + q_2^2 - l^2 = 0$ has two possible solutions for q_2. Specifically, $q_2 = \pm\sqrt{l^2 - q_1^2}$. The problem (5.45)-(5.52) will tend to give a solution that is closest to the initial estimate $q_2^{(0)}$.

Using $q_2^{(0)} = 0.1$ gives the consistent initial conditions $q_1 = \sqrt{2}/2$, $q_2 = \sqrt{2}/2$, $\dot{f}_2 = 9.8$. The initial conditions for all the remaining variables are zero. On the other hand, using $q_2^{(0)} = -0.1$ gives the consistent initial conditions $q_1 = \sqrt{2}/2$, $q_2 = -\sqrt{2}/2$, $\dot{f}_2 = 9.8$. The initial conditions for all the remaining variables are zero.

5.3.5 An implicit Runge-Kutta method for IDEs

This section describes an implementation of an implicit Runge-Kutta method for implicit differential equations (IDEs)

$$\Phi(y, \dot{y}, t) = 0, \tag{5.53}$$

where $y(t) \in \mathcal{R}^{n_y}$, $\dot{y}(t) \in \mathcal{R}^{n_y}$, $\Phi \in \mathcal{R}^{n_y}$, and the initial conditions $y(t_0)$, $\dot{y}(t_0)$ at time $t = t_0$ are consistent. From the presentation given above it should be clear that the formulation (5.53) includes ordinary differential equations as well as the Lagrangian DAEs.

In the algorithm described below we will use the Radau IIA, $s = 3$ coefficients. The development here follows that of Section 5.2.4 with specific modifications to deal with implicit differential equations.

To describe this numerical integration technique let $(y^{(k)}, \dot{y}^{(k)})$ be the solution to (5.53) at time $t^{(k)}$. Given a step size, h, the solution at time $t^{(k+1)} = t^{(k)} + h$ can be obtained using an implicit Runge-Kutta method as described by (5.37). That is

$$y^{(k+1)} = y^{(k)} + h \sum_{i=1}^{s} b_i Y_i',$$

$$0 = \Phi(Y_i, Y_i', \tau_i), \; i = 1, 2, \cdots, s$$
$$Y_i = y^{(k)} + h \sum_{j=1}^{s} a_{ij} Y_j',$$
$$\tau_i = t^{(k)} + c_i h,$$

where, $Y_i' \in \mathcal{R}^n$ is the stage derivative, and $Y_i \in \mathcal{R}^n$ is the stage value. Moreover, A, b and c are the coefficients associated with the Radau IIA method.

If we define the stage increment as $Z_i = Y_i - y^{(k)}$, and use the fact that A is nonsingular we get that the stage derivatives must satisfy

$$Y_i' = \frac{1}{h} \sum_{j=1}^{s} w_{ij} Z_j,$$

where w_{ij} is the (i, j)-th element of $W = A^{-1}$. Then, the increments, Z_i, $i = 1, 2, \cdots, s$ are determined by solving the equations

$$\Phi(y^{(k)} + Z_i, \frac{1}{h} \sum_{j=1}^{s} w_{ij} Z_j, \tau_i) = 0, \; i = 1, 2, \cdots, s. \tag{5.54}$$

As in section 5.2.3 we will employ the simplified Newton's method to solve this system of equations. Specifically, given an initial estimate $Z_i^{(0)}$ we perform the iterations

```
for q = 0, 1, ··· do
    Solve the linear system
```

$$D\bar{\Phi}\, \Delta Z^{(q)} = -\bar{\Phi}(Z^{(q)}) \tag{5.55}$$

```
    to find the correction ΔZ^(q).
    Set Z^(q+1) = Z^(q) + ΔZ^(q)
    if ||ΔZ^(q)||, or ||Φ̄(Z^(q+1))|| is sufficiently small then STOP end if
end for
```

Here,

$$Z^{(q)} = \begin{bmatrix} Z_1^{(q)} \\ Z_2^{(q)} \\ \vdots \\ Z_s^{(q)} \end{bmatrix}, \; \bar{\Phi}(Z^{(q)}) = \begin{bmatrix} \Phi(y^{(k)} + Z_1^{(q)}, \frac{1}{h} \sum_{j=1}^{s} w_{1j} Z_j^{(q)}, \tau_1) \\ \Phi(y^{(k)} + Z_2^{(q)}, \frac{1}{h} \sum_{j=1}^{s} w_{2j} Z_j^{(q)}, \tau_2) \\ \vdots \\ \Phi(y^{(k)} + Z_s^{(q)}, \frac{1}{h} \sum_{j=1}^{s} w_{sj} Z_j^{(q)}, \tau_s) \end{bmatrix},$$

$$\Delta Z^{(q)} = \begin{bmatrix} \Delta Z_1^{(q)} \\ \Delta Z_2^{(q)} \\ \vdots \\ \Delta Z_s^{(q)} \end{bmatrix}, \quad D\bar{\Phi} = I \otimes J + \frac{1}{h}(W \otimes M),$$

$J = \partial\Phi(y^{(k)}, \dot{y}^{(k)}, t^{(k)})/\partial y$, and $M = \partial\Phi(y^{(k)}, \dot{y}^{(k)}, t^{(k)})/\partial \dot{y}$.

In practice we have found that the performance of the implicit Runge-Kutta algorithm for IDEs depends on; (i) the initial estimate, $Z^{(0)}$, used in the simplified Newton's method, (ii) the efficient solution of the linear system (5.55), (iii) the criteria used to terminate the Newton's method, (iv) the accuracy of the local error estimation, and (v) the step size control technique. Each of these algorithm details will be discussed next.

The initial estimate $Z^{(0)}$.

At the first step of the integration, i.e., $k = 0$, we use $Z^{(0)} = 0$. For steps $k > 0$ we use a polynomial extrapolation technique to estimate the increments Z_i, $i = 1, 2, \cdots, s$. This procedure is described as follows. At the end of the k-th step we have available the following data; $y^{(k)}$, the solution to IDEs at time $t^{(k)}$, and the stage values Y_i, which are the solutions to the IDEs at times $\tau_i = t^{(k)} + c_i h$, $i = 1, 2, \cdots, s$. (Note that for the Radau IIA method $Y_s = y^{(k+1)}$.) Hence, we can approximate the solution to the IDEs using the Lagrange polynomial

$$\hat{y}(t) = \sum_{i=0}^{s} L_i(t) Y_i, \quad L_i(t) = \prod_{j=0, j\neq i}^{s} \frac{t - t_j}{t_i - t_j}, \tag{5.56}$$

where $Y_0 = y^{(k)}$, $t_0 = t^{(k)}$, $t_i = \tau_i$, $i = 1, 2, \cdots, s$.

Now, given a step size $\bar{h}$ we would like to find the solution to the IDEs at time $t^{(k+2)} = t^{(k+1)} + \bar{h}$. Using (5.56) we can get estimates for the stage values at $\bar{\tau}_i = t^{(k+1)} + c_i\bar{h}$, $i = 1, 2, \cdots, s$. That is, the stage values in the interval $t^{(k+1)} \leq t \leq t^{(k+2)}$ can be approximated by $\bar{Y}_i = \hat{y}(\bar{\tau}_i)$. Hence, to advance the solution from $t^{(k+1)}$ to $t^{(k+2)}$ we take the initial estimate of the increments, to be used in the simplified Newton iteration, as

$$Z_i^{(0)} = \bar{Y}_i - y^{(k+1)} = \hat{y}(\bar{\tau}_i) - y^{(k+1)}.$$

It should be noted that the polynomial (5.56) can also be used to approximate the solution to the IDEs in the interval $t^{(k)} \leq t \leq t^{(k+1)}$.

Solution of the linear system (5.55).

The corrections, $\Delta Z^{(q)}$, in the simplified Newton's method are determined by solving the linear system of equations (5.55). From Section 5.2.4 we know that there is a nonsingular matrix T such that $T\Gamma = AT$, where A is the coefficient matrix for the Radau IIA method, and Γ is matrix in Jordan canonical form. Therefore, we can obtain $T^{-1}A^{-1}T = \Gamma^{-1}$. If we define $\Delta V^{(q)}$ such that

$$(T \otimes I)\Delta V^{(q)} = \Delta Z^{(q)}, \tag{5.57}$$

then (5.55) can be written as

$$(I \otimes J + \frac{1}{h}(W \otimes M))(T \otimes I)\Delta V^{(q)} = -\bar{\Phi}(Z^{(q)}).$$

Multiplying both sides of this equation by $(T^{-1} \otimes I)$ gives

$$(I \otimes J + \frac{1}{h}(\Gamma^{-1} \otimes M))\Delta V^{(q)} = -(T^{-1} \otimes I)\bar{\Phi}(Z^{(q)}) = -U. \tag{5.58}$$

But, in the case of the Radau IIA ($s = 3$) method, the matrix Γ^{-1} has the form

$$\Gamma^{-1} = \begin{bmatrix} \mu_1 & 0 & 0 \\ 0 & \mu_2 & \mu_3 \\ 0 & -\mu_3 & \mu_2 \end{bmatrix}.$$

Let us partition $\Delta V^{(q)}$ as $\Delta V^{(q)} = [v_1^T, v_2^T, v_3^T]^T$, and U as $U = [u_1^T, u_2^T, u_3^T]^T$, where $v_i \in \mathcal{R}^{n_y}$ and $u_i \in \mathcal{R}^{n_y}$, $i = 1, 2, 3$. Then, the linear system (5.58) can be written as

$$(J + \frac{\mu_1}{h}M)v_1 = -u_1, \tag{5.59}$$

$$(J + \frac{\mu_2 - i\mu_3}{h}M)(v_2 + iv_3) = -(u_2 + iu_3), \tag{5.60}$$

where $i = \sqrt{-1}$. Therefore, the real linear system (5.58) with a coefficient matrix of dimension $3n_y \times 3n_y$ is reduced to a real system (5.59) of dimension $3n_y$ and complex system (5.60) of dimension $3n_y$. Once $\Delta V^{(q)}$ is known we can find the correction $\Delta Z^{(q)}$ from (5.57). Solving the systems (5.59) and (5.60) to find the correction $\Delta Z^{(q)}$ requires fewer operations than solving the system (5.55) directly.

Termination of the simplified Newton's method.

In our implementation of the simplified Newton's method the iterations terminate successfully if

$$\frac{\Theta}{1 - \Theta}\|\Delta Z^{(q)}\|_w \leq \texttt{ctol}_1, \tag{5.61}$$

or

$$\|\bar{\Phi}(Z^{(q+1)})\| \leq \texttt{ctol}_2, \tag{5.62}$$

where $\Theta = \Delta Z^{(q)}/\Delta Z^{(q-1)}$. The weighted norm in (5.61) is computed as

$$\|\Delta Z^{(q)}\|_w = \sqrt{\frac{1}{3n_y}\sum_{j=1}^{3}\sum_{i=1}^{n_y}\left(\frac{|\Delta Z_{j,i}^{(q)}|}{\mathtt{atol}_i + \mathtt{rtol}_i|y_i^{(k)}|}\right)^2}, \qquad (5.63)$$

where $\mathtt{atol}_i$ and $\mathtt{rtol}_i$, $i = 1, 2, \cdots, n_y$ are specified absolute and relative tolerances, respectively. Also, $\Delta Z_{j,i}^{(q)}$ is the i-th element of $\Delta Z_j^{(q)}$.

We select the termination tolerance to be

$$\mathtt{ctol}_1 = \max(10\epsilon_M/\hat{\mathtt{rtol}}, \min(0.01, \sqrt{\hat{\mathtt{rtol}}})),$$

where $\hat{\mathtt{rtol}} = \min(\mathtt{rtol}_i)$, $i = 1, 2, \cdots, n_y$, and ϵ_M is the machine precision. We also select $\mathtt{ctol}_2 = 100\epsilon_M$.

The iterations are terminated unsuccessfully if (i) the iterations diverge, i.e., $\Theta \geq 1$, (ii) q exceeds the maximum number of iterations allowed, i.e., $q = \mathtt{q\_MAX}$, or (iii) the rate of convergence is too slow, i.e.,

$$\frac{\Theta^{\mathtt{q\_MAX}-q}}{1-\Theta}\|\Delta Z^{(q)}\|_w > \mathtt{ctol}_1.$$

Local error estimation.

To estimate the local error at time $t^{(k+1)}$ we consider the approximate solution to the IDEs given by

$$\hat{y}^{(k+1)} = y^{(k)} + h\sum_{i=1}^{s}\hat{b}_i Y_i' + h\hat{b}_0\dot{y}^{(k)} + h\hat{\gamma}\dot{\hat{y}}^{(k+1)}. \qquad (5.64)$$

Note that the last two terms in (5.64) distinguishes $\hat{y}^{(k+1)}$ from $y^{(k+1)}$ given in (5.37). Moreover, the coefficients $\hat{\gamma}$ and $\hat{b}_i$, $i = 0, 1, \cdots, s$ are selected so that $\hat{y}^{(k+1)}$ has a local error of $O(h^{s+1})$, for the differential variables in the IDEs. This implies that the coefficients in (5.64) must satisfy the order conditions (5.28). That is,

$$\hat{C}\hat{b} = \left[1 - \hat{b}_0, \frac{1}{2}, \cdots, \frac{1}{s}\right]^T - \hat{\gamma}\mathbf{1},$$

where $\hat{C} \in \mathcal{R}^{s\times s}$, the i, j-th element of the matrix $\hat{C}$ is given by $\hat{C}_{ij} = c_j^{i-1}$, $j = 1, 2, \cdots, s$, $\hat{b} = [\hat{b}_1, \hat{b}_2, \cdots, \hat{b}_s]^T$, and $\mathbf{1} = [1, 1, \cdots, 1]^T \in \mathcal{R}^s$.

As in section 5.2.4 we will treat $\hat{b}_0$ and $\hat{\gamma}$ as free parameters. In the case of the Radau IIA ($s = 3$) method, it is convenient to select $\hat{\gamma} = \gamma_1$, since this eliminates the need for an additional matrix factorization. Also, computational experience has shown that using $\hat{b}_0 = 0.02$ yields reliable error estimates for a wide range of problems. With $\hat{b}_0$ and $\hat{\gamma}$ known we can compute the remaining coefficients from (5.28).

From (5.64) we get

$$\dot{\hat{y}}^{(k+1)} = \frac{1}{h\hat{\gamma}}(\hat{y}^{(k+1)} - (y^{(k)} + h\sum_{i=1}^{s}\hat{b}_i Y_i' + h\hat{b}_0\dot{y}^{(k)})).$$

Putting this in $\Phi(\hat{y}^{(k+1)}, \dot{\hat{y}}^{(k+1)}, t^{(k+1)}) = 0$, and solving for $\hat{y}^{(k+1)}$ using the simplified Newton's method gives the iteration

$$\begin{aligned} D\Phi\Delta\hat{y} &= -\Phi(\hat{y}^{(k+1)}, \dot{\hat{y}}^{(k+1)}, t^{(k+1)}), \\ \hat{y}^{(k+1)} &\leftarrow \hat{y}^{(k+1)} + \Delta\hat{y}, \end{aligned}$$

where $D\Phi = (J + (\mu_1/h)M)$. If we perform one iteration of this method with a starting value $\hat{y}^{(k+1)} = y^{(k+1)}$ we get an estimate of the local error as

$$\begin{aligned} \eta^{(k+1)} &= y^{(k+1)} - \hat{y}^{(k+1)} \\ &= -\Delta\hat{y} \\ &= -(J + (\mu_1/h)M)^{-1}\Phi(y^{(k+1)}, \dot{\hat{y}}^{(k+1)}, t^{(k+1)}), \end{aligned} \tag{5.65}$$

where

$$\begin{aligned} \dot{\hat{y}}^{(k+1)} &= (\mu_1/h)(y^{(k+1)} - (y^{(k)} + h\sum_{i=1}^{s}\hat{b}_i Y_i' + h\hat{b}_0\dot{y}^{(k)})) \\ &= \mu_1\left(\sum_{i=1}^{s}(b_i - \hat{b}_i)Y_i' - \hat{b}_0\dot{y}^{(k)})\right) \end{aligned}$$

From these equations it can be seen that using $\hat{\gamma} = \gamma_1 = 1/\mu_1$ allows us to reuse the factorization of the matrix $(J + (\mu_1/h)M)$ in the computation of $\eta^{(k+1)}$. (Recall that the factorization of this matrix is used to solve the linear system (5.59)).

Step size control.

As with other numerical integration methods we attempt to adjust the step size, h, so that the local error is within a desired tolerance. Assume that the local error, with current step size h, behaves according to the formula $\|\eta^{(k+1)}\| = Ch^{s+1}$, where $C > 0$ is some constant. We would like the new step size, $\bar{h}$, to be such that the local error satisfies a desired tolerance, ϵ. That is, we would like $\|\bar{\eta}^{(k+1)}\| = C\bar{h}^{s+1} = \epsilon$. Therefore, using $\|\eta^{(k+1)}\|$ and $\|\bar{\eta}^{(k+1)}\|$ we see that

$$\frac{\|\bar{\eta}^{(k+1)}\|}{\|\eta^{(k+1)}\|} = \left(\frac{\bar{h}}{h}\right)^{s+1} = \frac{\epsilon}{\|\eta^{(k+1)}\|}.$$

Hence,

$$\bar{h} = h\rho_1, \tag{5.66}$$

where $\rho_1 = (\|\eta^{(k+1)}\|/\epsilon)^{-1/(s+1)}$. A practical step size adjustment technique is to select $\bar{h}$ so that

$$\bar{h} = h \min(\texttt{fac}_1, \max(\texttt{fac}_0, \beta\rho_1)), \tag{5.67}$$

where $0 < \texttt{fac}_0 < \texttt{fac}_1$, and $0 < \beta < 1$. Typical values for these parameters are $\beta = 0.9$, $\texttt{fac}_0 = 0.2$ and $\texttt{fac}_1 = 5$, which ensures that $0.2h \leq \bar{h} \leq 5h$.

Another useful step size control method is given by Gustafsson, et al. (1988). This technique is based on the assumption that at the k-th step the local error satisfies

$$\|\eta^{(k+1)}\| = C^{(k)} h_{(k)}^{s+1},$$

where $h_{(k)}$ is the step size at the k-th step, and the non-negative coefficient $C^{(k)}$ is such that

$$\frac{C^{(k+1)}}{C^{(k)}} \approx \frac{C^{(k)}}{C^{(k-1)}}.$$

If we desire that the new step size, $\bar{h}$, satisfy the condition $C^{(k+1)}\bar{h}^{s+1} = \epsilon$ then, using $C^{(k)} = \|\eta^{(k+1)}\|/h_{(k)}^{s+1}$ and $C^{(k-1)} = \|\eta^{(k)}\|/h_{(k-1)}^{s+1}$ we get

$$\frac{C^{(k+1)}}{C^{(k)}} = \frac{C^{(k)}}{C^{(k-1)}} = \frac{\epsilon}{\|\eta^{(k+1)}\|}\left(\frac{h_{(k)}}{\bar{h}}\right)^{s+1} = \frac{\|\eta^{(k+1)}\|}{\|\eta^{(k)}\|}\left(\frac{h_{(k-1)}}{h_{(k)}}\right)^{s+1}.$$

This gives

$$\bar{h} = h_{(k)}\rho_2, \tag{5.68}$$

where

$$\rho_2 = \left(\frac{\epsilon}{\|\eta^{(k+1)}\|}\right)^{1/(s+1)} \left(\frac{h_{(k)}}{h_{(k-1)}}\right) \left(\frac{\|\eta^{(k)}\|}{\|\eta^{(k+1)}\|}\right)^{1/(s+1)}.$$

Therefore, if we know the step sizes from the past two steps, $(h_{(k)},\ h_{(k-1)})$, and the corresponding local errors, $(\eta^{(k+1)},\ \eta^{(k)})$, we can compute a new step size using (5.68).

In developing the step size control techniques described above we have assumed that the local error due to the formula (5.64) is of order h^{s+1}. Unfortunately, for variables in $y(t)$ that have differentiation index greater than 1, this assumption does not hold. In fact, for variables in $y(t)$ with differentiation index greater than 1, this formula suffers from order reduction. One approach to dealing with this loss in accuracy is to scale $\eta_i^{(k+1)}$ by $h^{1-\texttt{ind}_i}$, if $\texttt{ind}_i > 1$, where $\texttt{ind}_i$ is the differentiation index of the i-th variable in $y(t)$. Alternatively, we can ignore variables with differentiation index greater than 1 when computing $\|\eta^{(k+1)}\|$. In the latter case we compute the weighted norm

$$\|\eta^{(k+1)}\| = \sqrt{\frac{1}{n_{\mathcal{D}}}\sum_{i\in\mathcal{D}}\left(\frac{|\eta_i^{(k+1)}|}{\mathtt{atol}_i + \mathtt{rtol}_i|y_i^{(k)}|}\right)^2}, \tag{5.69}$$

where $\mathcal{D} = \{i \mid \mathtt{ind}_i \leq 1\}$, and $n_{\mathcal{D}}$ is the cardinality of $\mathcal{D}$.

An implicit Runge-Kutta algorithm for IDEs.

Using the techniques described in the previous sections we can state an algorithm for the integration of implicit differential equations that is based on the implicit Runge-Kutta method. This scheme is similar to Algorithm 5.2.3 but with the following modifications. In step 3 we perform the simplified Newton's iteration (5.55). In step 5 we compute the local error using (5.65). In step 6 we use $\sigma = \min(\rho_1, \rho_2)$, where ρ_1 is computed in (5.67), and ρ_2 is computed in (5.68). In both cases we use the scaled norm (5.69). A MATLAB implementation of this algorithm is given in the Chapter 6. The function is called `ride`, and is used to solve the system simulation problems in Chapter 6.

5.4 The Multistep BDF Method

Heretofore we have emphasized the single step Runge-Kutta methods for the solution of ODEs and DAEs. In this section we will outline a multi-step method based on the backward differentiation formula (BDF). Multistep methods utilize one or more of the solutions at the previous time steps to find the solution at the next time step. (Whereas, single step methods use only the most recent solution to find the solution at the next time step.) Implementations of the BDF method have been shown to be very effective at solving stiff differential equations and differential-algebraic equations.

The multistep BDF method is described as follows. Given the differential equation

$$\dot{y} = f(y, t), \tag{a}$$

where $y(t) \in \mathcal{R}^n$, $f(y,t) \in \mathcal{R}^n$ and t is the (independent) time variable. Suppose we know the exact solution to (a) at the times, $t^{(j)}$, $t^{(j-1)}$, $\cdots$, $t^{(j-m)}$. That is, we know $y(t^{(j)})$, $y(t^{(j-1)})$, $\cdots$, $y(t^{(j-m)})$. Now we would like to find the solution to (a) at time $t^{(j+1)} = t^{(j)} + h$, where h is the step size.

To do so, let us construct a polynomial $y_c(t)$ that (i) interpolates the exact solution at times $t^{(j-i)}$, $i = 0, 1, \cdots, k-1$, and (ii) satisfies the differential equation at time $t^{(j+1)}$. Therefore, $y_c(t)$ must satisfy

$$\begin{aligned} y_c(t^{(j-i)}) &= y(t^{(j-i)}),\ i = 0, 1, \cdots, k-1, \\ \dot{y}_c(t^{(j+1)}) &= f(y_c(t^{(j+1)}), t^{(j+1)}). \end{aligned}$$

Thus, $y_c(t^{(j+1)})$ represents an approximate solution to the differential equation at time $t^{(j+1)}$.

We can define $y_c(t)$ as a Lagrange polynomial

$$y_c(t) = \sum_{i=0}^{k} L_i(t) y(t^{(j+1-i)}), \quad L_i(t) = \prod_{q=0,\ q\neq i}^{k} \frac{t - t^{(j+1-q)}}{t^{(j+1-i)} - t^{(j+1-q)}}$$

Using this equation we see that

$$\dot{y}_c(t^{(j+1)}) = \alpha(t^{(j+1)}) y_c(t^{(j+1)}) + \gamma(t^{(j+1)}), \tag{b}$$

where

$$\alpha(t^{(j+1)}) = \dot{L}_0(t^{(j+1)}), \quad \gamma(t^{(j+1)}) = \sum_{i=1}^{k} \dot{L}_i(t^{(j+1)}) y(t^{(j+1-i)}).$$

Equation (b) is called the backward differentiation formula, and using this in the differential equations (a) gives

$$\Psi(y_c(t^{(j+1)})) = \alpha y_c(t^{(j+1)}) + \gamma - f(y_c(t^{(j+1)}, t^{(j+1)})) = 0,$$

where we have used $\alpha = \alpha(t^{(j+1)})$ and $\gamma = \gamma(t^{(j+1)})$. Now, $\Psi(y_c(t^{(j+1)})) = 0$ is a system of algebraic equations that must be solved to determine the approximate solution at time $t^{(j+1)}$, i.e., $y_c(t^{(j+1)})$. This system is solved using a simplified Newton's iteration.

It can be shown that the local error in these BDF methods is $O(h^{k+1})$. Moreover, the methods are stable for formulas with $1 \leq k \leq 6$. Further details on the properties and implementation on specific BDF algorithms can be found in Brayton, Gustavson and Hachtel, (1972), Brenan, Campbell and Petzold, (1996), Shampine and Reichelt, (1997), Ascher and Petzold, (1998), and Shampine, (2002).

References

1. U. M. Ascher, and L. R. Petzold, *Computer methods for ordinary differential equations and differential-algebraic equations*, SIAM, 1998.
2. J. C. Butcher, "Implicit Runge-Kutta Processes," *Mathematics of Computation*, **18**, 50–64, 1964.
3. J. C. Butcher, *Numerical Analysis of Ordinary Differential Equations*, John Wiley and Sons, New York, 1987.
4. R. K. Brayton, F. G. Gustavson and G. D. Hachtel, "A new efficient algorithm for solving differential-algebraic systems using implicit backward differentiation formulas," *Proceedings of the IEEE*, **60**, 98–108, 1972.

5. K. E. Brenan, S. L. Campbell and L. R. Petzold, *Numerical solution of initial-value problems in differential-algebraic equations*, 2nd Edition, SIAM, 1996.
6. S. D. Conte and C. de Boor, *Elementary Numerical Analysis*, 3rd-Edition, McGraw-Hill, 1980.
7. J. E. Dennis and R. B. Schnabel, *Numerical Methods for Unconstrained Optimization and Nonlinear Equations*, Prentice-Hall, New Jersey, 1983.
8. P. Dueflhard and F. Bornemann, *Scientific computing with ordinary differential equations*, Springer, 2002.
9. C. W. Gear, G. K. Gupta and B. Leimkuhler, "Automatic integration of Euler Lagrange equations with constraints," *J. Comput. Appl. Math.*, **12, 13**, 77–90, 1985.
10. K. Gustafsson, M. Lundh and G. Söderlind, "A PI stepsize control for the numerical integration of ordinary differential equations," *JBIT*, **28**, 270–287, 1988.
11. E. Hairer, C. Lubich and M. Roche, *The numerical solution of differential-algebraic systems by Runge-Kutta methods*, Lecture Notes in Mathematics, 1409, Springer-Verlag, 1989.
12. E. Hairer, P. Norsett and G. Wanner, *Solving ordinary differential equations I: Nonstiff Problems*, 2nd. Edition, Springer, 1993.
13. E. Hairer and G. Wanner, *Solving ordinary differential equations II: Stiff and Differential-Algebraic Problems*, 2nd. Edition, Springer, 1996.
14. A. C. Hindmarsh and L. R. Petzold, "Algorithms and software for ordinary differential equations and differential-algebraic equations, Part II: Higher-order methods and software packages," *Computers in Physics*, **9**, 148–155, 1995.
15. Ch. Lubich and M. Roche, "Rosenbrock methods for differential-algebraic systems with solution-dependent singular matrix multiplying the derivative," *Computing*, **43**, 325–342, 1990.
16. H. Olsson and G. Söderlind, "Stage value predictors and efficient Newton iteration in implicit Runge-Kutta methods," *SIAM Journal Scientific Computing*, **20**, 185–202, 1990.
17. L. F. Shampine, "Solving $0 = F(t, y(t), y'(t))$ in Matlab," *J. Numer. Math.*, **10**, 291–310, 2002.
18. L. F. Shampine and M. W. Reichelt, "The MATLAB ODE Suite," *SIAM Journal on Scientific Computing*, **18**, 1–22, 1997.
19. J. Stoer and R. Bulirsch, *Introduction to Numerical Analysis*, Springer-Verlag, New York, (2002)
20. J. de Swart and G. Söderlind, "On the construction of error estimators for implicit Runge-Kutta methods," *Journal of Computational and Applied Mathematics*, **87**, 347–358, 1997.

Problems

1. Apply the explicit and implicit Euler methods to solve the differential equations
$$\begin{bmatrix} \dot{y}_1 \\ \dot{y}_2 \end{bmatrix} = \begin{bmatrix} -1 & 1 \\ 0 & -1000 \end{bmatrix} \begin{bmatrix} y_1 \\ y_2 \end{bmatrix},$$
with initial conditions $y(0) = [1, -2]^T$, in the interval $0 \leq t \leq 5$. Use step sizes (i) $h = 0.1$, (ii) $h = 0.01$ and (iii) $h = 0.001$. Compare the numerical results with the analytical solution in each case.
2. Use the explicit and implicit Euler methods to solve the ODE
$$\dot{y} = -50(y - \cos t), \; y(0) = 0.$$
Use step sizes (i) $h = 1.974/50$ and (ii) $h = 1.875/50$ for $t = 0$ to $t = 1.5$. Plot $y(t)$ in each case (4 plots).
3. Show that the stability function for the p-th order Taylor series formula, and the p-th order explicit Runge-Kutta formula is
$$R(z) = 1 + z + \frac{z^2}{2} + \frac{z^3}{3!} + \cdots + \frac{z^p}{p!}.$$
In the case of the Runge-Kutta formula assume that $p \leq 4$.
4. Plot the stability region for the explicit Runge-Kutta methods of order $p = 1, 2, 3$ and 4. *Hint: The boundary of the stability region is determined by the roots of the complex variable equation,* $R(z) - e^{i\theta} = 0$, *for* $0 \leq \theta \leq 2\pi$. Are these methods A-stable?
5. Derive the order conditions for the three stage explicit Runge-Kutta method
$$\begin{aligned} y^{(k+1)} &= y^{(k)} + h\left[b_1 k_1 + b_2 k_2 + b_3 k_3\right], \\ k_1 &= f(y^{(k)}, t^{(k)}), \\ k_2 &= f(y^{(k)} + h a_{21} k_1, t^{(k)} + c_2 h), \\ k_3 &= f(y^{(k)} + h a_{31} k_1 + h a_{32} k_2, t^{(k)} + c_3 h). \end{aligned}$$
Use the simplifying assumptions $a_{21} = c_2$ and $a_{31} + a_{32} = c_3$. Show that by proper selection of the coefficients the local discretization error is of the form $\delta^{(k+1)} = Ch^4$ for some constant C that is independent of h.
6. Apply the explicit Euler method, the implicit Euler method and the classical 4-th order Runge-Kutta method to the following problems.

- $\dot{x} = -0.1x + 1$, $x(0) = 0$, and $0 \leq t \leq 1$.
- $\dot{x}_1 = x_2$, $\dot{x}_2 = -2\xi\omega_n x_2 - \omega_n^2 x_1$, $x_1(0) = 1$, $x_2(0) = 0$, $0 \leq t \leq 1$, $\xi = 0.1$, and $\omega_n = 2\pi$.

For each of the numerical methods use the following step sizes; $h = 10^{-1},\ 10^{-2},\ 10^{-3},\ 10^{-4}$. Compare the numerical solutions with the exact solution in each case.

7. Use the classical 4-th order Runge-Kutta method to solve the differential equation

$$\begin{aligned}\dot{y}_1 &= y_2,\\ \dot{y}_2 &= \mu(1 - y_1^2)y_2 - y_1,\end{aligned}$$

with initial conditions $y(0) = [2,\ 4]^T$. Solve this problem for the cases (i) $\mu = 1$, final time $t = 20$, and (ii) $\mu = 1000$, final time $t = 6000$.

8. Apply the implicit Euler method to the following problems.

- $\dot{x} = f,\ 0 = -cf + F,\ x(0) = 0,\ f(0) = 0,\ c = 0.1,\ F(t) = \sin 4\pi t,\ 0 \le t \le 1.$
- $\dot{x} = f,\ 0 = -kx + F,\ x(0) = 0,\ f(0) = 0,\ k = \pi^2,\ F(t) = \sin 4\pi t,\ 0 \le t \le 1.$

Use the following step sizes; $h = 10^{-1},\ 10^{-2},\ 10^{-3},\ 10^{-4}$, and compare the numerical solutions with the exact solution in each case.

9. Use the implicit Euler method to solve the following problems for $0 \le t \le 2.5$.

(a)The ODEs

$$\begin{aligned}\dot{x}_1 &= x_2,\\ \dot{x}_2 &= (g/l)\sin x_1,\end{aligned}$$

where $g = 9$, $l = 1$, $x_1(0) = \pi/2$ and $x_2(0) = 0$. Plot x_1 and x_2.

(b)The index-3 DAE

$$\begin{aligned}\dot{x}_1 &= x_3,\\ \dot{x}_2 &= x_4,\\ \dot{x}_3 &= -2x_1x_5,\\ \dot{x}_4 &= -2x_2x_5 + g,\\ 0 &= x_1^2 + x_2^2 - 1.\end{aligned}$$

(c)The index-2 DAE

$$\begin{aligned}\dot{x}_1 &= x_3,\\ \dot{x}_2 &= x_4,\\ \dot{x}_3 &= -2x_1x_5,\\ \dot{x}_4 &= -2x_2x_5 + g,\\ 0 &= x_1x_3 + x_2x_4.\end{aligned}$$

(d)The index-1 DAE

$$\begin{aligned}
\dot{x}_1 &= x_3,\\
\dot{x}_2 &= x_4,\\
\dot{x}_3 &= -2x_1x_5,\\
\dot{x}_4 &= -2x_2x_5 + g,\\
0 &= x_3^2 + x_4^2 - 2x_5 + x_2g.
\end{aligned}$$

For problems (b), (c) and (d) use $x_1(0) = 1$, $x_2(0) = x_3(0) = x_4(0) = x_5(0) = 0$. Plot x_1 versus x_2, and $x_1^2 + x_2^2 - 1$ in each case.

10. Plot the stability region of the implicit trapezoidal method. (See Example 5.2.)
11. Show that the implicit trapezoidal method and the implicit midpoint method,

$$\begin{array}{c|c} 1/2 & 1/2 \\ \hline & 1 \end{array},$$

have the same stability function.
12. • Show that the implicit Euler method is L-stable.
    • Use the implicit Euler method and the implicit trapezoidal method to solve the problem

$$\dot{y} = -2000(y - \cos t), \quad y(0) = 0, \quad t_i = 0, \quad t_f = 1.5.$$

Use the fixed step size $h = 1.4/40$.
    • Can you draw any conclusions regarding the L-stable method. (From Example 5.2 we know that the implicit trapezoidal method is not L-stable.)
13. Consider the autonomous ODE $\dot{y} = f(y(t))$, $y(t_i) = y_i$. A 2 stage Rosenbrock Wanner method applied to this equation has the form

$$\begin{aligned}
y^{(k+1)} &= y^{(k)} + b_1k_1 + b_2k_2,\\
(I - h\gamma J)k_1 &= hf(y^{(k)}),\\
(I - h\gamma J)k_2 &= hf(y^{(k)} + \alpha_{21}k_1) + h\gamma_{21}Jk_1,
\end{aligned}$$

where $J = df/dy$. The coefficients of the methods are $b_1, b_2, \gamma, \gamma_{21}$ and α_{21}. Show that this method is second order if the following order conditions are satisfied.

$$\begin{aligned}
b_1 + b_2 &= 1,\\
b_2(\alpha_{21} + \gamma_{21}) &= \frac{1}{2} - \gamma.
\end{aligned}$$

14. Derive an expression for the Jacobian of the nonlinear equations (5.22) in the implicit Runge-Kutta method. Compare this exact Jacobian with the Jacobian approximation (5.25). Where are the computational savings?
15. Reduce the index-3 DAE (5.36) to a system of ordinary differential equations by taking three time derivatives of $g(x) = 0$, and solving for $\dot{u}$.
16. Develop expressions for the s-stage implicit Runge-Kutta method applied directly to Hessenberg index-2 and Hessenberg index-3 DAEs.
17. Apply the implicit Euler method to the GGL index-2 formulation for a simple pendulum. Compare the simulation results with that obtained from Problem 9.

Chapter 6
Dynamic System Analysis and Simulation

This chapter presents the analysis and numerical simulation of some of the systems modeled in the previous chapters. Section 6.1 presents some basic concepts associated with dynamic systems. In particular, the ideas of equilibrium, and stability, in the sense of Lyapunov, are discussed. In Section 6.2 the numerical simulations of the dynamic systems are performed using the computer code `ride`. The program `ride` is a MATLAB/Octave implementation of an implicit Runge-Kutta method for the solution of implicit differential equations. Here, we simulate the behavior of several dynamic systems described by Lagrangian differential-algebraic equations.

6.1 System Analysis

The equations of motion for many the dynamic systems modeled in this text can be written as

$$\dot{y}(t) = f(y(t), e^s(t), t), \;\; t_i \leq t \leq t_f, \tag{6.1}$$

where t is the time variable, $y(t) \in \mathcal{R}^{n_y}$ is the state vector, $e^s(t) \in \mathcal{R}^{n_e}$ is the input to the system, t_i is an initial time, and t_f is a final time. From the results in Chapter 3 and 4 it can be seen that the state is constructed from the displacements and flows, whereas the input represents the efforts applied to the system. For systems represented by Lagrangian differential-algebraic equations, i.e., equation (4.7), we can find a representation like (6.1) by reducing the system to the underlying ODEs (see Section 4.5.1).

Throughout this text it is assumed that the function $f(y(t), e^s(t), t)$ has continuous derivatives with respect to $y(t)$ and $e^s(t)$, in the time interval $t_i \leq t \leq t_f$. This assumption ensures that, given an initial condition $y(t_i)$ and an input $e^s(t)$, a solution to (6.1) exists and is unique (See Boyce and DiPrima, p. 70).

B. Fabien, *Analytical System Dynamics: Modeling and Simulation*,
DOI 10.1007/978-0-387-85605-6_6,

6.1.1 Equilibrium and stability

Suppose $e^s(t)$ is a known function of time, then (6.1) can be rewritten as

$$\dot{y}(t) = f(y(t), t). \tag{6.2}$$

Moreover, if (6.2) does not depend explicitly on the time variable it can be written as

$$\dot{y}(t) = f(y(t)). \tag{6.3}$$

Since the system (6.3) does not contain the time explicitly it is called an *autonomous* system. While the systems (6.1) and (6.2) are called *nonautonomous.* We note that any nonautonomous system can be written as an autonomous system by replacing t with τ, and appending the differential equation $\dot{\tau} = 1$ to the system.

The point y^* is called an *equilibrium* point of (6.3) if

$$f(y^*) = 0.$$

This implies that if $y(t^*) = y^*$, then $y(t) = y^*$ for all $t \geq t^*$. We will find it convenient to rewrite the equation (6.3) so that the equilibrium point is translated to the origin. This can be accomplished using the transformation $y(t) = x(t) + y^*$ hence, $\dot{y}(t) = \dot{x}(t)$, and (6.3) becomes

$$\dot{x}(t) = f(x(t) + y^*),$$

which has an equilibrium at the origin $x = 0$.

Example 6.1. The system

$$\begin{aligned}\dot{y}_1 &= y_2,\\ \dot{y}_2 &= -y_1^2 - y_2 + 1,\end{aligned}$$

has equilibrium points at $(1, 0)$ and $(-1, 0)$.

Using the transformation $y_1(t) = x_1(t) + 1$, $y_2(t) = x_2(t)$ these differential equations become

$$\begin{aligned}\dot{x}_1 &= x_2,\\ \dot{x}_2 &= -x_1^2 - 2x_1 - x_2.\end{aligned}$$

This system has an equilibrium at $x = 0$ which corresponds to the equilibrium point $y_1 = 1,\ y_2 = 0$.

Similarly, the transformation $y_1(t) = x_1(t) - 1$, $y_2(t) = x_2(t)$ yields the system

$$\dot{x}_1 = x_2,$$

$$\dot{x}_2 = -x_1^2 + 2x_1 - x_2.$$

This system has an equilibrium at $x = 0$ which corresponds to the equilibrium point $y_1 = -1,\ y_2 = 0$.

The stability of the dynamic system is defined by how the system behaves in the vicinity of the equilibrium point. In particular, for autonomous systems, *stability in the sense of Lyapunov* is defined as follows.

- Suppose, $y^* = 0$ is an equilibrium of the system (6.3). Then, y^* is stable in the sense of Lyapunov if, for each $\epsilon > 0$ there is a $\delta > 0$ such that, if $\|y(t_i)\| < \delta$ then $\|y(t)\| < \epsilon$, for all $t \geq t_i$.
- The equilibrium y^* is called *unstable* if it is not stable.
- The equilibrium y^* is *asymptotically stable* if it is stable, and $\lim_{t\to\infty} y(t) = y^*$.

The definition of stability can be interpreted as follows. Let S_ϵ represent an n_y-dimensional spherical region, radius ϵ, centered at $y^* = 0$. Let S_δ represent an n_y-dimensional spherical region, radius δ, centered at $y^* = 0$. Then, $y^* = 0$ is stable in the sense of Lyapunov if all trajectories starting in S_δ remain within S_ϵ for all $t \geq t_i$.

Instability is simply the absence of stability, whereas asymptotic stability implies that any initial condition within S_δ eventually returns to the equilibrium y^*.

6.1.2 Lyapunov indirect method

A technique for evaluating the stability of an equilibrium point is called the Lyapunov indirect method, or the linearization method. In this approach the nonlinear equations of motion are linearized about the equilibrium point, and the behavior of the resultant linear system is examined. For example, suppose $y^* = 0$ is an equilibrium of the system (6.3). Let $y(t) = y^* + x(t)$, where $x(t)$ represents a 'small' variation from the equilibrium. Then, writing $f(y^* + x(t))$ in a Taylor series centered at y^* gives

$$\begin{aligned} \dot{x}(t) &= f(y(t)) = f(y^* + x(t)) \\ \dot{x}(t) &= f(y^*) + \frac{\partial f(y^*)}{\partial x} x + \text{higher order terms} \\ \dot{x}(t) &= Ax(t). \end{aligned} \tag{6.4}$$

To get (6.4) we use $\dot{y}(t) = \dot{x}(t)$, $f(y^*) = 0$,

$$A = \frac{\partial f(y^*)}{\partial x} \in \mathcal{R}^{n_y \times n_y},$$

and we assume that $\|x(t)\|$ is so small that the higher-order terms in x can been neglected. The linear system (6.4) is called *time-invariant* because A does not depend of t. Now, the stability of the equilibrium y^* depends on the the solution of the linear time-invariant system (6.4). If the system is perturbed and the solution to (6.4) remains sufficiently close to the origin, then we can conclude that the equilibrium, y^*, is stable. Otherwise, the equilibrium is unstable.

The definition of equilibrium can be extended to dynamic systems of the form

$$\dot{y}(t) = f(y(t), e^s(t)). \tag{6.5}$$

(Note that this is the autonomous version of (6.1).) In this case the point $y(t) = y^*$, $e^s(t) = e^*$ is said to be an equilibrium of (6.5) if,

$$f(y^*, e^*) = 0.$$

A linearization of the system (6.5) about the equilibrium y^*, e^* proceeds as follows. Let $y(t) = y^* + x(t)$ and $e(t) = e^* + u(t)$ where $x(t)$ and $u(t)$ are small variations in $y(t)$ and $e^s(t)$, respectively. Then, by neglecting the higher-order terms in a Taylor series expansion of $f(y(t), e(t))$ about the point y^*, e^* we get

$$\dot{x}(t) = Ax(t) + Bu(t), \tag{6.6}$$

where

$$A = \frac{\partial f(y^*, e^*)}{\partial y} \in \mathcal{R}^{n_y \times n_y}, \; B = \frac{\partial f(y^*, e^*)}{\partial e^s} \in \mathcal{R}^{n_y \times n_e}.$$

Example 6.2. Consider the nonlinear system

$$\begin{aligned} \dot{y}_1 &= y_2 \\ \dot{y}_2 &= -y_1^2 - y_2 + e^s, \end{aligned}$$

where the nominal value of the input is $e^s(t) = 0$. Then, this system has an equilibrium point at $y = y^* = [0, \; 0]^T$, $e^s = e^* = 0$. Therefore, the linearized equations of motion about this equilibrium is

$$\dot{x} = Ax + Bu,$$

where

$$A = \frac{\partial f(y^*, e^*)}{\partial y} = \begin{bmatrix} \partial f_1/\partial y_1 & \partial f_1/\partial y_2 \\ \partial f_2/\partial y_1 & \partial f_2/\partial y_2 \end{bmatrix} = \begin{bmatrix} 0 & 1 \\ 0 & -1 \end{bmatrix},$$

$$B = \frac{\partial f(y^*, e^*)}{\partial e^s} = \begin{bmatrix} \partial f_1/\partial e^s \\ \partial f_2/\partial e^s \end{bmatrix} = \begin{bmatrix} 0 \\ 1 \end{bmatrix}.$$

Linear system analysis

In this section we examine the solution to the linear invariant systems (6.4) and (6.6). In addition to being an important problems in their own right, we see from the previous section that the solution to these problems have important implications regarding the stability of the nonlinear system (6.3) and (6.5).

-Homogeneous Equations

First, consider the linear, time-invariant, initial value problem

$$\dot{x}(t) = Ax(t), \tag{6.7}$$

with initial condition $x(0) = x_i$. Here, $x(t) \in \mathcal{R}^n$ is the state, the coefficient matrix $A \in \mathcal{R}^{n \times n}$ is constant, and the initial time $t_i = 0$. Note that since this system is time-invariant any problem with a nonzero initial time, i.e., $t_i \neq 0$, can be recast in the form of (6.7) by defining a new time variable, say $\tau = t - t_i$.

To develop a solution to this problem, assume that

$$x(t) = \sum_{k=0}^{\infty} a_k t^k, \tag{a}$$

where $a_k \in \mathcal{R}^n$, $k = 0, 1, \cdots$. Hence, (a) presents the solution to (6.7) as a power series in t, with vector coefficients a_k. Using (a) it can be seen that

$$\dot{x}(t) = \sum_{k=1}^{\infty} k a_k t^{k-1}. \tag{b}$$

Putting (a) and (b) into (6.7) gives

$$\sum_{k=1}^{\infty} k a_k t^{k-1} = \sum_{k=0}^{\infty} A a_k t^k.$$

Next, equating the like powers of t gives

$$\begin{aligned}
t^0 &: a_1 = Aa_0, \\
t^1 &: 2a_2 = Aa_1 = A^2 a_0, \\
t^2 &: 3a_3 = Aa_2 = \frac{1}{2} A^3 a_0, \\
&\vdots \\
t^k &: ka_k = Aa_{k-1} = \frac{1}{(k-1)!} A^k a_0.
\end{aligned}$$

Hence,

$$x(t) = \left(I + At + \frac{1}{2!}A^2t^2 + \frac{1}{3!}A^3t^3 + \cdots\right) a_0 = \sum_{k=0}^{\infty} \frac{t^k}{k!} A^k a_0 = e^{At} a_0,$$

where

$$e^{At} = \left(I + At + \frac{1}{2!}A^2t^2 + \frac{1}{3!}A^3t^3 + \cdots\right) = \sum_{k=0}^{\infty} \frac{t^k}{k!} A^k$$

is called the *matrix exponential.* The vector a_0 is defined by the initial condition. Specifically, using (a) it can be seen that at $t = 0$, $x(0) = x_i = a_0$. Therefore, the solution the initial value problem (6.7) is

$$x(t) = e^{At} x_i. \tag{6.8}$$

-Some properties of e^{At}:

It can be shown that the matrix exponential has the following properties;

- $$\frac{d}{dt} e^{At} = A e^{At} = e^{At} A.$$
- $$e^{-At} = e^{A^{-1}t}.$$
- If A is a diagonal matrix, i.e., $A = \text{diag}(\lambda_1,\ \lambda_2, \cdots, \lambda_n)$, where $\lambda_i,\ i = 1, 2, \cdots, n$ are scalar variables. Then,

$$e^{At} = \begin{bmatrix} e^{\lambda_1 t} & 0 & \cdots & 0 \\ 0 & e^{\lambda_2 t} & \cdots & 0 \\ & & \ddots & \\ 0 & 0 & \cdots & e^{\lambda_n t} \end{bmatrix}.$$

- If A is block diagonal, i.e., $A = \text{diag}(A_1,\ A_2,\ \cdots, A_n)$, where $A_i,\ i = 1, 2, \cdots, n$ are square matrices. Then,

$$e^{At} = \begin{bmatrix} e^{A_1 t} & 0 & \cdots & 0 \\ 0 & e^{A_2 t} & \cdots & 0 \\ & & \ddots & \\ 0 & 0 & \cdots & e^{A_n t} \end{bmatrix}.$$

- If $A = \sigma I + B$, where σ is a scalar, I is the identity matrix, and B is a square matrix. Then,

$$e^{At} = e^{\sigma t} e^{Bt}.$$

- If $A \in \mathcal{R}^{n\times n}$ is of the form

$$A = \begin{bmatrix} 0 & 1 & 0 & \cdots & 0 \\ 0 & 0 & 1 & \cdots & 0 \\ & & & \ddots & \\ 0 & 0 & 0 & \cdots & 1 \\ 0 & 0 & 0 & \cdots & 0 \end{bmatrix}$$

then,

$$e^{At} = \begin{bmatrix} 1 & t & \frac{t^2}{2} & \cdots & \frac{t^{n-1}}{(n-1)!} \\ 0 & 1 & t & \cdots & \frac{t^{n-2}}{(n-2)!} \\ & & & \ddots & \\ 0 & 0 & 0 & \cdots & t \\ 0 & 0 & 0 & \cdots & 1 \end{bmatrix}.$$

- If

$$A = \begin{bmatrix} 0 & \omega \\ -\omega & 0 \end{bmatrix} \text{ then, } e^{At} = \begin{bmatrix} \cos\omega t & \sin\omega t \\ -\sin\omega t & \cos\omega t \end{bmatrix}.$$

- If

$$A = \begin{bmatrix} \sigma & \omega \\ -\omega & \sigma \end{bmatrix} \text{ then, } e^{At} = e^{\sigma t}\begin{bmatrix} \cos\omega t & \sin\omega t \\ -\sin\omega t & \cos\omega t \end{bmatrix}.$$

- If the matrix $A \in \mathcal{R}^{n\times n}$ has distinct eigenvalues, $\lambda_1,\ \lambda_2,\ \cdots,\ \lambda_n$, and the n by n matrix Q represents the corresponding eigenvectors then,

$$QAQ^{-1} = \begin{bmatrix} \lambda_1 & 0 & \cdots & 0 \\ 0 & \lambda_2 & \cdots & 0 \\ & & \ddots & \\ 0 & 0 & \cdots & \lambda_n \end{bmatrix},$$

and

$$e^{At} = Q^{-1}\begin{bmatrix} e^{\lambda_1 t} & 0 & \cdots & 0 \\ 0 & e^{\lambda_2 t} & \cdots & 0 \\ & & \ddots & \\ 0 & 0 & \cdots & e^{\lambda_n t} \end{bmatrix}Q.$$

- If the matrix $A \in \mathcal{R}^{n\times n}$ has repeated eigenvalues then there is an n by n matrix Q that transforms A into the Jordan form, i.e.,

$$QAQ^{-1} = \begin{bmatrix} J_1 & 0 & \cdots & 0 \\ 0 & J_2 & \cdots & 0 \\ & & \ddots & \\ 0 & 0 & \cdots & J_m \end{bmatrix},$$

where the n_i by n_i block diagonal matrices J_i, $i = 1, 2, \cdots, m$ are of the form

$$J_i = \begin{bmatrix} \lambda_i & 1 & 0 & \cdots & 0 \\ 0 & \lambda_i & 1 & \cdots & 0 \\ & & & \ddots & \\ 0 & 0 & 0 & \cdots & 1 \\ 0 & 0 & 0 & \cdots & \lambda_i \end{bmatrix}.$$

In this case

$$e^{At} = Q^{-1} \begin{bmatrix} e^{J_1 t} & 0 & \cdots & 0 \\ 0 & e^{J_2 t} & \cdots & 0 \\ & & \ddots & \\ 0 & 0 & \cdots & e^{J_m t} \end{bmatrix} Q.$$

(See Nobel and Daniel (1977) and Strang (1980).)

-Nonhomogeneous Equations

Next, consider the linear nonhomogeneous system of differential equations

$$\dot{x}(t) = Ax(t) + Bu(t), \tag{6.9}$$

where $x(t) \in \mathcal{R}^n$ is the state, and $u(t) \in \mathcal{R}^m$ is the input to the system. Here, $u(t)$ is a known function of time. In (6.9) the matrices $A \in \mathcal{R}^{n \times n}$, and $B \in \mathcal{R}^{n \times m}$ are constant. Moreover, the system has an initial condition $x(0) = x_i$.

To develop a solution to (6.9) we can use the property that

$$\begin{aligned} \frac{d}{dt}\left(e^{-At}x(t)\right) &= e^{-At}\dot{x}(t) - e^{-At}Ax(t) \\ &= e^{-At}\left(\dot{x}(t) - Ax(t)\right). \end{aligned}$$

Hence,

$$e^{-At}\left(\dot{x}(t) - Ax(t)\right) = e^{-At}Bu(t).$$

Integrating both sides of this expression gives

$$\begin{aligned} e^{-A\tau}x(\tau)|_0^t &= \int_0^t e^{-A\tau}Bu(\tau)\,d\tau \\ e^{-At}x(t) - x(0) &= \int_0^t e^{-A\tau}Bu(\tau)\,d\tau \\ x(t) &= e^{At}x_i + e^{At}\int_0^t e^{-A\tau}Bu(\tau)\,d\tau. \end{aligned} \tag{6.10}$$

Example 6.3.
(a) Consider the homogeneous scalar linear differential equation

$$\dot{x} = \lambda x(t),$$

with initial condition $x(0) = x_i$. Note that $x^* = 0$ is the equilibrium point for the system. From (6.8) it can be seen that the solution to this differential equation is

$$x(t) = e^{\lambda t} x_i.$$

Clearly, the system response depends on the initial condition x_i, and the parameter λ.

If $x_i \neq 0$ and $\lambda < 0$ then, $\lim_{t\to\infty} x(t) = 0$. Thus, if $\lambda < 0$ the equilibrium point $x^* = 0$ is asymptotically stable. Since, the system eventually returns to the equilibrium x^* starting from any initial condition.

On the other hand if $\lambda > 0$, then as $t \to \infty$ the response $x(t)$ becomes unbounded. This behavior occurs for any nonzero initial condition, no matter how small. Thus, if $\lambda > 0$ the equilibrium $x^* = 0$ is unstable.

(b) The solution to the nonhomogeneous scalar linear differential equation

$$\dot{x} = \lambda x(t) + bu(t),$$

with initial condition $x(0) = x_i$, is given by (6.10) as

$$x(t) = e^{\lambda t} x_i + \int_0^t e^{\lambda(t-\tau)} bu(\tau)\, d\tau.$$

If $\lambda < 0$ and $u(t) = 1,\ t \geq 0$ then, the system response is

$$x(t) = e^{\lambda t} x_i + \frac{b}{\lambda}(e^{\lambda t} - 1).$$

The first term in this expression is called the initial condition response and it is the solution to the homogeneous differential equation, i.e., the system with $u(t) = 0$. The second term is called the forced response and is due to the input $u(t) = 1$. Since $\lambda < 0$ it can be seen that

$$\lim_{t\to\infty} x(t) = -\frac{b}{\lambda},$$

which is called the steady-state solution.

Example 6.4.
Consider the second-order homogeneous differential equation

$$\dot{x}(t) = Ax(t),$$

where $x(t) \in \mathcal{R}^2$, $A \in \mathcal{R}^{2\times 2}$ and initial condition $x(0) = x_i$.

Depending on the eigenvalues of A there is a nonsingular matrix $Q \in \mathcal{R}^{2\times 2}$ that can transform A into one of the following forms.

1. A diagonal matrix
$$QAQ^{-1} = \Lambda = \begin{bmatrix} \lambda_1 & 0 \\ 0 & \lambda_2 \end{bmatrix}, \tag{a}$$
where λ_1 and λ_2 are the real eigenvalues of A.
2. A Jordan block
$$QAQ^{-1} = \Lambda = \begin{bmatrix} \lambda & 1 \\ 0 & \lambda \end{bmatrix}, \tag{b}$$
where λ is the repeated real eigenvalue of A.
3. A matrix
$$QAQ^{-1} = \Lambda = \begin{bmatrix} \sigma & \omega \\ -\omega & \sigma \end{bmatrix}, \tag{c}$$
where $\lambda_1 = \sigma + i\omega$ and $\lambda_2 = \sigma - i\omega$ are the complex conjugate eigenvalues of A.

If we define a new vector $z(t) = Qx(t)$ then, the linear system becomes

$$\dot{z}(t) = \Lambda z(t),$$

with initial condition $z(0) = Qx(0) = z_0$.

In the case where Q diagonalizes A, i.e., (a) holds, the transformed differential equations are

$$\dot{z}_1(t) = \lambda_1 z_1(t),$$
$$\dot{z}_2(t) = \lambda_2 z_2(t).$$

The solution to this system is

$$z_1(t) = e^{\lambda_1 t} z_{10},$$
$$z_2(t) = e^{\lambda_2 t} z_{20},$$

where $z_1(0) = z_{10}$ and $z_2(0) = z_{20}$. If $\lambda_1 < 0$ and $\lambda_2 < 0$ then the equilibrium point $z^* = x^* = 0$ is asymptotically stable, since $\lim_{t\to\infty} z(t) = 0$.

If on the other hand $\lambda_1 > 0$ and/or $\lambda_2 > 0$ then the equilibrium point $z^* = 0$ is unstable because $z_1(t)$ and/or $z_2(t)$ become unbounded as $t \to \infty$, for any nonzero initial condition.

In the case where Q transforms A to the Jordan form, i.e., (b) holds, the differential equations become

$$\dot{z}_1(t) = \lambda z_1(t) + z_2(t),$$
$$\dot{z}_2(t) = \lambda z_2(t).$$

The solution to this system is

$$z_1(t) = e^{\lambda t} z_{10} + t e^{\lambda t} z_{20},$$
$$z_2(t) = e^{\lambda t} z_{20}.$$

Again, the stability of the equilibrium depends on the sign of λ. If $\lambda < 0$ then the equilibrium point $z^* = 0$ is asymptotically stable, since $\lim_{t\to\infty} z(t) = 0$. If $\lambda > 0$ the equilibrium is unstable.

If the eigenvalues are A are the complex conjugate pair $\lambda = \sigma \pm i\omega$, i.e., (c) holds, then the system response is

$$z(t) = e^{\sigma t} \begin{bmatrix} \cos \omega t & \sin \omega t \\ -\sin \omega t & \cos \omega t \end{bmatrix} \begin{bmatrix} z_{10} \\ z_{20} \end{bmatrix}.$$

If $\sigma < 0$ then the equilibrium is asymptotically stable, since $z(t) \to 0$ as $t \to \infty$. If $\sigma > 0$ then $z(t)$ becomes unbounded as $t \to \infty$, and the equilibrium is therefore unstable. If $\sigma = 0$ then $z(t)$ is a response that oscillates about the equilibrium. In fact the trajectory $(z_1(t),\, z_2(t))$ forms a closed path around the equilibrium. In this case the equilibrium is said to be *marginally stable.*

The stability results obtained in last two examples can be extended to linear systems in general. In particular, we can state the following. Let λ_j, $j = 1, 2, \cdots, n$ be the eigenvalues of the matrix A in the linear system (6.7). Let $\mathrm{Re}(\lambda_j)$ denote the real part of the j-th eigenvalue of A. Then, sufficient conditions for the stability of the linear system are as follows.

(i) The equilibrium is asymptotically stable if $\mathrm{Re}(\lambda_j) < 0$, $j = 1, 2, \cdots, n$. That is, all the eigenvalues of A are strictly in the left-hand complex plane.
(ii) The equilibrium is unstable if $\mathrm{Re}(\lambda_j) > 0$ for any j. That is, at least one eigenvalue of A is in the right-hand complex plane.
(iii) The equilibrium is marginally stable if $\mathrm{Re}(\lambda_j) \leq 0$, $j = 1, 2, \cdots, n$, and there are no repeated eigenvalues on the imaginary axis. That is, at least one eigenvalue of A is on the imaginary axis, while the others are in the left-hand complex plane.

In the case where (6.7) is obtained by the linearization of a nonlinear system then, $\mathrm{Re}(\lambda_j) < 0$, $j = 1, 2, \cdots, n$ implies that the equilibrium of the nonlinear system is asymptotically stable. Whereas, if $\mathrm{Re}(\lambda_j) > 0$ for any j then, the equilibrium of the nonlinear system is unstable. No conclusion can be made regarding the stability of the equilibrium of the nonlinear system if some of the eigenvalues of A lie on the imaginary axis and the rest are in the left-hand complex plane.

6.1.3 Lyapunov direct method

In the linearization method described above the stability of the dynamic system is determined by examining the eigenvalues of the linearized equations of motion. In this section the stability of the equilibrium is determined by examining a positive definite scalar 'energy function' for the system. To motivate why such an analysis may be fruitful, consider an unforced dynamic system consisting of ideal inductors, capacitors and resistors. In such a system any initial energy (due to perturbation of the system) will be dissipated by the resistors, and the system will eventually return to an equilibrium point. The Lyapunov direct method involves the construction and analysis of suitable scalar energy-like function for the system.

Before stating the Lyapunov stability theorems we need some definitions. Let $x \in \mathcal{R}^n$ be some vector, and Ω be region in $\mathcal{R}^n$ that contains the origin. Then;

- The scalar function $\mathcal{V}(x)$ is said to be locally positive definite if $\mathcal{V}(0) = 0$ and $\mathcal{V}(x) > 0$ for all $x \neq 0$ in Ω.
- The scalar function $\mathcal{V}(x)$ is said to be locally negative definite if $\mathcal{V}(0) = 0$ and $\mathcal{V}(x) < 0$ for all $x \neq 0$ in Ω.
- The scalar function $\mathcal{V}(x)$ is said to be locally positive semi-definite if $\mathcal{V}(0) = 0$ and $\mathcal{V}(x) \geq 0$ for all $x \neq 0$ in Ω.
- The scalar function $\mathcal{V}(x)$ is said to be locally negative semi-definite if $\mathcal{V}(0) = 0$ and $\mathcal{V}(x) \leq 0$ for all $x \neq$ in Ω.
- If $\mathcal{V}(x) \geq 0$ for some values of $x \in \Omega$, and $\mathcal{V}(x) \leq 0$ for other values of $x \in \Omega$ then, $\mathcal{V}(x)$ is said to be indefinite.

If Ω is the entire space $\mathcal{R}^n$ then these definitions apply *globally* instead of locally.

Lyapunov Stability Theorems

Consider the autonomous dynamic system

$$\dot{x}(t) = f(x(t)), \tag{6.11}$$

where the state $x(t) \in \mathcal{R}^n$, and $x^* = 0$ is an equilibrium. Let Ω be region in $\mathcal{R}^n$ that contains the origin, $x^* = 0$. Then, the following theorems give sufficient conditions for $x^* = 0$ to be stable, or asymptotically stable equilibrium.

I Lyapunov stability theorem.
If for $x \in \Omega$ there is a continuous, scalar, positive-definite function $\mathcal{V}(x)$ such that

$$\frac{d\mathcal{V}(x)}{dt} = \dot{\mathcal{V}}(x) \leq 0.$$

Then, the equilibrium $x^* = 0$ is stable in the sense of Lyapunov.

II Lyapunov asymptotic stability theorem.

If for $x \in \Omega$ there is a continuous, scalar, positive-definite function $\mathcal{V}(x)$ such that

$$\dot{\mathcal{V}}(x) < 0.$$

Then, the equilibrium $x^* = 0$ is asymptotically stable in the sense of Lyapunov.

III Krasovsky asymptotic stability theorem.

If for $x \in \Omega$ there is a continuous, scalar, positive-definite function $\mathcal{V}(x)$ such that

$$\dot{\mathcal{V}}(x) < 0, \ x \notin \Omega_0,$$
$$\dot{\mathcal{V}}(x) = 0, \ x \in \Omega_0,$$

where $\Omega_0 \subset \Omega$, (excluding the origin) is defined by the scalar function $F(x) = 0$. Then, the equilibrium $x^* = 0$ is asymptotically stable in the sense of Lyapunov if

$$\mathcal{U}(x) = \left(\frac{dF(x)}{dx}\right)^T f(x) \neq 0, \ x \in \Omega_0.$$

If the conditions in Theorem I or Theorem II or Theorem III are satisfied then, $\mathcal{V}(x)$ is called a *Lyapunov function.*

Theorem I indicates that if we can find a scalar function $\mathcal{V}(x)$ that is positive definite and $\dot{\mathcal{V}}(x)$ is negative semi-definite then, $x^* = 0$ is a stable equilibrium point.

Theorem II indicates that if we can find a scalar function $\mathcal{V}(x)$ that is positive definite and $\dot{\mathcal{V}}(x)$ is negative definite then, $x^* = 0$ is an asymptotically stable equilibrium point.

Theorem III indicates that $x^* = 0$ is an asymptotically stable equilibrium point if the following conditions hold;

- The scalar function $\mathcal{V}(x)$ is positive definite.
- $\dot{\mathcal{V}}(x) = 0$ in some subset of Ω, say Ω_0 which is defined by the surface $F(x) = 0$.
- $\dot{\mathcal{V}}(x)$ is negative definite outside of Ω_0.
- $\mathcal{U}(x) \neq 0$, $x \in \Omega_0$. This condition implies that the system trajectory does not stay in Ω_0 forever. Note that Ω_0 excludes the origin.

Since the system does not remain in Ω_0 for all time, and $\dot{\mathcal{V}}(x)$ is negative definite outside of Ω_0, it can be inferred that the systems 'energy', $\mathcal{V}(x)$, will decrease until it reaches the equilibrium.

In Theorem II asymptotic stability requires a monotonic decrease in the function $\mathcal{V}(x)$, while Theorem III allows $\mathcal{V}(x)$ to decrease piecewise fashion. See Merkin (1997) for the proofs of these theorems.

Note that these stability theorems are all local and apply only for $x \in \Omega$. However, similar statements hold globally if in addition to the conditions stated above, we have $\mathcal{V}(x) \to \infty$ as $\|x\| \to \infty$. The utility of these theorems will be demonstrated via the examples below.

Example 6.5.
The equations of motion for the unforced linear mass-spring system shown here are

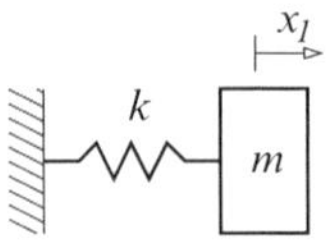

$$\dot{x}_1 = x_2$$
$$\dot{x}_2 = -\frac{k}{m}x_1,$$

where x_1 is the position, x_2 the velocity, m is the mass, and k is the spring stiffness. The equilibrium for this system is $x_1^* = 0$, $x_2^* = 0$.

Now, select

$$\mathcal{V}(x) = \frac{1}{2}(kx_1^2 + mx_2^2)$$

as a Lyapunov function *candidate*. (The function $\mathcal{V}(x)$ is called a Lyapunov function only if its satisfies one of the stability theorems given above.)

First, note that $\mathcal{V}(x)$ is a scalar positive definite function, and it is the sum of the kinetic coenergy and the potential energy of the system.

Next,

$$\begin{aligned}\dot{\mathcal{V}}(x) &= kx_1\dot{x}_1 + mx_2\dot{x}_2 \\ &= kx_1x_2 + mx_2(-\frac{k}{m}x_1) \\ &= 0.\end{aligned}$$

Since, $\mathcal{V}(x)$ is positive definite and $\dot{\mathcal{V}}(x) = 0$, the Lyapunov stability Theorem I, indicates that the equilibrium is stable. The same result can be obtained using the linearization method.

Example 6.6.
For the dynamic system

$$\dot{x} = \lambda x,$$

where $\lambda < 0$, the point $x^* = 0$ is the equilibrium.

Let the scalar positive definite function

$$\mathcal{V}(x) = \frac{1}{2}x^2$$

be a Lyapunov function candidate. Then,

$$\begin{aligned}\dot{\mathcal{V}}(x) &= x\dot{x} \\ &= \lambda x^2 < 0,\end{aligned}$$

for all $x \neq 0$. Therefore, according to the stability Theorem II we can conclude that the equilibrium, $x^* = 0$, is asymptotically stable.

Example 6.7.
The equations of motion for the unforced linear mass-spring-damper system shown here are

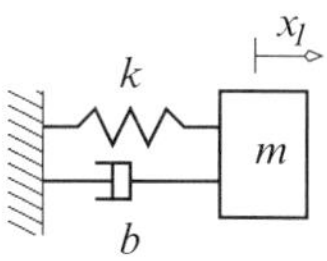

$$\dot{x}_1 = x_2,$$
$$\dot{x}_2 = -\frac{k}{m}x_1 - \frac{b}{m}x_2,$$

where x_1 is the position, x_2 the velocity, m is the mass, k is the spring stiffness, and b is the damping coefficient. The equilibrium for this system is $x_1^* = 0$, $x_2^* = 0$.

Let

$$\mathcal{V}(x) = \frac{1}{2}(kx_1^2 + mx_2^2)$$

be the Lyapunov function candidate. Then,

$$\begin{aligned}\dot{\mathcal{V}}(x) &= kx_1\dot{x}_1 + mx_2\dot{x}_2 \\ &= kx_1x_2 + mx_2\left(-\frac{k}{m}x_1 - \frac{b}{m}x_2\right) \\ &= -bx_2^2.\end{aligned}$$

Since, $\dot{\mathcal{V}}$ is negative semi-definite we can conclude from the stability Theorem I that the equilibrium is stable. However, using the stability Theorem III it can be shown that, in fact, the equilibrium is asymptotically stable. To do so let $\Omega_0 = \{x \in \mathcal{R}^2 \mid x_2 = 0\}$, excluding the origin. That is, Ω_0 is all points in $\mathcal{R}^2$ where $x_2 = 0$, except the origin. Noting this exception, it can be seen that $\dot{\mathcal{V}}(x) = 0$, for $x \in \Omega_0$, and we can define Ω_0 using the function

$$F(x) = x_2 = 0.$$

Therefore,

$$\begin{aligned}\mathcal{U}(x) &= \left(\frac{dF(x)}{dx}\right)^T f(x) \\ &= \begin{bmatrix} 0 & 1 \end{bmatrix}\begin{bmatrix} x_2 \\ -\frac{k}{m}x_1 - \frac{b}{m}x_2 \end{bmatrix} \\ &= -\frac{k}{m}x_1 - \frac{b}{m}x_2 \\ &\neq 0, \ x \in \Omega_0.\end{aligned}$$

That is, $\mathcal{U}(x)$ is not identically zero for all points in Ω_0. Thus, the system trajectory is such that x does not remain in Ω_0 for all time, and since $\dot{\mathcal{V}}(x)$ is negative definite outside of Ω_0 the system eventually approaches the equilibrium. Hence, from the stability Theorem III we can conclude that the equilibrium is asymptotically stable.

6.1.4 Lagrange's stability theorem

Let $q(t) \in \mathcal{R}^n$ denote the generalized displacements of a dynamic system, and let $\dot{q}(t)$ be the corresponding flows. The system is called a *natural dynamic system* if the kinetic coenergy is of the form $T^*(q, \dot{q}) = \frac{1}{2}\dot{q}(t)^T M(q(t))\dot{q}(t)$, where $M(q(t)) \in \mathcal{R}^{n \times n}$ is a symmetric positive definite matrix.

If the natural dynamic system is conservative then, the equilibrium and stability of the system can be determined by evaluating the potential energy function. In this case all the applied efforts, including gravity forces, must be included in the potential energy function.

Equilibrium

Let

$$e_i^C = -\frac{\partial V(q)}{\partial q_i}, \quad i = 1, 2, \cdots, n,$$

be the efforts due to the capacitors and gravity. Here, $V(q(t))$ is the potential energy function. Let

$$e_i^R = -\frac{\partial D(\dot{q})}{\partial \dot{q}_i}, \quad i = 1, 2, \cdots, n,$$

be the efforts due to ideal resistors, where $D(\dot{q})$ is the dissipation function. Let e_i^s, $i = 1, 2, \cdots, n$ be the efforts due to sources. If the system is conservative then the efforts due to the resistors and sources are zero, i.e., $e_i^R = 0$, and $e_i^s = 0$, $i = 1, 2, \cdots, n$.

In the equilibrium position the the virtual work done by all the efforts applied to the system must sum to zero, i.e.,

$$\begin{aligned}
\delta\mathcal{W}(q) &= \sum_{i=1}^{n} (e_i^C + e_i^R + e_i^s)\delta q_i \\
&= \sum_{i=1}^{n} -\frac{\partial V}{\partial q_i}\delta q_i \\
&= 0.
\end{aligned}$$

Since, the displacements are independent this expression will hold if and only if

$$\partial V/\partial q_i = 0, \;\; i = 1, 2, \cdots, n.$$

Thus, a necessary condition for q^* to be an equilibrium of a conservative natural system is that the potential energy function must be stationary at q^*.

Stability

Writing the potential energy function as a Taylor series centered at q^* gives

$$V(q) = V(q^*) + \frac{\partial V(q^*)}{\partial q}(q - q^*) + \frac{1}{2}(q - q^*)^T H(q^*)(q - q^*) \\ + \text{ higher order terms,}$$

where

$$H(q) = \frac{\partial^2 V(q)}{\partial q^2}$$

is the Hessian of the potential energy function. At the equilibrium, q^*, we have $\partial V(q^*)/\partial q = 0$. Also, we can assume that $V(q^*) = 0$, since a constant can be added to the potential energy function without affecting the efforts applied to the system. Then, in a neighborhood of the equilibrium, the potential energy function can be approximated by

$$V(q) = \frac{1}{2}(q - q^*)^T H(q^*)(q - q^*).$$

Now, suppose that the the Hessian matrix $H(q^*)$ is positive definite at the equilibrium $q = q^*$ then, the potential energy function is a minimum at q^*. Moreover, $V(q)$ is positive definite in a neighborhood of the equilibrium.

To evaluate the stability of the equilibrium, consider the Lyapunov function candidate

$$\begin{aligned} \mathcal{V}(q, \dot{q}) &= \sum_{i=1}^{n} \frac{\partial T^*}{\partial \dot{q}_i}\dot{q}_i - T^* + V \\ &= \frac{1}{2}\dot{q}^T M(q)\dot{q} + V(q). \end{aligned}$$

Since, $M(q)$ is positive definite, and $V(q^*)$ is a minimum then, there is a neighborhood of the point $q = q^*$, $\dot{q} = 0$ where $\mathcal{V}(q, \dot{q})$ is positive definite. Also,

$$\dot{\mathcal{V}} = \frac{d}{dt}\left[\sum_{i=1}^{n} \frac{\partial T^*}{\partial \dot{q}_i}\dot{q}_i - T^* + V\right]$$

$$
\begin{aligned}
&= \sum_{i=1}^{n} \left(\frac{d}{dt} \frac{\partial T^*}{\partial \dot{q}_i} \right) \dot{q}_i + \sum_{i=1}^{n} \frac{\partial T^*}{\partial \dot{q}_i} \ddot{q}_i - \sum_{i=1}^{n} \frac{\partial T^*}{\partial q_i} \dot{q}_i - \sum_{i=1}^{n} \frac{\partial T^*}{\partial \dot{q}_i} \ddot{q}_i + \sum_{i=1}^{n} \frac{\partial V}{\partial q_i} \dot{q}_i \\
&= \sum_{i=1}^{n} \left[\frac{d}{dt} \frac{\partial T^*}{\partial \dot{q}_i} - \frac{\partial T^*}{\partial q_i} + \frac{\partial V}{\partial q_i} \right] \dot{q}_i \\
&= 0.
\end{aligned}
$$

Thus, according to the Lyapunov stability Theorem I, the equilibrium $q = q^*$, $\dot{q} = 0$ is stable. Note that to obtain the result $\dot{\mathcal{V}} = 0$, we have used the fact that

$$
\frac{d}{dt} \frac{\partial T^*}{\partial \dot{q}_i} - \frac{\partial T^*}{\partial q_i} + \frac{\partial V}{\partial q_i} = -\frac{\partial D}{\partial \dot{q}_i} + e_i^s = 0, \;\; i = 1, 2, \cdots, n.
$$

The result developed above can be summarized as follows.

Stability Theorem of Lagrange

If the potential energy function of a conservative, natural system, has a minimum at an equilibrium, then the equilibrium is stable in the sense of Lyapunov.

The stability theorem of Lagrange provides sufficient conditions for the equilibrium to be stable. The inverse of this result is given by two theorems attributed to Lyapunov (Meirovitch (1970)). Specifically, for a conservative, natural system, sufficient conditions for the equilibrium to be unstable are given by the following results.

- Lyapunov's theorem.
  The equilibrium q^* is unstable if the potential energy function is a maximum at q^*.
- Extended Lyapunov's theorem.
  If the potential energy function has no minimum at the equilibrium point, the equilibrium is unstable.

Example 6.8.
In the system shown on the right the mass m_1 is attached to a spring with stiffness $k > 0$. The displacement of the spring, as measured from its free length is x. The mass m_2 is attached to m_1 via a inertialess rod with length l. The angular displacement of the rod is given by the angle θ, and g is the acceleration due to gravity.
Using x and θ as the generalized coordinates the potential energy for the system is

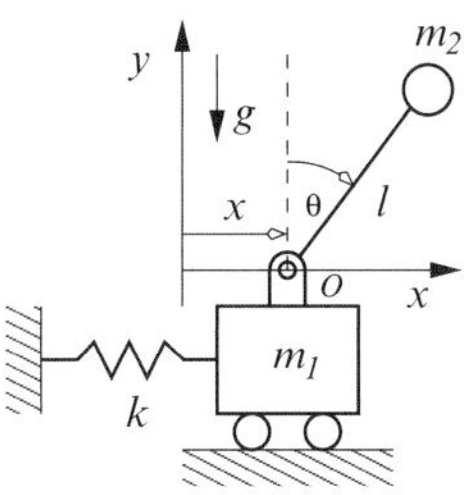

$$V = \frac{1}{2}kx^2 + m_2 gl\cos\theta.$$

The first term is the potential energy due to the spring force, and the second term is the potential energy due to the force of gravity on m_2.

At the equilibrium the potential energy function is stationary, hence we require

$$\frac{\partial V}{\partial x} = kx = 0,$$
$$\frac{\partial V}{\partial \theta} = -m_2 gl\sin\theta = 0.$$

These equations are satisfied at the points

$$\begin{bmatrix} x^* \\ \theta^* \end{bmatrix} = \begin{bmatrix} 0 \\ \pm n\pi \end{bmatrix}, \quad n = 0, 1, 2, \cdots.$$

We will examine the two unique equilibrium positions, i.e., (a) $x^* = 0$, $\theta^* = 0$, and (b) $x^* = 0$, $\theta^* = \pi$. In either case the Hessian of the potential energy function is

$$H(x,\theta) = \begin{bmatrix} \dfrac{\partial^2 V}{\partial x^2} & \dfrac{\partial^2 V}{\partial x \partial \theta} \\ \dfrac{\partial^2 V}{\partial \theta \partial x} & \dfrac{\partial^2 V}{\partial \theta^2} \end{bmatrix} = \begin{bmatrix} k & 0 \\ 0 & -m_2 gl\cos\theta \end{bmatrix}.$$

Therefore, at the first equilibrium point the Hessian is

$$H(0,0) = \begin{bmatrix} k & 0 \\ 0 & -m_2 gl \end{bmatrix}.$$

Since $k > 0$, $m_2 > 0$, $g > 0$ and $l > 0$, the Hessian matrix, $H(0,0)$, is indefinite, i.e., $x^* = 0$, $\theta^* = 0$ is a saddle point of the potential energy function. By the extended Lyapunov theorem it can be concluded that this equilibrium is unstable. Intuitively this can be shown to be true since, at

$\theta = 0$ any small perturbation in the angle θ will cause the rod to tip over, and thus move 'far' away from the equilibrium.

At the second equilibrium point the Hessian is

$$H(0, \pi) = \begin{bmatrix} k & 0 \\ 0 & m_2 g l \end{bmatrix},$$

which is a positive definite matrix. In this case we can use the Lagrange stability theorem to argue that the equilibrium point $x^* = 0$, $\theta^* = \pi$ is stable.

6.2 System Simulation

One of the attractive features of the modeling technique described in this book is that the process can be readily automated on computer systems, so as to obtain the equations of motion and simulate the behavior of the model. In this section we describe two computer programs that facilitate the numerical solution of the Lagrangian DAEs. We also present several simulation examples using these programs.

The program `ldaetrans`, which is described in section 6.2.1, takes a system described by q, f, e^s, e, s, T^*, V, D, ϕ, ψ, Γ, and Σ, and generates the LDAEs (4.8) via symbolic differentiation of the appropriate terms. This program also produces a file that can be used to simulate the equations of motion using MATLAB or Octave.

The function `ride`, which is described in section 6.2.2, is an implementation of an implicit Runge-Kutta method for the solution of implicit differential equations (IDEs). The solution technique is based on the 3-stage Radau IIA implicit Runge-Kutta method as presented in Chapter 5. The function is written in the MATLAB/Octave programming language.

Both computer codes can be found on the web site:

`http://abs-5.me.washington.edu/pub/fabien/asd`

6.2.1 The translator `ldaetrans`

The program `ldaetrans` translates an ASCII input file description of the model into the LDAEs (4.8), and it also produces files suitable for the numerical integration of the model using MATLAB or Octave. (Here, we used the function `ride` to integrate the LDAEs.) The program `ldaetrans` is written in ANSI C, and has been compiled and executed on several UNIX based operating systems including Linux, Solaris, Mac OSX, OpenBSD and FreeBSD.

The input to this translator takes advantage of the fact that the modeling approach developed in this book leads to a set of highly structured equations. The ASCII input file uses keywords to enter the system variables, scalar energy functions, constraints, etc. A list of the keywords recognized by the program are shown in Table 6.1. These keywords are used as follows.

DisplacementVariables	FlowVariables	EffortVariables
DynamicVariables	KineticCoEnergy	PotentialEnergy
DissipationFunction	GeneralizedEfforts	DisplacementConstraints
FlowConstraints	EffortConstraints	DynamicConstraints
InitialDisplacements	InitialFlows	InitialEfforts
InitialDynamic	InitialTime	FinalTime
OutputPoints		

Table 6.1 `ldaetrans` keywords

- `DisplacementVariables`: This keyword is used to input the displacement variables in the model. For example, the displacement variables $q = [q_1,\ q_2,\ q_3]^T$ are entered using the statement

  ```
  DisplacementVariables = [q1,q2,q3];
  ```

  Note that, all vector inputs to the program are enclosed in square brackets, and the elements of the vector are separated by commas. In addition, all input lines to the program end with a semicolon.
  Similar input statements are used to enter the flow variables f, the effort variables e, and the dynamic variables s, via the keywords `FlowVariables`, `EffortVariables` and `DynamicVariables`, respectively.
- KineticCoEnergy: This keyword is used to enter the kinetic coenergy for the system. For example, $T^* = \frac{1}{2}mf_1^2$ can be entered using the statement

  ```
  KineticCoEnergy = (m*f1*f1)/2;
  ```

  Similar statements are used to enter the potential energy V, and the dissipation function D via the keywords `PotentialEnergy` and `DissipationFunction`, respectively.
- `GeneralizedEfforts`: The efforts $e^s = [e_1^s,\ e_2^s,\ \cdots,\ e_n^s]^T$ are entered using this keyword. For example, consider a system with displacements $q = [q_1,\ q_2,\ q_3]^T$, and the virtual work due to the applied efforts is $\delta\mathcal{W} = \tau\,\delta q_1 + F\,\delta q_3$. In this case $e^s = [\tau,\ 0,\ F]^T$, and can be described using the statement

  ```
  GeneralizedEfforts = [tau,0,F];
  ```

  Note that the efforts e^s are distinct from the efforts associated with the effort constraints Γ, i.e., the effort variables e. Also, in a strict sense these are not 'generalized efforts' since, the model may have more displacement variables than degrees of freedom. Here, we simply use the term `GeneralizedEfforts` to distinguish e^s from e.

- `DisplacementConstraints`: The displacement constraints ϕ are entered using this keyword. For example, the constraints $\phi_1 = q1^2 + q_2^2 - L_1^2$, and $\phi_2 = (q_1 - q_3)^2 + (q_2 - q_4)^2 - L_2^2$ can be entered using the statement

```
DisplacementConstraints = [q1*q1 + q2*q2 - L1*L1,
    (q1-q3)*(q1-q3)+(q2-q4)*(q1-q3)-L2*L2];
```

  The flow constraints ψ, the effort constraints Γ, and the dynamic constraints Σ can be entered using the keywords `FlowConstraints`, `EffortConstraints`, and `DynamicConstraints`, respectively.
- InitialDisplacements: This keyword is used to specify the initial displacement to be used in the integration of the equations of motion. For example, the initial conditions $q(t_0) = [0, 1, 3]^T$ can be entered using the statement

```
InitialDisplacements = [0,1,3];
```

  The initial flows $f(t_0)$, the initial effort variables $e(t_0)$, and the initial dynamic variables $s(t_0)$, are entered using the keywords `InitialFlows`, `InitialEfforts`, and `InitialDynamic`, respectively.
- `InitialTime`: This keyword is used to specify the initial time to be used in the numerical integration of the equations of motion. For example an initial time $t_0 = 1.2$ is entered using the statement

```
InitialTime = 1.2;
```

  Similarly, the final time for the numerical integration is entered using the keyword `FinalTime`.
- `OutputPoints`: This keyword is used to specify the number of points that are output form the numerical solver in the interval `InitialTime`$\leq t \leq$`FinalTime`.

To illustrate an input file for the program `ldaetrans` let us reconsider Example 4.10.

Example 6.9.

This example models a DC motor with proportional plus integral feedback control. The `ldaetrans` input file for the model is shown in Listing 1 below.

```
Kt = 1;
Kb = 1;
L = 0.001;
I = 0.1;
R = 100;
b = 3;
```

```
7  theta_d = 1;
8  Kp = -100;
9  Ki = -10;
10
11 DisplacementVariables = [q,theta];
12 FlowVariables = [qdot,thetadot];
13 EffortVariables = [v,vb,tau];
14 DynamicVariables = [s];
15
16 KineticCoEnergy = (L*qdot*qdot + I*thetadot*thetadot)/2;
17
18 DissipationFunction = (R*qdot*qdot + b*thetadot*thetadot)/2;
19
20 EffortConstraints = [vb-Kb*thetadot,
21                      tau-Kt*qdot,
22                      v-Kp*(theta-theta_d)-Ki*s];
23
24 DynamicConstraints = [theta-theta_d];
25
26 GeneralizedEfforts = [(v-vb),tau];
27
28 InitialDisplacements = [0,0];
29 InitialFlows = [0,0];
30 InitialEfforts = [0,0,0];
31 InitialDynamic = [0];
32
33 InitialTime = 0;
34 FinalTime = 50;
35 OutputPoints = 1000;
```

Listing 1. The DC motor `ldaetrans` input file.

Lines 1 through 9 specify the various model parameters. The other lines in this input file have obvious meaning when compared with Example 4.10.

Using this file as the input, the program `ldaetrans` generates three ASCII files. One file gives a description of the LDAEs, and the other two files are inputs by the program `ride`, which is used to obtain a numerical solution to the problem.

The LDAEs description generated by `ldaetrans` is shown in Listing 2.

```
LDAEs

Kt = 1
Kb = 1
L = 0.001
I = 0.1
```

```
R = 100
b = 3
theta_d = 1
Kp = -100
Ki = -10

0 = q_ddt-qdot
0 = theta_ddt-thetadot
0 = 0.5*(L+L)*qdot_ddt+0.5*(R*qdot+R*qdot)-((v-vb))
0 = 0.5*(I+I)*thetadot_ddt+0.5*(b*thetadot+b*thetadot)-(tau)
0 = s_ddt-(theta-theta_d)
0 = vb-Kb*thetadot
0 = tau-Kt*qdot
0 = v-Kp*(theta-theta_d)-Ki*s
```

Listing 2. Output from `ldaetrans` for Example 6.9.

In this listing, the postscript `_ddt` is used to indicate differentiation with respect to time. Therefore, the term `q_ddt` is equivalent to $\frac{dq}{dt}$. This result should be compared with that obtained in Example 4.10. (Notice that `ldaetrans` computes derivatives accurately, but does not simplify expressions.)

The other two files generated by `ldaetrans` are of the form `IDEFUN` and `IDEJAC` for this system model. The structure of these files is discussed in the next section.

6.2.2 The solver `ride`

The function `ride` is used to simulate the behavior of the dynamic systems described in the remainder of this chapter. In this section we will explain the basic usage of the function.

The function `ride` is a MATLAB/Octave function which can be called using one of the following statements;

```
[TOUT,YOUT,INFO] = ride('IDEFUN','IDEJAC',TSPAN,Y0,YP0)  or
[TOUT,YOUT,INFO] = ride('IDEFUN','IDEJAC',TSPAN,Y0,YP0,...
                                                OPTIONS)  or
[TOUT,YOUT,INFO] = ride('IDEFUN','',TSPAN,Y0,YP0)  or
[TOUT,YOUT,INFO] = ride('IDEFUN','',TSPAN,Y0,YP0,OPTIONS)
```

Here,

`TOUT` is is a $(N \times 1)$ vector of output times at which the solution is computed.

`YOUT` is a $(N \times n)$ matrix of solutions to the IDEs, where n is the dimension of the IDEs.

`INFO` is a structure that provides some statistics related to the solution algorithm. In particular,

`INFO.nfun` is the number of function evaluations.
`INFO.njac` is the number of Jacobian evaluations.
`INFO.naccept` is the number of successful steps.
`INFO.nreject` is the number of failed steps.

`IDEFUN` is a function that returns the $(n \times 1)$ vector `res = Phi(y,yp,t)`, where `y` is the $(n \times 1)$ state vector of the IDEs, `yp` = `dy/dt` is the $(n \times 1)$ state derivative, `t` is the time, and `Phi` is the $(n \times 1)$ system of IDEs. This function has the form

```
function res = IDEFUN(y,yp,t)
```

In this description `Phi(y, yp, t)` $= \Phi(y, \dot{y}, t)$ are the IDEs, `y` $= y(t)$ and `yp` $= \dot{y}$.

`IDEJAC` is a function that computes the $(n \times n)$ derivatives (i.e., the Jacobians) `J` = `Phi/dy` $= d\Phi/dy$, and $(n \times n)$ matrix `M` = `dPhi/dyp` $= d\Phi/\dot{y}$. This function has the form

```
function [J,M] = IDEJAC(y,yp,t)
```

where

$$\mathtt{J} = \begin{bmatrix} d\Phi_1/dy_1 & d\Phi_1/dy_2 & \cdots & d\Phi_1/dy_n \\ d\Phi_2/dy_1 & d\Phi_2/dy_2 & \cdots & d\Phi_2/dy_n \\ \vdots & \vdots & \ddots & \vdots \\ d\Phi_n/dy_1 & d\Phi_n/dy_2 & \cdots & d\Phi_n/dy_n \end{bmatrix}$$

and

$$\mathtt{M} = \begin{bmatrix} d\Phi_1/d\dot{y}_1 & d\Phi_1/d\dot{y}_2 & \cdots & d\Phi_1/d\dot{y}_n \\ d\Phi_2/d\dot{y}_1 & d\Phi_2/d\dot{y}_2 & \cdots & d\Phi_2/d\dot{y}_n \\ \vdots & \vdots & \ddots & \vdots \\ d\Phi_n/d\dot{y}_1 & d\Phi_n/d\dot{y}_2 & \cdots & d\Phi_n/d\dot{y}_n \end{bmatrix}.$$

If `IDEJAC` is not included in the argument list for `ride` then `J` and `M` are computed via a finite difference approximation.

`TSPAN` This vector defines the interval of integration, and time points where the solutions to the DAEs are computed. The `TSPAN` is an $(S \times 1)$ vector of the form,

```
TSPAN = [t_1;t_2;...;t_S]
```

where the times `t_1, t_2, ..., t_S` is a monotone increasing or decreasing) sequence of time points. The solver integrates from `t_1` to `t_S`, and computes the solution to the IDEs at all points `t_1, t_2, ..., t_S`.

`Y0` This $(n \times 1)$ vector gives the initial state for the IDEs, i.e., `Y0` $= y(t_1)$.

YP0 This $(n \times 1)$ vector gives the initial state derivative for the IDEs, i.e., YP0 $= \dot{y}(t_1)$. Moreover, it is assumed that these initial conditions are consistent, i.e., `res` $=$ `Phi(Y0,YP0,TSPAN(1))` $= 0$. Note that the initial time t_1 $=$ `t_1` in the TSPAN vector.

OPTIONS This is a structure that provides various options for the numerical solution algorithm. The default values for the members of this structure can be ascertained using the function

```
OPTIONS = ride_options()
```

The function options can be assigned using the function

```
OPTIONS = ride_set_option(OPTIONS,'OPTION',value)
```

The options for the function ride and the default values are as follows;

OPTIONS.ATOL is the absolute error tolerance. (Default 10^{-6}).

OPTIONS.RTOL is the relative error tolerance. (Default 10^{-6}). Note that ATOL and RTOL can also be $(n \times 1)$ vectors.

OPTIONS.INITIAL_STEP_SIZE is the initial step size. (Default 10^{-8}).

OPTIONS.MAX_STEPS is maximum number of steps that is performed by the function. (Default 1000).

OPTIONS.DIFF_INDEX is a $(n\times 1)$ vector that indicates the differentiation index of the i-th state variable, y_i. (Default [], i.e., all state variables are assumed to be index 0).

OPTIONS.T_EVENT is a $(P \times 1)$ vector of time points of the form

```
OPTIONS.T_EVENT = [te_1;te_1;te_2;...;te_P]
```

where the times `te_1, te_2, ..., te_P` is a monotone increasing (or decreasing) sequence. In the algorithm the step sizes are selected such that the event times, `T_EVENT(k)`, are at the end (or start) of the integration interval. This is useful for modeling discontinuous time varying inputs or, discontinuous implicit differential equations.

Example 6.10.
The program `ldaetrans` produced m-files of the form IDEFUN and IDEJAC for the DC motor model described in Example 4.10 and Example 6.10. These files are shown below. The IDEFUN file has the name `lex14_ride_ldae.m`, and the IDEJAC file has the name `lex14_ride_jacobian.m`.

```
function [res_] = lex14_ride_ldae(x_,xdot_,t)

q = x_(1);
q_ddt = xdot_(1);
theta = x_(2);
theta_ddt = xdot_(2);
qdot = x_(3);
```

```
qdot_ddt = xdot_(3);
thetadot = x_(4);
thetadot_ddt = xdot_(4);
s = x_(5);
s_ddt = xdot_(5);
v = x_(6);
vb = x_(7);
tau = x_(8);
Kt = 1;
Kb = 1;
L = 0.001;
I = 0.1;
R = 100;
b = 3;
theta_d = 1;
Kp = -100;
Ki = -10;

res_ = zeros(size(x_));
res_ = [
q_ddt-qdot;
theta_ddt-thetadot;
0.5*(L+L)*qdot_ddt+0.5*(R*qdot+R*qdot)-((v-vb));
0.5*(I+I)*thetadot_ddt+0.5*(b*thetadot+b*thetadot)-(tau);
s_ddt-(theta-theta_d);
vb-Kb*thetadot;
tau-Kt*qdot;
v-Kp*(theta-theta_d)-Ki*s;
];
```

The `IDEFUN` file generated by `ldaetrans` for Example 6.9.

```
function [dfdx_,dfdxdot_] = lex14_ride_jacobian(x_,xdot_,t)

q = x_(1);
q_ddt = xdot_(1);
theta = x_(2);
theta_ddt = xdot_(2);
qdot = x_(3);
qdot_ddt = xdot_(3);
thetadot = x_(4);
thetadot_ddt = xdot_(4);
s = x_(5);
s_ddt = xdot_(5);
v = x_(6);
vb = x_(7);
```

```
tau = x_(8);
Kt = 1;
Kb = 1;
L = 0.001;
I = 0.1;
R = 100;
b = 3;
theta_d = 1;
Kp = -100;
Ki = -10;

dfdx_ = zeros(8,8);
dfdxdot_ = zeros(8,8);
dfdxdot_(1,1) = 1.0;
dfdxdot_(3,3) = 0.5*(L+L);
dfdxdot_(2,2) = 1.0;
dfdxdot_(4,4) = 0.5*(I+I);
dfdxdot_(5,5) = 1.0;
dfdx_(1,3) = -1;
dfdx_(2,4) = -1;
dfdx_(3,3) = 0.5*(R+R);
dfdx_(3,6) = -1;
dfdx_(3,7) = 1;
dfdx_(4,4) = 0.5*(b+b);
dfdx_(4,8) = -1;
dfdx_(5,2) = -1;
dfdx_(6,4) = -Kb;
dfdx_(6,7) = 1;
dfdx_(7,3) = -Kt;
dfdx_(7,8) = 1;
dfdx_(8,2) = -Kp;
dfdx_(8,5) = -Ki;
dfdx_(8,6) = 1;
```

The IDEJAC file generated by `ldaetrans` for Example 6.9.

Note that one can use `ride` independent of the program `ldaetrans`. In fact in the first two examples in the next section we show how to construct the files IDEFUN and IDEJAC without using `ldaetrans`.

6.2.3 System simulation examples

Example 6.11.

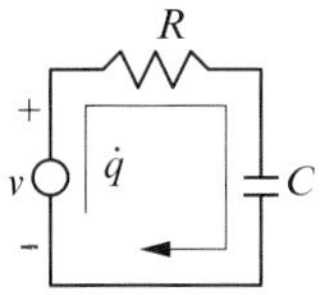

The system shown here is simple RC circuit, with resistance $R = 5 \times 10^3$ ohms, and capacitance $C = 10^{-6}$ farads. Also, in this example we will take the applied voltage to be a constant, $v(t) = 3$ volts, i.e., a step input. Using the charge, q, as the generalized displacement we get that Lagrange's equation of motion for the system is

$$R\dot{q} + \frac{q}{C} = v. \tag{a}$$

If we assume that the initial charge is $q(0) = q_0$, then the analytical solution to this linear first-order differential equation is

$$q(t) = Cv(1 - e^{-t/\tau}) + q_0 e^{-t/\tau}, \tag{b}$$

where $\tau = RC$ is called the time constant.

To solve the differential equation (a) using the function `ride` we use the following procedure.

- Construct an 'm-file' that describes the differential equation (a) as an implicit differential equation (IDE). Here, m-files are MATLAB/Octave script files. The m-file `ex1_ide.m`, shown in Listing 1, provides the IDEs for this problem. Line 1 in this listing declares the function, its output and input variables. Lines 3, 4, 5 initializes the model parameters. Line 7 computes the 'residual' of the IDEs at time t, i.e., `Phi` $= \Phi(y, \dot{y}, t)$. For this problem `y(1)` $= y = q$, and `yp(1)` $= \dot{y} = \dot{q}$.
- Construct an m-file that computes the Jacobian of the IDEs. This m-file is called `ex1_jacobian.m`, and its content is shown in Listing 2. Line 1 of this listing gives the declaration of the function. The Jacobian `J` $= d\Phi/dy = 1/C$ is on line 7, and the term `M` $= d\Phi/d\dot{y} = R$ is on line 8.
- Construct an m-file that establishes the initial conditions, interval of integration, and options required by the function `ride`. Listing 3 shows the commands used to solve the IDEs using `ride`. The command on line 1 of this listing clears all variables and functions from the MATLAB/Octave interpreter.
  The model parameters are initialized on lines 2, 3, and 4.
  The integration interval is defined in the vector `tspan` on line 5. Here, `tspan` is defined as a vector of 100 equally spaced points from 0 to 0.05. Thus, we seek the solution to the IDE in the interval $0 \le t \le 0.05$ seconds.
  The initial charge is given by `y0` $= q(0) = 0$ on line 6.
  The initial current is given by `yp0` $= \dot{q}(0) = v/R$ on line 7. It is easy to verify that these initial conditions are consistent, i.e., $\Phi(y(0), \dot{y}(0), 0) = 0$.

The command on line 8 calls the solver, and the commands on line 9, 10 and 11 are used to plot the result.

The results of this simulation is shown in Fig. 6.1. The plot (i) shows the system response for the charge $q(t)$ that is computed by the function `ride`. The plot (ii) shows the absolute value of the error between the numerical solution and the analytical solution given by equation (b). As can be seen the error is much less than the default error tolerance, (10^{-6}), used by the solver.

```
function Phi = ex1_ide(y,yp,t)

R = 5.0e3;
C = 1.0e-6;
v = 3.0;

Phi = R*yp(1) + y(1)/C - v;

return;
```

Listing 1. The file `ex1_ide.m`.

```
function [J,M] = ex1_jacobian(y,yp,t)

R = 5.0e3;
C = 1.0e-6;
v = 3.0;

J = [1.0/C];
M = [R];

return;
```

Listing 2. The file `ex1_jacobian.m`.

```
1  clear all
2  R = 5.0e3;
3  C = 1.0e-6;
4  v = 3.0;
5  tspan = linspace(0;0.05,100)';
6  y0 = [0];
7  yp0 = [v/R];
8  [T,Y] = ride('ex1_ide','ex1_jacobian',tspan,y0,yp0);
9  plot(T,Y(:,1));
10 xlabel('time');
11 ylabel('charge, q');
```

Listing 3. The main script.

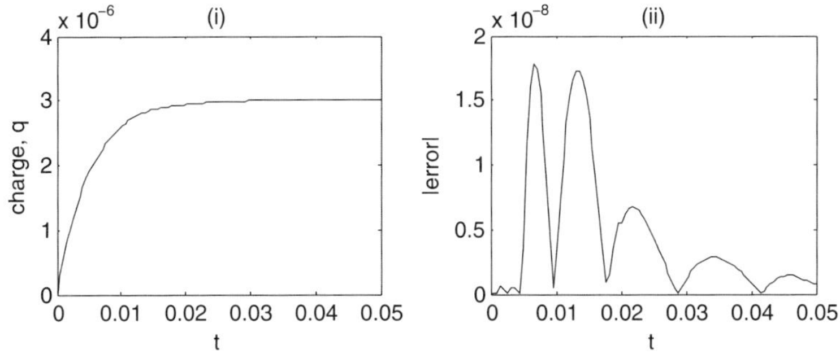

Fig. 6.1 Example 6.11

Example 6.12.

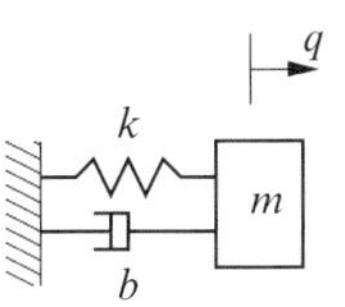

In this example we will consider the initial condition response of the linear mass-spring-damper system shown here. If we use q as the generalized displacement for the system then, Lagrange's equation of motion can be written as

$$m\ddot{q} + b\dot{q} + kq = 0. \tag{a}$$

If we define the state variables $y_1 = q$ and $y_2 = \dot{q}$, equation (a) can be rewritten as the implicit differential equations

$$\begin{aligned} 0 &= \dot{y}_1 - y_2, \\ 0 &= m\dot{y}_2 + by_2 + ky_1. \end{aligned} \tag{b}$$

The parameters for this model are $m = 0.3$ kg, $k = 45$ N/m, and $b = 0.75$ N-s/m. We will use the function `ride` to compute the response of this system with the initial condition $y_1(0) = 10^{-2}$ m, and $y_2 = 0$ m/s. The m-files for this model are shown in the listing below. Here, the function `ex2_ide` computes the implicit differential equations (b), and the file `ex2_jacobian` computes the Jacobian of the IDEs. The file `ex2.m` sets the initial conditions, the solver options, and calls the function `ride`. This problem is solved in the interval $0 \leq t \leq 3$ seconds.

```
%
% file: ex2_ide.m
%
function Phi = ex2_ide(y,yp,t)


m=0.3;
k=45.0;
```

```
b=0.75;

Phi = zeros(2,1);
Phi(1,1) = yp(1)-y(2);
Phi(2,1) = m*yp(2) + b*y(2) + k*y(1);

return;
% end file: ex2_ide.m

%
% file: ex2_jacobian.m
%
function [J,M] = ex2_jacobian(y,yp,t)

m=0.3;
k=45.0;
b=0.75;

M = zeros(2,2);
J = zeros(2,2);

M(1,1) = 1;
M(2,2) = m;
J(1,2) = -1;
J(2,1) = k;
J(2,2) = b;

return;
% end file: ex2_jacobian.m

%
% file: ex2.m
%
clear all
m=0.3;
k=45.0;
b=0.75;
tspan = linspace(0,3,100)';
y0 = [1.0e-2;0];
yp0 = [y0(2);-k*y0(1)/m];
[T,Y] = ride('ex2_ide','ex2_jacobian',tspan,y0,yp0);
% end file: ex2.m
```

The differential equation (b) can be written as

$$\ddot{y}_1 + 2\zeta\omega\dot{y}_1 + \omega^2 y_1 = 0,$$

where $\omega^2 = k/m$, $\zeta = (b/m)/(2\omega)$, $\omega_d^2 = \omega_n^2(1 - \zeta^2)$. If $\zeta < 1$ then, it can be shown that the analytical solution to this differential equation is

$$y_1(t) = y_1(0)e^{\zeta\omega t}\left(\cos\omega_d t + \frac{\zeta}{\sqrt{1-\zeta^2}}\sin\omega_d t\right) + \frac{\dot{y}_1(0)}{\omega_d}e^{-\zeta\omega t}\sin\omega_d t.$$

Recall that the condition $\zeta < 1$ implies that the system response is 'underdamped'. If $\zeta = 1$ then the system response is said to be 'critically damped'. If $\zeta > 1$ the system response is said to be 'overdamped'.

For the model parameters used here it can be seen that the system is underdamped. The plots (i) and (ii) in Fig. 6.2 show the numerically computed displacement and velocity of the system. The plots (iii) and (iv) in Fig. 6.2 show the absolute value of the error between the analytical solution and the computed results. These figures indicate that the error in the numerical solution is well below the specified error tolerance.

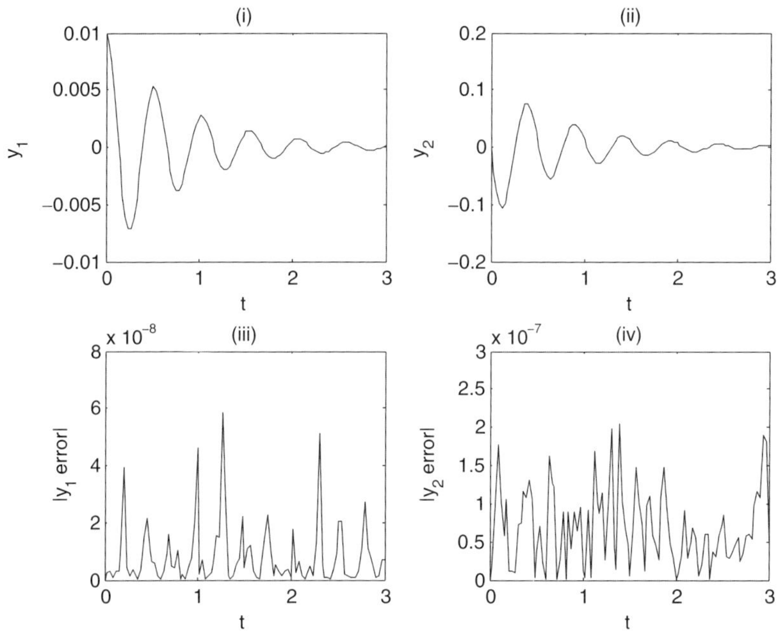

Fig. 6.2 Example 6.12

Example 6.13.

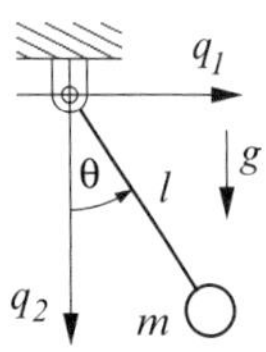

This example simulates the behavior of the simple pendulum shown here using two different models of the system. In the first model the angle θ is taken as the generalized displacement. In the second model the variables q_1 and q_2 are used as the displacement coordinates for the system. Using θ as the generalized displacement we can show that Lagrange's equation of motion is

$$ml^2\ddot{\theta} + mgl\sin\theta = 0. \qquad (a)$$

If we define the state variables $y_1 = \theta$ and $y_2 = \dot{\theta}$ then, (a) can be rewritten as the implicit differential equations

$$\begin{aligned} 0 &= \dot{y}_1 - y_2, \\ 0 &= \dot{y}_2 + \frac{g}{l}\sin y_1. \end{aligned} \qquad \text{(b)}$$

The parameters for this model are $m = 0.25$ kg, $l = 0.3$ m, and $g = 9.8$ m/s^2. The initial conditions are $\theta = y_1 = \pi/2$ and $\dot{\theta} = y_2 = 0$.

The m-files required to simulate this version model are shown below.

```
%
% file: ex3a_ide.m
%
function Phi = ex3a_ide(y,yp,t)
m = 0.25;
L = 0.3;
g = 9.8;
Phi = zeros(2,1);
Phi(1,1) = yp(1) - y(2);
Phi(2,1) = yp(2) + (g/L)*sin(y(1));
return;
```

```
%
% file: ex3a_jacobian.m
%
function [J,M] = ex3a_jacobian(y,yp,t)
m = 0.25;
L = 0.3;
g = 9.8;
M = zeros(2,2);
J = zeros(2,2);
M(1,1) = 1;
M(2,2) = 1;
J(1,2) = -1;
J(2,1) = (g/L)*cos(y(1));
return;
```

```
%
% file ex3a.m
%
m = 0.25;
L = 0.3;
g = 9.8;
tspan = linspace(0,3,100)';
y0 = [pi/2.0;0];
yp0 = [y0(1);-(g/L)*sin(y0(1))];
[T,Y] = ride('ex3a_ide','ex3a_jacobian',tspan,y0,yp0);
```

If we use q_1 and q_2 as the displacements, the Lagrange's equations of motion are

$$\begin{aligned} m\ddot{q}_1 + 2\lambda q_1 &= 0, \\ m\ddot{q}_2 + 2\lambda q_2 - mg &= 0, \\ q_1^2 + q_2^2 - l^2 &= 0. \end{aligned} \tag{b}$$

The system (b) can be rewritten in state variable form using $y_1 = q_1$, $y_2 = q_2$, $y_3 = \dot{q}_1$, $y_4 = \dot{q}_2$ and $y_5 = \lambda$. This yields the index-3 DAEs

$$\begin{aligned} 0 &= \dot{y}_1 - y_3, \\ 0 &= \dot{y}_2 - y_4, \\ 0 &= m\dot{y}_3 + 2y_5 y_1, \\ 0 &= m\dot{y}_4 + 2y_5 y_2 - mg, \\ 0 &= y_1^2 + y_2^2 - l^2. \end{aligned} \tag{c}$$

Finally, we can put (c) in the GGL stabilized index-2 form (see Section 5.3.3). This is accomplished by adding the multiplier y_6 and adjoining the the derivative of the displacement constraint to get

$$\begin{aligned} 0 &= \dot{y}_1 - y_3 + 2y_1 y_6, \\ 0 &= \dot{y}_2 - y_4 + 2y_2 y_6, \\ 0 &= m\dot{y}_3 + 2y_1 y_5, \\ 0 &= m\dot{y}_4 + 2y_2 y_5 - mg, \\ 0 &= y_1^2 + y_2^2 - l^2, \\ 0 &= 2y_1 y_3 + 2y_2 y_4. \end{aligned} \tag{d}$$

The m-files used to solve the system (d) are shown below. Note that in the file `ex3b.m` we use the optional input `DIFF_INDEX` to specify the differentiation index of the variables in the system. In particular we assign differentiation index 2 to the multipliers y_5 and y_6. All the other variables are assigned differentiation index 0. Also, the initial step size is set to 10^{-6} in this case, since this adjustment leads to fewer errors in the simplified Newton's iteration.

```
%
% file ex3b_ide.m
%
```

```
function Phi = ex3b_ide(y,yp,t)
m = 0.25;
L = 0.3;
g = 9.8;
Phi = zeros(6,1);
Phi(1) = yp(1)-y(3)+2*y(1)*y(6);
Phi(2) = yp(2)-y(4)+2*y(2)*y(6);
Phi(3) = m*yp(3)+2*y(1)*y(5);
Phi(4) = m*yp(4)+2*y(2)*y(5)-m*g;
Phi(5) = y(1)^2+y(2)^2-L^2;
Phi(6) = y(1)*y(3)+y(2)*y(4);
return;

%
% file ex3b_jacobian.m
%
function [J,M] = ex3b_jacobian(y,yp,t)
m = 0.25;
L = 0.3;
g = 9.8;
M = zeros(6,6);
J = zeros(6,6);
M(1,1) = 1; M(2,2) = 1; M(3,3) = m; M(4,4) = m;
J(1,2) = 2*y(6); J(1,3) = -1; J(1,6) = 2*y(1);
J(2,2) = 2*y(6); J(2,4) = -1; J(2,6) = 2*y(2);
J(3,1) = 2*y(5); J(3,5) = 2*y(1);
J(4,2) = 2*y(5); J(4,5) = 2*y(2);
J(5,1) = 2*y(1); J(5,2) = 2*y(2);
J(6,1) = y(3); J(6,2) = y(4); J(6,3) = y(1); J(6,4) = y(2);
return;

%
% file ex3b.m
%
m = 0.25;
L = 0.3;
g = 9.8;
yb0 = [L;0;0;0;0;0];
ypb0 = [yb0(3);yb0(4);0;g;0;0];
options = ride_options();
options.DIFF_INDEX = [0;0;0;0;2;2];
options.INITIAL_STEP_SIZE = 1.0e-6;
[T,Yb] = ride('ex3b_ide','ex3b_jacobian',...
                    tspan,yb0,ypb0,options);
```

Some of the results from these simulations are shown in Fig. 6.3 and Fig. 6.4. Plots (i) and (ii) in Fig.6.3 show $\theta(t)$ and $\dot{\theta}(t)$, which are are computed using the model ex3a_ide. Plots (iii) through (vi) in Fig 6.3 show $q_1(t)$, $q_2(t)$, $\dot{q}_1(t)$, $\dot{q}_2(t)$, $y_5(t) = \lambda(t)$ and $y_6(t)$, which are are computed using the model ex3b_ide. Recall that y_6 is the multiplier associated with the GGL stabilized index-2 formulation, and this variable should be zero along the solution to the equations of motion.

The plots shown in Fig. 6.4 compare the solutions obtained using the models ex3a_ide and ex3b_ide. Here, we plot the errors:

$$\begin{aligned} e_1(t) &= |l \sin\theta(t) - q_1(t)|, \\ e_2(t) &= |l \cos\theta(t) - q_2(t)|, \quad \text{and} \\ e_3(t) &= |q_1(t)^2 + q_2(t)^2 - l^2|, \end{aligned}$$

where $\theta(t)$ is the result from the model ex3a_ide, and $q_1(t)$, $q_2(t)$ are the results from the model ex3b_ide. The small values obtained for $e_1(t)$ and $e_2(t)$ indicate that both models give essentially the same result. The small values obtained for e_3 shows that the displacement constraint is satisfied, to within the desired error tolerance (10^{-6}).

Example 6.14.

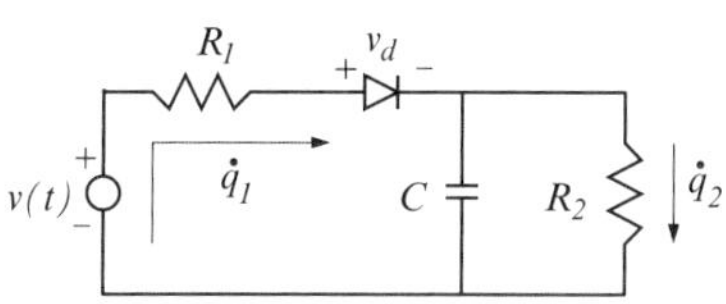

The power supply filter shown here has the following model parameters; $R_1 = 8$ ohm, $R_2 = 48$ ohm, $C = 3900 \times 10^{-6}$ farad, and $v(t) = 12 \sin 120\pi t$. Also, the current through the diode satisfies $\dot{q}_1 = I_s(e^{\alpha v_d} - 1)$, where the reverse saturation current is $I_s = 10^{-12}$ amp. The thermal voltage is $(1/\alpha) = 25 \times 10^{-3}$ volt, and v_d is the voltage across the diode. Using the charges q_1, q_2 as generalized displacements, and $f_1 = \dot{q}_1, f_2 = \dot{q}_2$ as the corresponding flows, Lagrange's equations of motion for this system can be written as

$$\begin{aligned} 0 &= \dot{q}_1 - f_1, \\ 0 &= \dot{q}_2 - f_2, \\ 0 &= R_1 f_1 + \frac{(q_1 - q_2)}{C} - v + v_d, \\ 0 &= R_2 f_2 - \frac{(q_1 - q_2)}{C}, \\ 0 &= f_1 - I_s(e^{\alpha v_d} - 1). \end{aligned} \tag{a}$$

We solve these equations of motion using the function ride with zero initial conditions for all the variables, in the time interval $0 \le t \le 1$. The results of this simulation are shown in Fig. 6.5.

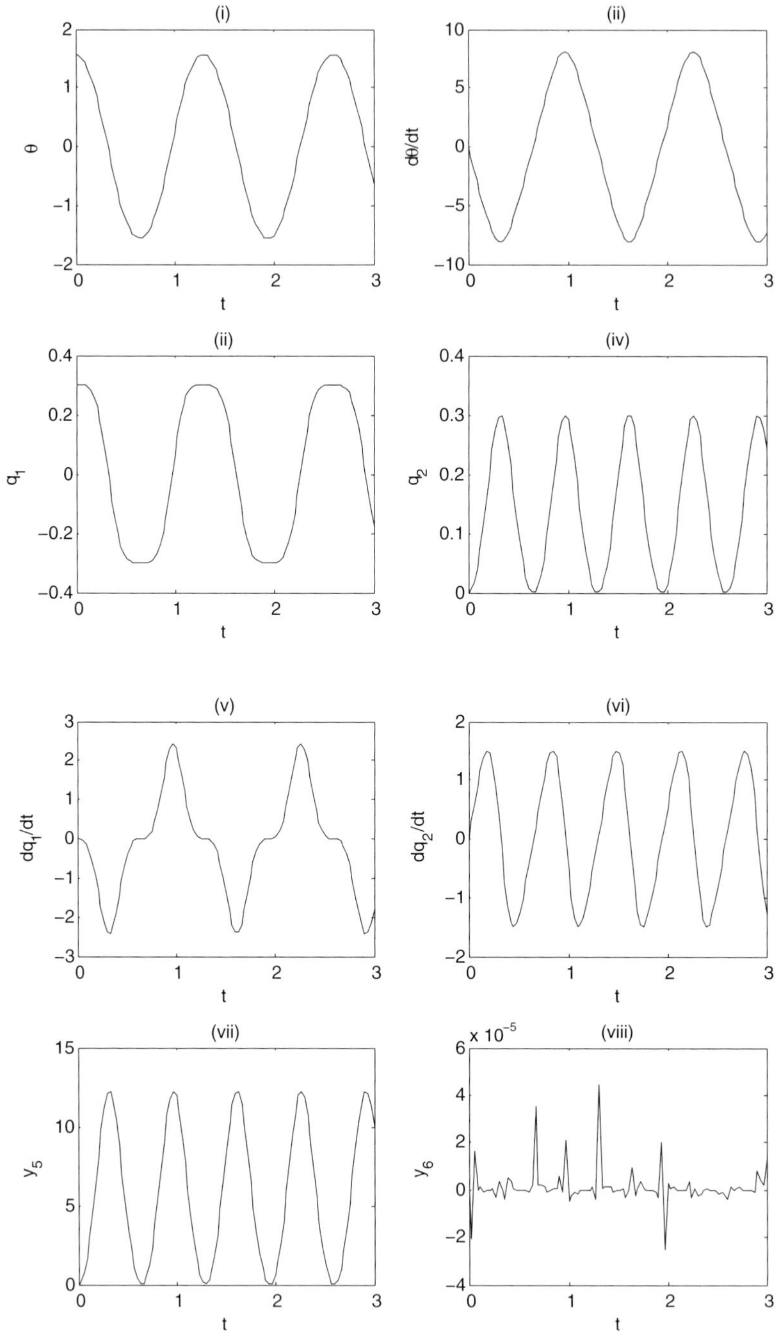

Fig. 6.3 Example 6.13

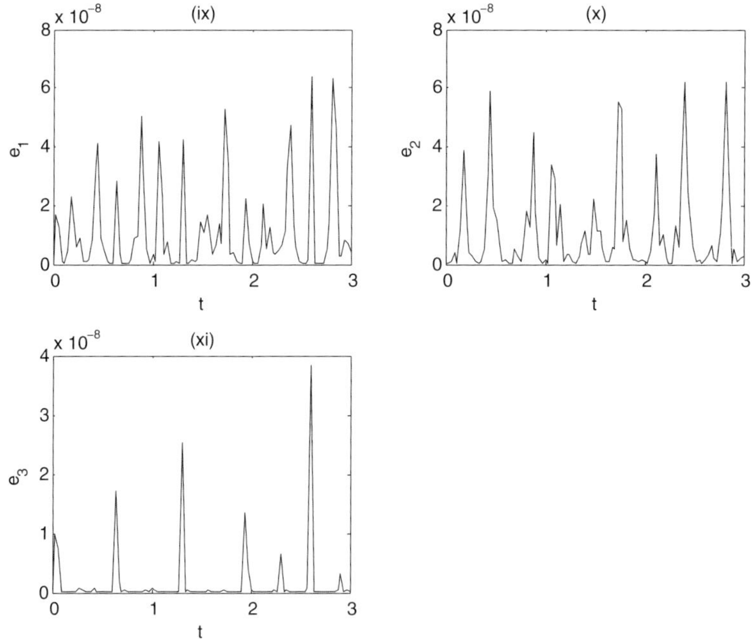

Fig. 6.4 Example 6.13, continued.

The plot (i) in Fig. 6.5 shows the voltage across the resistor R_2, i.e., $v_{R_2} = R_2 f_2$. This plot covers the entire solution interval, and shows that the system quickly approaches its steady-state response around $t = 0.3$ seconds. The plot (ii) in Fig. 6.5 shows the voltage v_{R_2} and the positive portion of the input voltage v in the interval $0.9 \leq t \leq 1$. This plot clearly shows the variation (or 'ripple') in the voltage v_{R_2}. The plot (iii) in Fig. 6.5 shows the input voltage, $v(t)$, and the voltage across the resistor R_1, i.e., $v_{R_1} = R_1 f_1$. Finally, the plot (iv) in Fig. 6.5 shows the input voltage, $v(t)$, and the voltage across the diode, i.e., $v_d(t)$.

These results clearly show the switching characteristics of the diode. In particular, it can be seen that when the diode voltage v_d attains a positive value the current $\dot{q}_1 > 0$, which yields a positive voltage v_{R_1}. On the other hand, when $v_d < 0$ we get $\dot{q}_1 = 0$, and hence, $v_{R_1} = 0$. The net effect is that the diode allows positive current flow when the voltage source $v(t)$ is positive, and no current flow when $v(t)$ is negative.

Example 6.15.
In this example we consider the modeling and simulation of the transistor

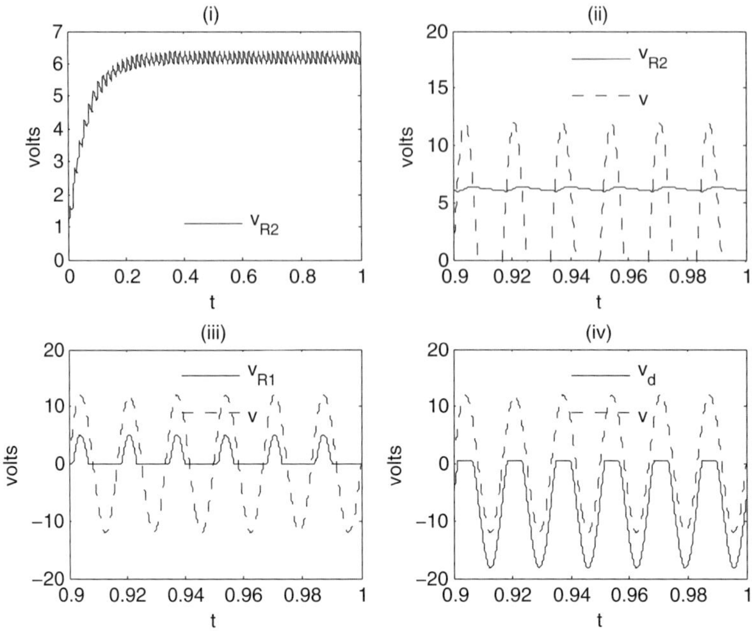

Fig. 6.5 Example 6.14

amplifier shown below. Here, the voltage source v_1 is constant, and provides power to the system. The voltage source $v_2(t)$ is an input signal that is amplified. The amplified signal drives the load resistor R_3.

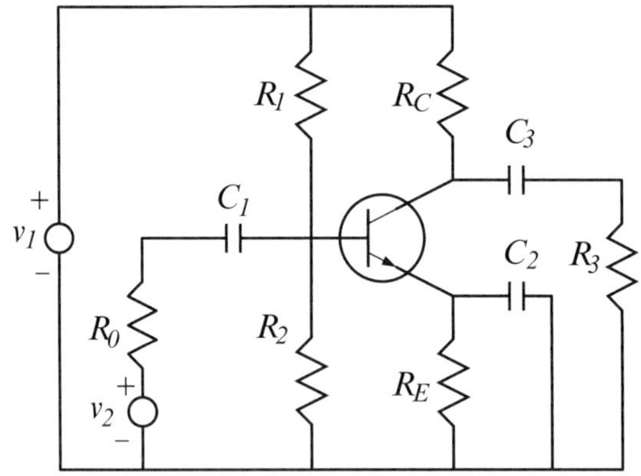

Amplifier circuit

Kinematic analysis:

The kinematic analysis of this circuit is performed using the node and flow assignment shown in the diagram below. Here, the variables $f_1,\ f_2,\ \cdots$, denote the flows (i.e., the currents) in the branches of the network. The displacements (i.e., the charges) corresponding to these flows are $q_1,\ q_2$, etc. This figure also shows an equivalent network model of the transistor. In this model the current at the collector, f_c, and emitter, f_e, are determined by the voltages v_{be} and v_{bc} as described by the Ebbers-Moll equations

$$\begin{bmatrix} f_e \\ f_c \end{bmatrix} - \begin{bmatrix} I_{ES} & -I_S \\ I_S & -I_{CS} \end{bmatrix} \begin{bmatrix} e^{\alpha v_{be}} - 1 \\ e^{\alpha v_{bc}} - 1 \end{bmatrix} = \begin{bmatrix} 0 \\ 0 \end{bmatrix}, \tag{a}$$

where I_{ES}, I_S and I_{CS} are constants for the transistor. (See Section 1.2.5.)

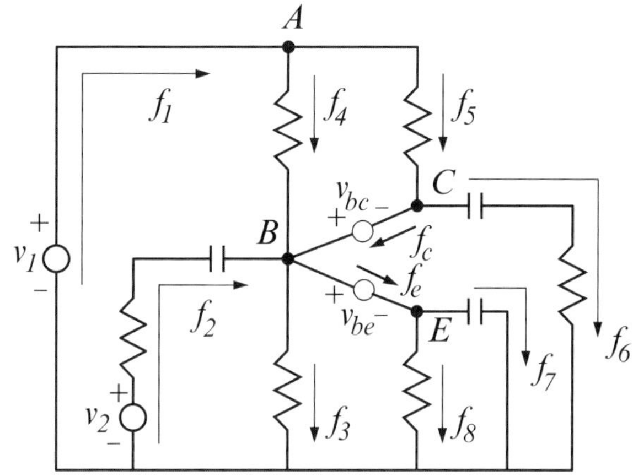

Node and flow assignment

This network has $B = 10$ branches, and $N = 5$ nodes $(A, B, C, E$ and the ground$)$. Hence, there are $6 = B - N + 1$ independent loops in the network. Since only 6 of the branch flows are independent we must construct 4 constraint equations to account for the 'extra' variables in the model. Applying Kirchhoff's current law at nodes A, B, C and E gives the following flow constraints;

$$\begin{aligned} \psi_1 &= f_1 - f_4 - f_5 &&= 0, \\ \psi_2 &= f_2 - f_3 + f_4 + f_c - f_e &&= 0, \\ \psi_3 &= f_5 - f_6 - f_c &&= 0, \\ \psi_4 &= f_e - f_7 - f_8 &&= 0. \end{aligned} \tag{b}$$

These constraints can be used to eliminate 4 of the flows from the model. Here however, we will retain all flow/displacement variables in the model.

Applied effort analysis:

The virtual work done by the applied efforts v_1 and v_2, and the implicit efforts v_{be} and v_{bc} is

$$\delta\mathcal{W} = v_1\delta q_1 + v_2\delta q_2 + v_{bc}\delta q_c - v_{be}\delta q_e.$$

Lagrange's equations:

The kinetic coenergy for this model is $T^* = 0$ because there are no inductors. The potential energy stored in the capacitors is

$$V = \frac{q_2^2}{2C_1} + \frac{q_6^2}{C_3} + \frac{q_7^2}{C_2}.$$

The dissipation function is

$$D = \frac{1}{2}\left(R_0 f_2^2 + R_1 f_4^2 + R_2 f_3^2 + R_C f_5^2 + R_E f_8^2 + R_3 f_6^2\right).$$

Therefore, the Lagrangian DAEs (4.8) for this system are

$$\begin{aligned}
&\dot{q}_1 - f_1 = 0, && \mu_1 - v_1 = 0, && f_1 - f_4 - f_5 = 0,\\
&\dot{q}_2 - f_2 = 0, && \frac{q_2}{C_1} + R_0 f_2 + \mu_2 - v_2 = 0, && f_2 - f_3 + f_4 + f_c - f_e = 0,\\
&\dot{q}_3 - f_3 = 0, && R_2 f_3 - \mu_2 = 0, && f_5 - f_6 - f_c = 0,\\
&\dot{q}_4 - f_4 = 0, && R_1 f_4 - \mu_1 + \mu_2 = 0, && f_e - f_7 - f_8 = 0,\\
&\dot{q}_5 - f_5 = 0, && R_C f_5 - \mu_1 + \mu_3 = 0, && f_e - I_{ES} I_F + I_s I_R = 0,\\
&\dot{q}_6 - f_6 = 0, && \frac{q_6}{C_3} + R_3 f_6 - \mu_3 = 0, && f_c - I_s I_F + I_{CS} I_R = 0,\\
&\dot{q}_7 - f_7 = 0, && \frac{q_7}{C_2} - \mu_4 = 0, && I_R - (e^{\alpha v_{bc}} - 1) = 0,\\
&\dot{q}_8 - f_8 = 0, && R_E f_8 - \mu_4 = 0, && I_F - (e^{\alpha v_{be}} - 1) = 0.\\
&\dot{q}_e - f_e = 0, && -\mu_2 + \mu_4 + v_{be} = 0, &&\\
&\dot{q}_c - f_c = 0, && \mu_2 - \mu_3 - v_{bc} = 0, &&
\end{aligned}$$

In these equations μ_1, μ_2, μ_3 and μ_4 are the Lagrange multipliers associated with the flow constraints (b). Also, the variables I_F and I_R are introduced solely to simplify the presentation of the equations.

System simulation:

The DAEs given above are solved using the code `ride` with the following model parameters; $R_0 = 10^3$ ohm, $R_1 = 30 \times 10^3$ ohm, $R_2 = 5.5 \times 10^3$ ohm, $R_3 = 10 \times 10^3$ ohm, $R_C = 2 \times 10^3$ ohm, $R_E = 0.5 \times 10^3$ ohm, $C_1 = 10 \times 10^{-6}$ farad, $C_2 = 100 \times 10^{-6}$ farad, $C_3 = 10 \times 10^{-6}$ farad, $v_1 = 12$ volt, $v_2 = 0.001 \sin(2000\pi t)$ volt, $(1/\alpha) = 0.025$ volt, $I_{ES} = 14.34 \times 10^{-15}$ amp, $I_S = \frac{170}{171} I_{ES}$ amp, $\alpha_R = 0.05$, $I_{CS} = I_s/\alpha_R$ amp.

The initial conditions used for this simulation are; $\mu_1(0) = v_1$, $f_4(0) = \mu_1(0)/R_1$, $f_5(0) = \mu_1(0)/R_C$, $f_1(0) = f_4(0) + f_5(0)$, $f_2(0) = -f_4(0)$, $f_6(0) =$

$f_5(0)$, $q_2(0) = -C_1R_0f_2(0)$, and $q_6(0) = -C_3R_3f_6(0)$. All other variables are initialized to zero.

The solution to the DAEs in the interval $0 \leq t \leq 1.5$ seconds is shown Fig. 6.6. Here, the plot (i) shows $v_{R_3}(t)$, the voltage across the resistor R_3. As can be seen the system response approaches a steady-state after about 1 second. The plot (ii) shows $v_{R_3}(t)$ in the interval $1.49 \leq t \leq 1$ second. This plot clearly shows the input signal is amplified with gain of approximately 80. The plot (iii) shows the voltage at node C, i.e., $v_C = v_1 - R_Cf_5$, and the plot (iv) shows the voltage at node B, i.e., $v_B = R_2f_3$, in the interval $1.49 \leq t \leq 1.5$. The results shown here are in good agreement with the solution presented in Lamey (pp. 101).

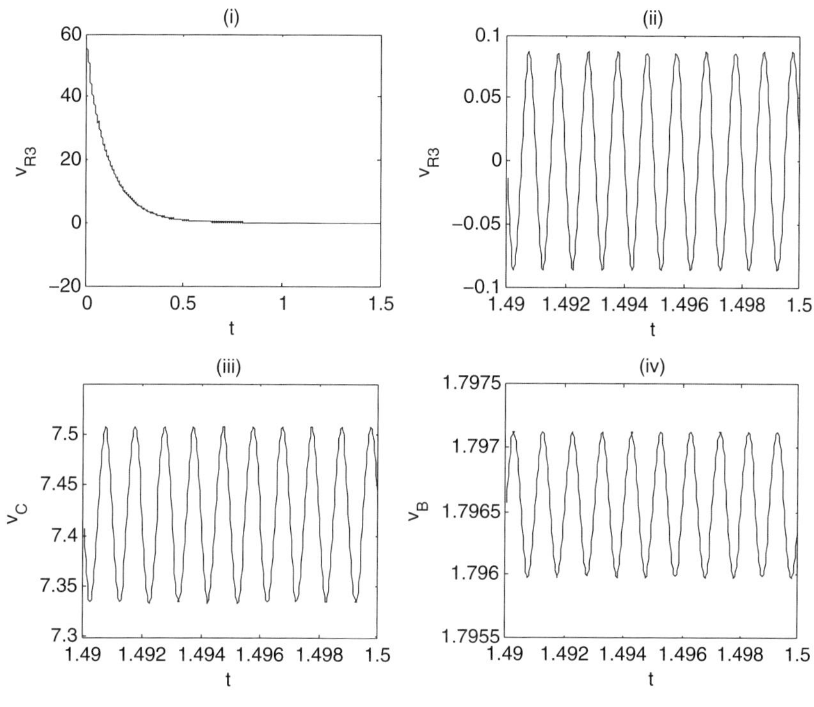

Fig. 6.6 Example 6.15

Example 6.16.

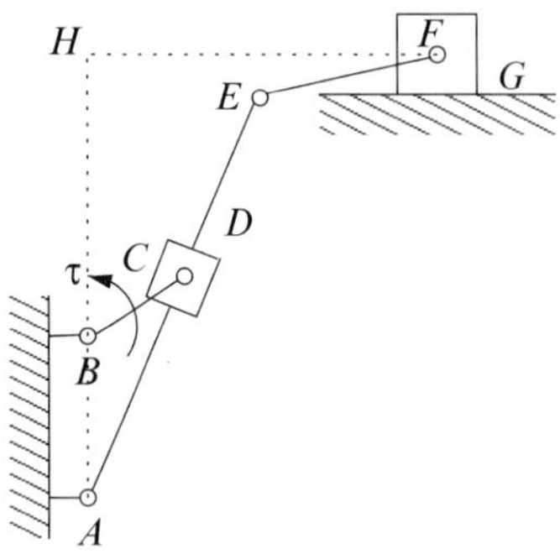

In the mechanism shown here an input torque τ is applied to the crank BC. This causes the slider at D to translate along the link AE, while the link AE simultaneously rotates about A. Also, the motion of the link EF causes the slider at G to translate in the horizontal direction. The motion of this device is such that the point C has a circular path, while the point E traverses a circular arc, and point F translates horizontally. In this device the links BC, AE and EF are treated as uniform bars. The sliders at D and G are treated as lumped masses.

Kinematic analysis:

It is easy to verify that this mechanism has one degree of freedom. However, it will be convenient to use the following displacements to model the system.

x_1, y_1: the location of the center of mass for link BC.
θ_1: the angle link BC makes with the x-axis.
x_2, y_2: the location of the center of mass for link AE.
θ_2: the angle link AE makes with the x-axis.
x_3, y_3: the location of the center of mass for link EF.
θ_3: the angle link EF makes with the x-axis.
r_2: the position of D along the link AE.
r_4: the position of G along the x-axis.

The angles θ_1, θ_2, θ_3, and the linear displacements r_2 and r_4 are shown in the figure below.

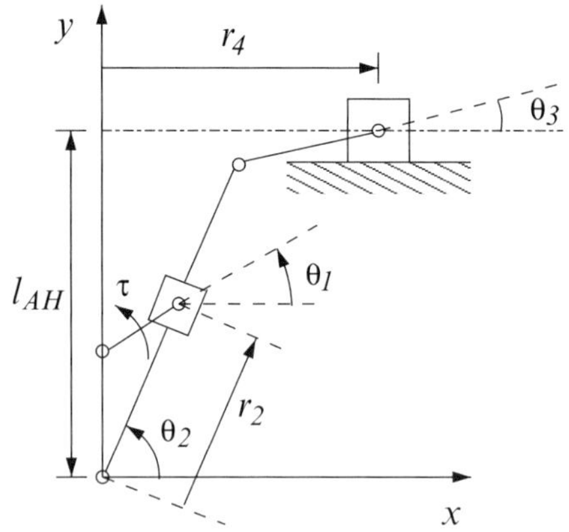

These 11 displacements are used to describe the 1 degree of freedom system. Hence, we must establish 10 independent constraint equations to account for the 'excess' variables. We can obtain 6 of these equations by simply stating the location of the mass centers for the links, i.e.,

$$\begin{aligned}
\phi_1 &= x_1 - \tfrac{l_{BC}}{2}\cos\theta_1 &&= 0,\\
\phi_2 &= y_1 - \tfrac{l_{BC}}{2}\sin\theta_1 - l_{AB} &&= 0,\\
\phi_3 &= x_2 - \tfrac{l_{AE}}{2}\cos\theta_2 &&= 0,\\
\phi_4 &= y_2 - \tfrac{l_{AE}}{2}\sin\theta_2 &&= 0,\\
\phi_5 &= x_3 - l_{AE}\cos\theta_2 - \tfrac{l_{EF}}{2}\cos\theta_3 &&= 0,\\
\phi_6 &= y_3 - l_{AE}\sin\theta_2 - \tfrac{l_{EF}}{2}\sin\theta_3 &&= 0.
\end{aligned}$$

Here, l_{AB} is the length of link AB, l_{BC} is the length of link BC, l_{AE} is the length of link AE, and l_{EF} is the length of link EF.

The other 4 constraint equations are obtained by considering the closed-loops $\bar{AB} + \bar{BC} - \bar{AC} = 0$, and $\bar{AE} + \bar{EF} - \bar{AH} - \bar{HF} = 0$. These vector equations give

$$\begin{aligned}
\phi_7 &= l_{BC}\cos\theta_1 - r_2\cos\theta_2 &&= 0,\\
\phi_8 &= l_{AB} + l_{BC}\sin\theta_1 - r_2\sin\theta_2 &&= 0,\\
\phi_9 &= l_{AE}\cos\theta_2 + l_{EF}\cos\theta_3 - r_4 &&= 0,\\
\phi_{10} &= l_{AE}\sin\theta_2 + l_{EF}\sin\theta_3 - l_{AH} &&= 0,
\end{aligned}$$

where l_{EF} is the length of link EF and l_{AH} is the distance from A to H.

Applied effort analysis:

The virtual work done by the applied torque is

$$\delta\mathcal{W} = \tau\delta\theta_1.$$

Lagrange's equations:

The kinetic coenergy for this device is

$$T^* = \frac{1}{2}(m_{BC}(\dot{x}_1^2 + \dot{y}_1^2) + I_{BC}\dot{\theta}_1^2 + m_{AE}(\dot{x}_2^2 + \dot{y}_2^2) + I_{AE}\dot{\theta}_2^2 \\ + m_{EF}(\dot{x}_3^2 + \dot{y}_3^2) + I_{EF}\dot{\theta}_3^2 + m_D l_{BC}^2\dot{\theta}_1^2 + m_G\dot{r}_4^2).$$

In this equation m_{BC} denotes the mass of link BC, and I_{BC} is the moment of inertia of link BC about its center of mass. The other parameters in the equation are similarly defined. Also, in this expression, the term $\frac{1}{2}(m_{BC}(\dot{x}_1^2 + \dot{y}_1^2) + I_{BC}\dot{\theta}_1^2)$ is the kinetic coenergy of the link BC, the term $\frac{1}{2}(m_{AE}(\dot{x}_2^2 + \dot{y}_2^2) + I_{AE}\dot{\theta}_2^2)$ is the kinetic coenergy of the link AE, the term $\frac{1}{2}m_D l_{BC}^2\dot{\theta}_1^2$ is the kinetic coenergy of the slider at D, and the term $\frac{1}{2}m_G\dot{r}_4^2$ is the kinetic coenergy of the slider at G.

The potential energy and dissipation function are $V = 0$ and $D = 0$, respectively.

Using these definitions the Lagrange's equations of motion of the system can be stated as

$$
\begin{aligned}
&m_{BC}\ddot{x}_1 + \lambda_1 &&= 0,\\
&m_{BC}\ddot{y}_1 + \lambda_2 &&= 0,\\
&(I_{BC} + m_D l_{BC}^2)\ddot{\theta}_1 + l_{BC}(\lambda_1/2 - \lambda_7)\sin\theta_1 &&\\
&\quad + l_{BC}(-\lambda_2/2 + \lambda_8)\cos\theta_1 &&= \tau,\\
&m_{AE}\ddot{x}_2 + \lambda_3 &&= 0,\\
&m_{AE}\ddot{y}_2 + \lambda_4 &&= 0,\\
&I_{AE}\ddot{\theta}_2 + l_{AE}(\lambda_3/2 + \lambda_5 - \lambda_9 + r_2\lambda_7/l_{AE})\sin\theta_2 &&\\
&\quad + l_{AE}(-\lambda_4/2 - \lambda_6 + \lambda_{10} - r_2\lambda_8/l_{AE})\cos\theta_2 &&= 0,\\
&m_{EF}\ddot{x}_3 + \lambda_5 &&= 0,\\
&m_{EF}\ddot{y}_3 + \lambda_6 &&= 0,\\
&I_{EF}\ddot{\theta}_3 + l_{EF}(\lambda_5/2 - \lambda_9)\sin\theta_3 + l_{EF}(-\lambda_6/2 + \lambda_{10})\cos\theta_3 &&= 0,\\
&-\lambda_7\cos\theta_2 - \lambda_8\sin\theta_2 &&= 0,\\
&m_G\ddot{r}_4 - \lambda_9 &&= 0,
\end{aligned}
\qquad (a)
$$

where $\lambda_1,\ \lambda_2,\ \cdots,\ \lambda_{10}$ are the Lagrange multipliers associated with the displacement constraints, $\phi_1,\ \phi_2,\ \cdots,\ \phi_{10}$.

System simulation:

The equations (a), along with the displacement constraints, define an index-3 differential-algebraic system. To obtain numerical solution to this problem we will restate these equations in the GGL stabilized index-2 form. To do so let us define the state displacement vector $q = [x_1,\ y_1,\ \theta_1,\ x_2,\ y_2,\ \theta_2,\ x_3,\ y_3,\ \theta_3,\ r_2,\ r_4]^T$, and let $f = dq/dt$ be the corresponding flow vector. Then, (5.42) gives

$$
\begin{aligned}
&\dot{q}_1 - f_1 + \nu_1 &&= 0,\\
&\dot{q}_2 - f_2 + \nu_2 &&= 0,\\
&\dot{q}_3 - f_3 + l_{BC}(\nu_1/2 - \nu_7)\sin q_3 + l_{BC}(-\nu_2/2 + \nu_8)\cos q_3 &&= 0,\\
&\dot{q}_4 - f_4 + \nu_3 &&= 0,\\
&\dot{q}_5 - f_5 + \nu_4 &&= 0,\\
&\dot{q}_6 - f_6 + l_{AE}(\nu_3/2 + \nu_5 - \nu_9 + q_{10}\nu_7/l_{AE})\sin q_6 &&\\
&\quad + l_{AE}(-\nu_4/2 - \nu_6 + \nu_{10} - q_{10}\nu_8/l_{AE})\cos q_6 &&= 0,\\
&\dot{q}_7 - f_7 + \nu_5 &&= 0,\\
&\dot{q}_8 - f_8 + \nu_6 &&= 0,\\
&\dot{q}_9 - f_9 + l_{EF}(\nu_5/2 - \nu_9)\sin q_9 + l_{EF}(-\nu_6/2 + \nu_{10})\cos q_9 &&= 0,\\
&\dot{q}_{10} - f_{10} - \nu_7\cos q_6 - \nu_8\sin q_6 &&= 0,\\
&\dot{q}_{11} - f_{11} - \nu_9 &&= 0,\\
&m_{BC}\dot{f}_1 + \lambda_1 &&= 0,\\
&m_{BC}\dot{f}_2 + \lambda_2 &&= 0,\\
&(I_{BC} + m_D l_{BC}^2)\dot{f}_3 + l_{BC}(\lambda_1/2 - \lambda_7)\sin q_3 &&\\
&\quad + l_{BC}(-\lambda_2/2 + \lambda_8)\cos q_3 &&= \tau,\\
&m_{AE}\dot{f}_4 + \lambda_3 &&= 0,\\
&m_{AE}\dot{f}_5 + \lambda_4 &&= 0,\\
&I_{AE}\dot{f}_6 + l_{AE}(\lambda_3/2 + \lambda_5 - \lambda_9 + q_{10}\lambda_7/l_{AE})\sin q_6 &&\\
&\quad + l_{AE}(-\lambda_4/2 - \lambda_6 + \lambda_{10} - q_{10}\lambda_8/l_{AE})\cos q_6 &&= 0,\\
&m_{EF}\dot{f}_7 + \lambda_5 &&= 0,\\
&m_{EF}\dot{f}_8 + \lambda_6 &&= 0,\\
&I_{EF}\dot{f}_9 + l_{EF}(\lambda_5/2 - \lambda_9)\sin q_9 + l_{EF}(-\lambda_6/2 + \lambda_{10})\cos q_9 &&= 0,\\
&-\lambda_7\cos q_6 - \lambda_8\sin q_6 &&= 0,\\
&m_G\dot{f}_{11} - \lambda_9 &&= 0,\\
&q_1 - \frac{l_{BC}}{2}\cos q_3 &&= 0,\\
&q_2 - \frac{l_{BC}}{2}\sin q_3 - l_{AB} &&= 0,\\
&q_4 - \frac{l_{AE}}{2}\cos q_6 &&= 0,\\
&q_5 - \frac{l_{AE}}{2}\sin q_6 &&= 0,
\end{aligned}
$$

$$
\begin{aligned}
q_7 - l_{AE}\cos q_6 - \frac{l_{EF}}{2}\cos q_9 &= 0,\\
q_8 - l_{AE}\sin q_6 - \frac{l_{EF}}{2}\sin q_9 &= 0,\\
l_{BC}\cos q_3 - q_{10}\cos q_6 &= 0,\\
l_{AB} + l_{BC}\sin q_3 - q_{10}\sin q_6 &= 0,\\
l_{AE}\cos q_6 + l_{EF}\cos q_9 - q_{11} &= 0,\\
l_{AE}\sin q_6 + l_{EF}\sin q_9 - l_{AH} &= 0,\\
f_1 + \frac{l_{BC}}{2} f_3 \sin q_3 &= 0,\\
f_2 - \frac{l_{BC}}{2} f_3 \cos q_3 &= 0,\\
f_4 + \frac{l_{AE}}{2} f_6 \sin q_6 &= 0,\\
f_5 - \frac{l_{AE}}{2} f_6 \cos q_6 &= 0,\\
f_7 + l_{AE} f_6 \sin q_6 + \frac{l_{EF}}{2} f_9 \sin q_9 &= 0,\\
f_8 - l_{AE} f_6 \cos q_6 - \frac{l_{EF}}{2} f_9 \cos q_9 &= 0,\\
-l_{BC} f_3 \sin q_3 - f_{10}\cos q_6 + q_{10} f_6 \sin f_6 &= 0,\\
l_{BC} f_3 \cos q_3 - f_{10}\sin q_6 - q_{10} f_6 \cos q_6 &= 0,\\
-l_{AE} f_6 \sin q_6 - l_{EF} f_9 \sin q_9 - f_{11} &= 0,\\
l_{AE} f_6 \cos q_6 + l_{EF} f_9 \cos q_9 &= 0.
\end{aligned}
$$

Here, $\nu_1,\ \nu_2, \cdots, \nu_{10}$ are the multiplies associated with adjoining the flow constraints $\frac{d}{dt}\phi_j = 0$, $j = 1, 2, \cdots, 10$, to the system.

These 42 differential-algebraic equations are solved to determine the behavior of the system. In this simulation we use the following model parameters; $AH = 8$ cm, $AB = 4$ cm, $BC = 2$ cm, $AE = 8$ cm, and $EF = 3$ cm, $m_{BC} = 0.049$ kg, $I_{BC} = 2.881 \times 10^{-6}$ kg-m^2, $m_{AE} = 0.198$ kg, $I_{AE} = 1.103 \times 10^{-4}$ kg-m^2, $m_{EF} = 0.074$ kg, $I_{EF} = 7.409 \times 10^{-6}$ kg-m^2 $m_D = 0.015$ kg, and $m_G = 0.15$ kg. The input torque is $\tau = 10(2 - \dot{\theta}_1)$ N-m

The nonzero initial conditions are

$$
\begin{aligned}
\theta_1(0) &= 0,\\
\theta_2(0) &= \tan^{-1}(l_{AB}/l_{BC}),\\
\theta_3(0) &= \sin^{-1}((l_{AH} - l_{AE}\sin\theta_2(0))/l_{EF}),\\
x_1(0) &= (l_{BC}/2)\cos\theta_1(0),\\
y_1(0) &= (l_{BC}/2)\sin\theta_1(0) + l_{AB},\\
x_2(0) &= (l_{AE}/2)\cos\theta_2(0),\\
y_2(0) &= (l_{AE}/2)\sin\theta_2(0),\\
x_3(0) &= l_{AE}\cos\theta_2(0) + (l_{EF}/2)/\cos\theta_3(0),\\
y_3(0) &= l_{AE}\sin\theta_2(0) + (l_{EF}/2)/\sin\theta_3(0),\\
r_2(0) &= (l_{AB} + l_{BC}\sin\theta_1(0))/\sin\theta_2(0),\\
r_4(0) &= l_{AE}\cos\theta_2(0) + l_{EF}\cos\theta_3(0),
\end{aligned}
$$

$$\ddot{\theta}_1(0) = 20/(I_{BC} + m_D l_{BC}^2).$$

In addition, we use ATOL=RTOL= 1.0×10^{-4}, and INITIAL_STEP_SIZE= 1.0×10^{-4}.

The results of this simulation are shown in Fig. 6.7 and Fig. 6.8. The plots (i) and (ii) in Fig. 6.7 show the crank angle θ_1 and the angular velocity $\dot{\theta}_1$. Note that the torque input $\tau = 10(2 - \dot{\theta}_1)$ keeps θ_1 close to 2 radians/second.

The plots (iii) through (x) show the displacements and flows θ_2, $\dot{\theta}_2$, θ_3, $\dot{\theta}_3$, r_2, $\dot{r}_2$, r_4, $\dot{r}_4$, respectively. Notice that the device takes about 2 seconds for the slider at G to travel from $r_4 = 0.068$ m to $r_4 = -0.012$ m. However, it only takes about 1 second for the slider to travel from $r_4 = -0.012$ m to $r_4 = 0.068$ m. For this reason this device is sometimes called a 'quick return' mechanism.

Finally, the plots (xi) and (xii) show the values of the constraints ϕ_7, ϕ_8, ϕ_9, and ϕ_{10}. As can be seen these constraints are small relative to the desired convergence tolerance (10^{-4}.

Example 6.17.

This example simulates the response of a planar R-R robot that is controlled to track a desired trajectory. A schematic of the robot is shown on the right. The angles q_1 and q_2 are the generalized displacements for the system. The input torque to the link AB is τ_1, and the input torque to link BC is τ_2. The dynamic equations of motion for this system are derived in Example 3.14. These equations can be written as

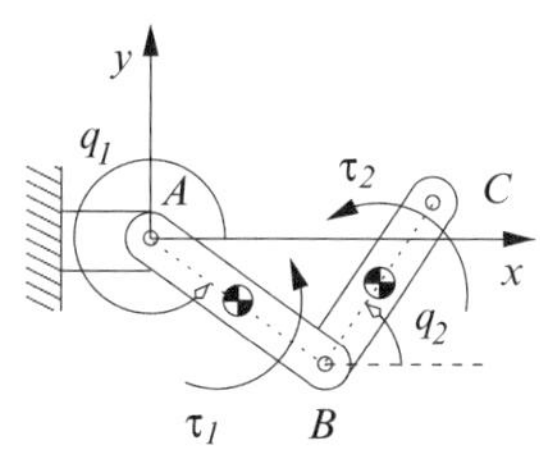

$$\begin{bmatrix} M_{11} & M_{12} \\ M_{21} & M_{22} \end{bmatrix} \begin{bmatrix} \ddot{q}_1 \\ \ddot{q}_2 \end{bmatrix} + \begin{bmatrix} f_1 \\ f_2 \end{bmatrix} = \begin{bmatrix} \tau_1 \\ \tau_2 \end{bmatrix}, \tag{a}$$

where

$$M_{11} = I_1 + m_1\frac{l_1^2}{4} + m_2 l_1^2,\ M_{12} = M_{21} = \frac{m_2 l_1 l_2}{2}\cos(q_1 - q_2),\ M_{22} = I_2 + m_2\frac{l_2^2}{4},$$

$$f_1 = \frac{m_2 l_1 l_2}{2}\dot{q}_2^2 \sin(q_1 - q_2), \quad f_2 = -\frac{m_2 l_1 l_2}{2}\dot{q}_1^2 \sin(q_1 - q_2),$$

and the model parameters; m_1, m_2, I_1, I_2, l_1 and l_2 are defined in Example 3.14.

Our goal here is to have the robot angles follow a desired path specified by angles $q_{1d}(t)$, and $q_{2d}(t)$. That is, we would like $q_1(t) = q_{1d}(t)$ and $q_2(t) = q_{2d}(t)$. One way to accomplish this is to specify the input torques such that

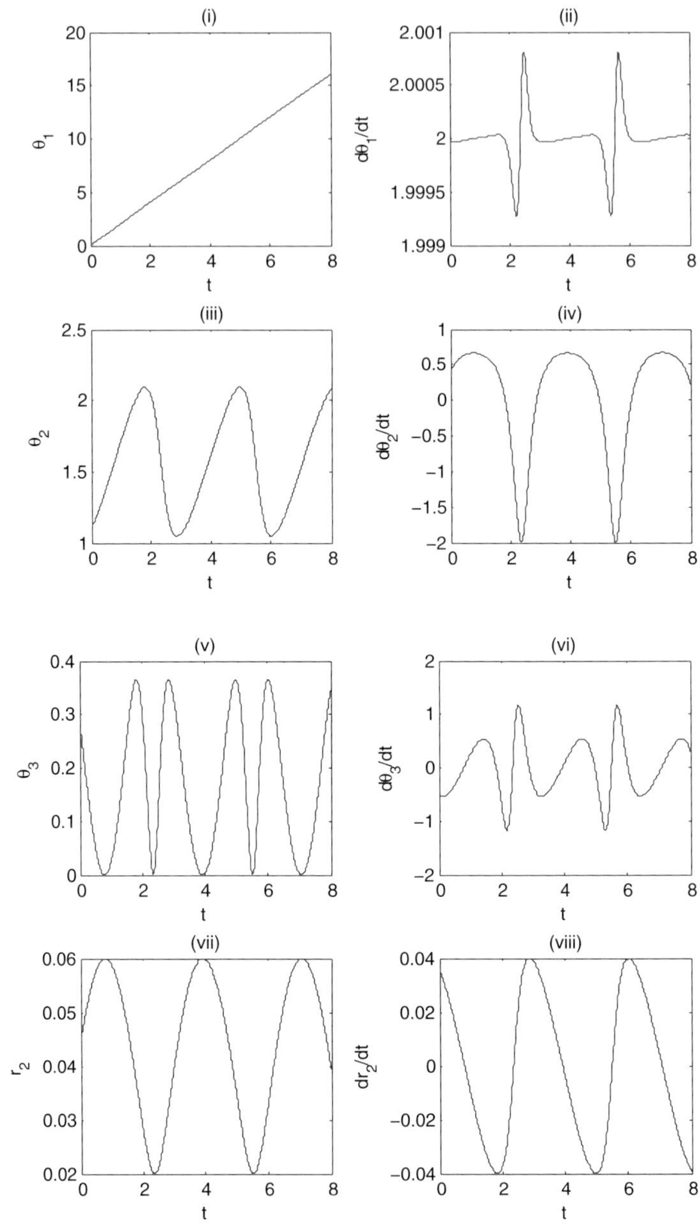

Fig. 6.7 Example 6.16

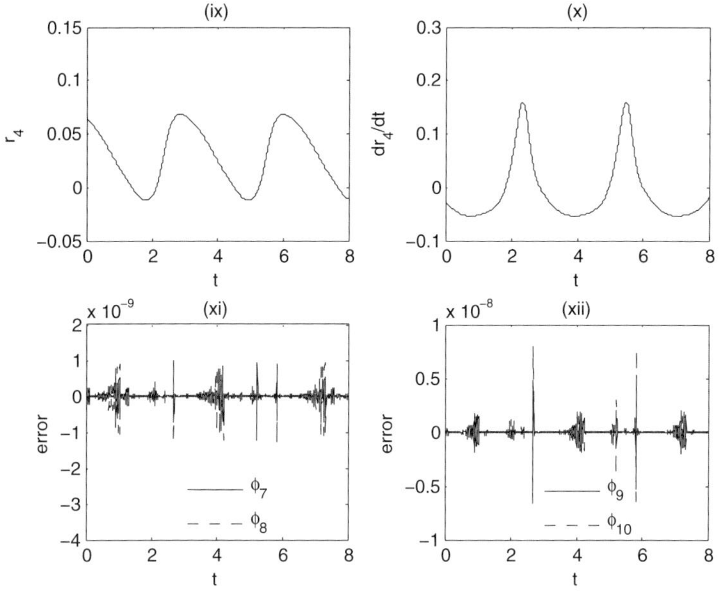

Fig. 6.8 Example 6.16, continued.

$$\begin{bmatrix} \tau_1 \\ \tau_2 \end{bmatrix} = \begin{bmatrix} M_{11} & M_{12} \\ M_{21} & M_{22} \end{bmatrix} \begin{bmatrix} \ddot{q}_{1d} \\ \ddot{q}_{2d} \end{bmatrix} + \begin{bmatrix} f_1 \\ f_2 \end{bmatrix} - 2\zeta\omega \begin{bmatrix} \dot{q}_1 - \dot{q}_{1d} \\ \dot{q}_2 - \dot{q}_{2d} \end{bmatrix} - \omega^2 \begin{bmatrix} q_1 - q_{1d} \\ q_2 - q_{2d} \end{bmatrix}, \quad \text{(b)}$$

where $\zeta > 0$ and $\omega > 0$ are constants. Then, using (b) in (a) gives

$$\ddot{e} + 2\zeta\omega\dot{e} + \omega^2 e = 0, \quad \text{(c)}$$

where $e = [(q_1 - q_{1d}),\ (q_2 - q_{2d})]^T$ is the tracking error. We see that this error satisfies a linear second-order differential equation, (c), with constant coefficients. Moreover, the coefficients of this error differential equation can be selected arbitrarily. If we set $\zeta > 1$ then the solution to (c) is called 'overdamped' and is of the form

$$e_i(t) = a_{i1}e^{\lambda_1 t} + a_{2i}e^{\lambda_2 t},\ i = 1, 2, \quad \text{(d)}$$

where $\lambda_1 = -\omega(\zeta + \sqrt{\zeta^2 - 1})$, $\lambda_1 = -\omega(\zeta - \sqrt{\zeta^2 - 1})$. The coefficients a_{1i}, and a_{2i}, $i = 1, 2$ depend on the initial conditions $e(0)$, and $\dot{e}(0)$. Since, $\lambda_1 < 0$ and $\lambda_2 < 0$ we see from (d) that

$$\lim_{t\to\infty} e(t) = 0.$$

Therefore, the torque input (b) is such that the tracking error vanishes asymptotically.

The system simulation uses the following model parameters; $l_1 = 0.65$ m, $l_2 = 0.5$ m, $m_1 = 1.6$ kg, $m_2 = 1.2$ kg, $I_1 = 0.06$ kg-m^2, $I_2 = 0.05$ kg-m^2. The desired trajectory of the robot is,

$$q_{1d}(t) = \begin{cases} \pi/4 & \text{if} \quad t < 2 \\ 0 & \text{if} \quad t \geq 2 \end{cases}, \quad q_{2d}(t) = \pi/2.$$

The parameters for the controller are selected as $\zeta = 1.1$ and $\omega = 5/\sqrt{\zeta^2 - 1}$.

Figure 6.9 shows the result of the numerical simulation. The plot (i) shows the response for the angle q_1, and the desired angle q_{1d}. The plot (ii) shows the response for the angle q_2, and the desired angle q_{2d}. In both cases we can see that the input torque allows the robot to follow the desired trajectory. The angular velocities $\dot{q}_1$ and $\dot{q}_2$ are shown in plots (iii) and (iv), respectively. Finally, plots (v) and (vi) show the torques τ_1 and τ_2 required to execute this maneuver.

It should be noted that the actual implementation of the control (b) is a nontrivial task. Since, it requires the measurements $q_1(t)$, $q_2(t)$, $\dot{q}_1(t)$ and $\dot{q}_2(t)$, as well as the model parameters. Other control techniques must be used if any of this information is unknown, or is imprecise. For examples of such control system designs see Craig (1986), and Slotine and Li (1991).

Example 6.18.

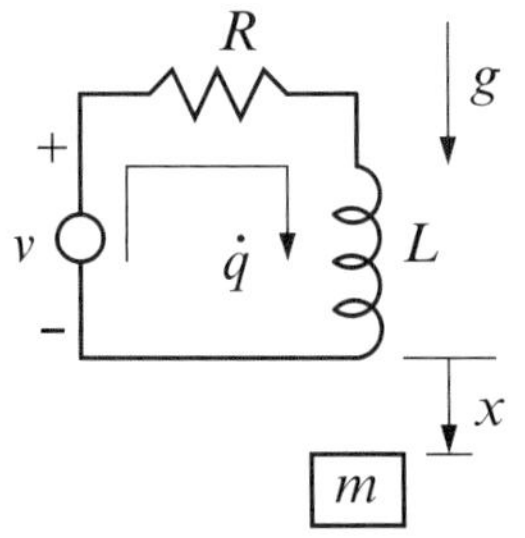

The schematic of an electromagnetic suspension is shown in the figure on the left. This device consists of a coil with inductance, L, and resistance, R, that is actuated by the voltage source $v(t)$. The mass, m, is made of a highly permeable magnetic material. For devices of this type it can be shown that the coil inductance satisfies an equation of the form

$$L(x) = \frac{\gamma_0}{\gamma_1 + x},$$

where $\gamma_0 > 0$ and $\gamma_1 > 0$ are parameters that depend on the material properties and the geometry of the system. In the system simulation given below uses the following model parameters, $m = 0.1$ kg, $R = 10$ ohm, $g = 9.8$ m/s^2, $\gamma_0 = 0.03$ volt-m-s/amp, and $\gamma_1 = 1 \times 10^{-5}$ m.

Kinematic analysis:

The network current $\dot{q}$ and the displacement of the mass x are taken as

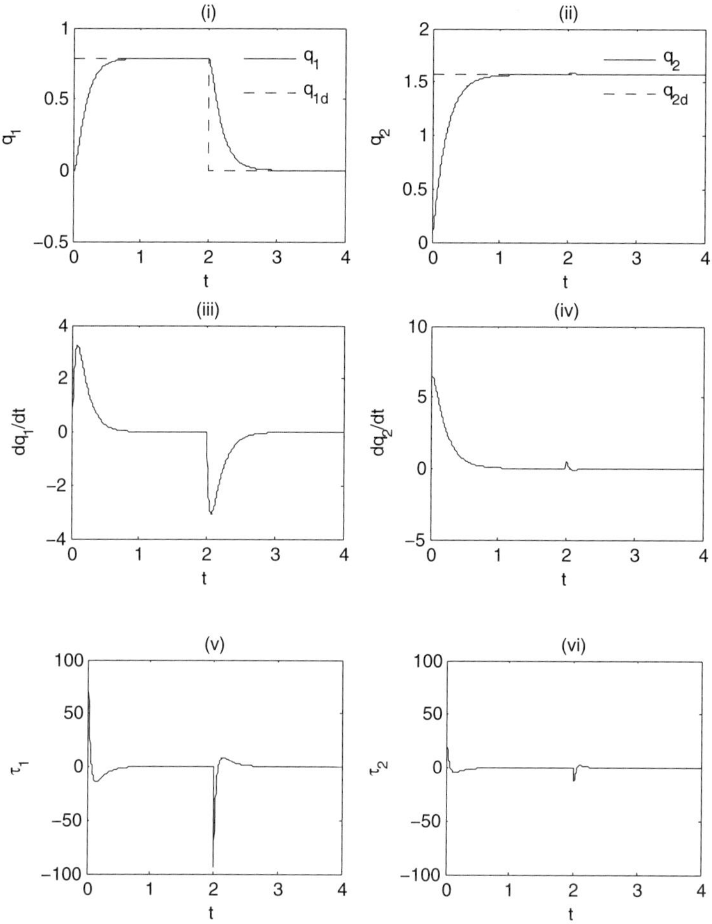

Fig. 6.9 Example 6.17

the generalized coordinates for the system. The variable x is also called the air gap in electromagnetic suspensions.

Applied effort analysis:

The virtual work done by the applied voltage, $v(t)$, and the weight, mg, is

$$\delta \mathcal{W} = v\,\delta q + mg\,\delta x.$$

Lagrange's equations:

The kinetic coenergy for the system is $T^* = L(x)\dot{q}^2/2 + m\dot{x}^2/2$. The potential energy and dissipation function are $V = 0$ and $D = R\dot{q}^2/2$, respectively. Hence, Lagrange's equations of motion for the system are

$$m\ddot{x} + \frac{\gamma_0 \dot{q}^2}{2(\gamma_1 + x)^2} = mg,$$

$$\frac{\gamma_0}{\gamma_1 + x}\ddot{q} - \frac{\gamma_0 \dot{q}\dot{x}}{(\gamma_1 + x)^2} + R\dot{q} = v.$$

System simulation:

By defining the state variables $q_1 = x$, $q_2 = \dot{x}$ and $q_3 = \dot{q}$, the equations of motion for the system can be rewritten as the implicit differential equations

$$0 = \dot{q}_1 - q_2,$$

$$0 = \dot{q}_2 + \frac{\gamma_0 q_3^2}{2m(\gamma_1 + q_1)^2} - g,$$

$$0 = \dot{q}_3 + \frac{\gamma_1 + q_1}{\gamma_0}(Rq_3 - v) - \frac{q_2 q_3}{\gamma_1 + q_1}.$$

Suppose the mass is suspended in an equilibrium position $q_1 = q_{10} = 1 \times 10^{-3}$ m and $q_2 = 0$ m/s. Then the corresponding current and input voltage can be obtained from the equations of motion by setting $\dot{q}_2 = \dot{q}_3 = 0$ to get $q_3 = q_{30} = \sqrt{2mg(\gamma_1 + q_{10})^2/\gamma_0}$ amp, and $v = v_0 = Rq_{30}$ volt. In this equilibrium position the force developed by the coil is

$$f_{\text{coil}} = \frac{\gamma_0 q_{30}^2}{2(\gamma_1 + q_{10})^2}.$$

This coil force just balances the weight mg.

To examine the response of the system, if it is perturbed from the equilibrium position, consider the initial conditions $q_1(0) = q_{10} + 1 \times 10^{-4}$ m, $q_2 = 0$ m/s, and $q_3(0) = q_{30}$ amp. In addition, the input voltage is set as $v(t) = v_0$ volt.

Figure 6.10 shows the response of the system to this initial condition. Here, plot (i) shows the air gap, plot(ii) shows the velocity of the mass, and plot (iii) shows the current in the coil. As can be seen the air gap, q_1, deviates quite far from the equilibrium position $q_{10} = 1 \times 10^{-3}$ m.

This type of motion illustrates an equilibrium position that is 'unstable'. In particular, an unstable equilibrium position is one for which a 'small' perturbation in the equilibrium position results in a trajectory that move 'far' away from the equilibrium position.

As described above the electromagnetic suspension is not very useful since, any small disturbance will cause the system to move far away from its equi-

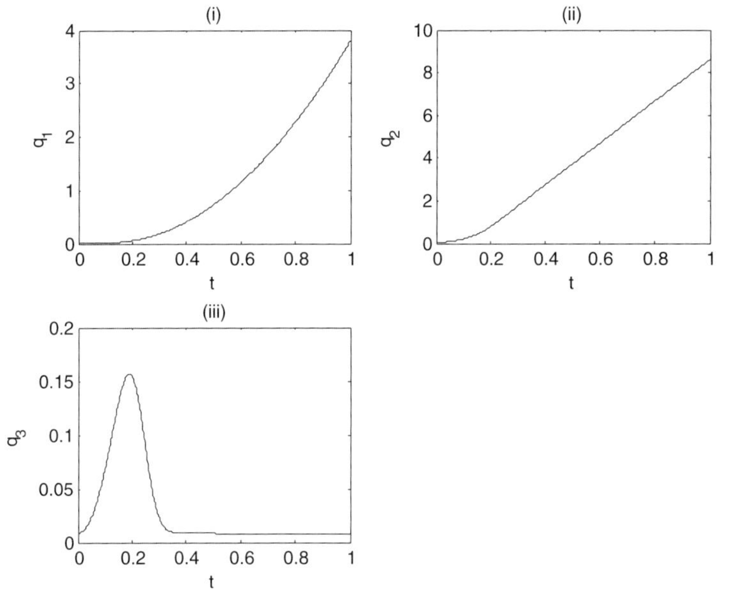

Fig. 6.10 Example 6.18 unstable response

librium. However, a number of techniques can be used to stabilize the equilibrium of the device. These techniques involve adjusting the voltage input, v, so that the air gap, q_1, remains close to q_{10} if the system is subject to small disturbances. One such method can be described as follows.

Let $z_1 = q_1 - q_{10} = x - q_{10}$ represent the change in the air gap, let $z_2 = q_2 = \dot{x}$ represent the velocity of the suspended mass, and let $z_3 = \dot{q}_2 = \ddot{x}$ represent the acceleration of the suspended mass. Then, the equations of motion can be written as

$$\begin{aligned} 0 &= \dot{z}_1 - z_2, \\ 0 &= \dot{z}_2 - z_3, \\ 0 &= \dot{z}_3 - u, \end{aligned}$$

where

$$u = \frac{q_3(Rq_3 - v)}{m(\gamma_1 + z_1 + q_{10})}.$$

Hence, by using the coordinate transformation from q_1, q_2, q_3 to z_1, z_2, z_3, and the input u we have made the dynamic equations of motion appear to be linear.

Next, using the input $u = -k_1 z_1 - k_2 z_2 - k_3 z_3$, where k_1, k_2 and k_3 are constants to be determined, the equations of motion become

$$\frac{d^3 z_1}{dt^3} + k_3 \frac{d^2 z_1}{dt^2} + k_2 \frac{dz_1}{dt} + k_1 z_1 = 0.$$

This third-order, linear ordinary differential equation determines the variation in the air gap $z_1(t)$. The general solution to this equation is

$$z_1(t) = a_1 e^{\lambda_1 t} + a_2 e^{\lambda_2 t} + a_3 e^{\lambda_3 t},$$

where a_1, a_2 and a_3 depend on the initial conditions $z_1(0)$, $z_2(0)$ and $z_3(0)$, and λ_1, λ_2 and λ_3 are the roots of the characteristic equation

$$\lambda^3 + k_3 \lambda^2 + k_2 \lambda + k_1 = 0.$$

In the system simulation below, the roots of the characteristic equation are placed at $\lambda_1 = -30$, $\lambda_2 = -20 + 10i$ and $\lambda_3 = -20 - 10i$, by selecting $k_1 = 15000$, $k_2 = 1700$, and $k_3 = 70$. Using these coefficients it can be seen that $\lim_{t \to \infty} z_1(t) = 0$, for any nonzero initial condition. That is, the air gap will eventually return to $x = q_{10}$ after some perturbation from the equilibrium position.

To implement this control technique the voltage input to the system is determined by

$$\begin{aligned} v &= Rq_3 - m(\gamma_1 + z_1 + q_{10})u/q_3 \\ &= Rq_3 + (m/q_3)(\gamma_1 + z_1 + q_{10})(k_1 z_1 + k_2 z_2 + k_3 z_3). \end{aligned}$$

Thus, in an actual device this controller requires the knowledge of the states z_1, z_2, z_3 and q_3, as well as the model parameters m, R and γ_1.

The simulation of the system with this nonlinear feedback controller is shown in Fig. 6.11. These plots show the response of the system with initial conditions $q_1(0) = q_{10} + 1 \times 10^{-4}$ m, $q_2 = 0$ m/s, and $q_3(0) = q_{30}$ amp. As can be seen the system is well behaved in this case, and the initial perturbation in the air gap eventually vanishes. In this case the equilibrium position $q_1 = q_{10}$, $q_2 = 0$, $q_3 = q_{30}$ is made asymptotically stable using the voltage control input. This result should be compared to the unstable system shown in Fig. 6.10.

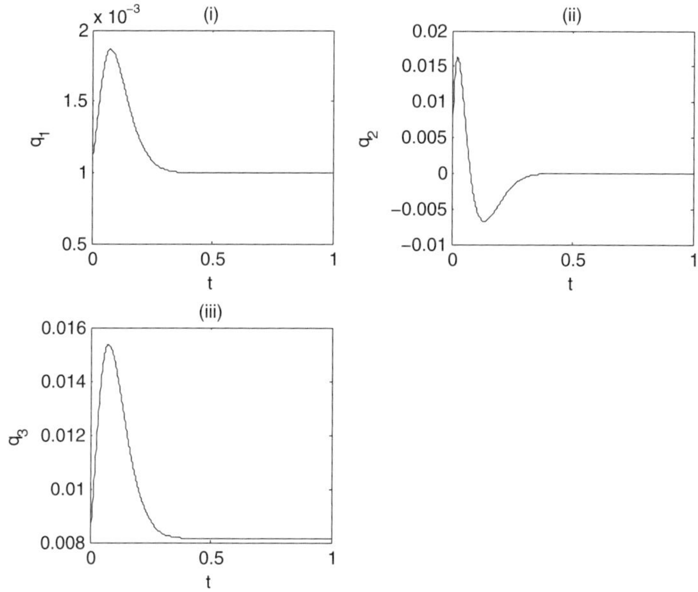

Fig. 6.11 Example 6.18 stable response

Example 6.19.

This example investigates the behavior of the inverted pendulum system shown on the right. The dynamic equations of motion for this system are derived in Example 3.10. Here, the system response is computed in following cases; (i) The initial condition response about the equilibrium position $\theta = \pi$. (ii) The initial condition response about the equilibrium position $\theta = \pi$, with $F = -c_1\dot{x} - c_2\dot{\theta}$, where c_1 and c_2 are positive constants. (iii) The initial condition response about the equilibrium position $\theta = 0$. (iv) The initial condition response about the equilibrium position $\theta = 0$, and F determined using linear state feedback control.

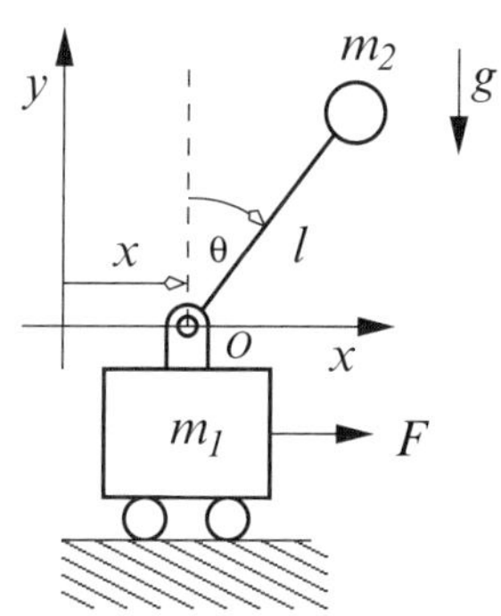

Lagrange's equations:

From Example 3.10 Lagrange's equations of motion for the system are

$$(m_1 + m_2)\ddot{x} + m_2 l(\ddot{\theta}\cos\theta - \dot{\theta}^2 \sin\theta) = F,$$

$$m_2(l^2\ddot{\theta} + l\ddot{x}\cos\theta) = m_2 gl \sin\theta.$$

These equations can be written as a system of first-order ordinary differential equations using the state variables $q_1 = x$, $q_2 = \theta$, $q_3 = \dot{x}$ and $q_4 = \dot{\theta}$. With this definition the equations of motion become

$$\begin{bmatrix} 1 & 0 & 0 & 0 \\ 0 & 1 & 0 & 0 \\ 0 & 0 & m_1 + m_2 & m_2 l \cos q_2 \\ 0 & 0 & m_2 l \cos q_2 & m_2 l^2 \end{bmatrix} \begin{bmatrix} \dot{q}_1 \\ \dot{q}_2 \\ \dot{q}_3 \\ \dot{q}_4 \end{bmatrix} - \begin{bmatrix} q_3 \\ q_4 \\ m_2 l q_4^2 \sin q_2 + F \\ m_2 gl \sin q_2 \end{bmatrix} = 0. \quad (a)$$

System simulation:

The following model parameters are used in the simulations described below; $m_1 = 0.5$ kg, $m_2 = 0.1$ kg, $l = 0.25$ m, and $g = 9.8$ m/s^2.

Case 1: $\theta(0) = \pi$ and $F = 0$.
This simulation examines the motion of the system perturbed from the *equilibrium* point,

$$q^* = [q_1^*, q_2^*, q_3^*, q_4^*]^T = [0,\ \pi,\ 0,\ 0]^T.$$

The point q^* is called an equilibrium point because the equations of motion are satisfied identically with $q(t) = q^*$, $\dot{q}(t) = 0$, and $F = 0$. (Here, $q(t) = [q_1(t),\ q_2(t),\ q_3(t),\ q_4(t)]^T$.)

Now consider the response of the system with initial condition $q(0) = [0, \pi, 0, 0.1]$. That is, the equilibrium position q^* is perturbed by an angular velocity $\dot{\theta} = q_4 = 0.1$. Figure 6.12 shows plots of the system trajectory. Here, the plot (i) shows that q_1 moves to the left, while the plot (ii) shows that $q_2 = \theta$ oscillates about the equilibrium $q_2 = \pi$. The velocities $\dot{q}_1 = q_3$ and $\dot{q}_2 = q_4$ are shown in plots (iii) and (iv), respectively.

Another view of the system response is illustrated in the state portraits shown in plots (v) and (vi) of Fig. 6.12. Plot (v) shows q_1 versus q_3, and plot (ii) shows q_2 versus q_4. Notice that q_2-q_4 form a closed orbit around the equilibrium position q^*.

Case 2: $\theta(0) = \pi$ and $F = -c_1\dot{x} - c_2\dot{\theta}$.
Here the force applied to the mass is $F = -c_1\dot{x} - c_2\dot{\theta}$, where c_1 and c_2 are positive constants. This is equivalent to adding a linear damper, with damping coefficient c_1 to the mass m_1, and a torsional damper with damping coefficient c_2 to the link l. As a result this force input can also be represented by the dissipation function $D = c_1\dot{x}^2/2 + c_2\dot{\theta}^2/2 = c_1 q_3^2/2 + c_2 q_4^2$. The simulation shown below uses $c_1 = 2$ N-s/m and $c_2 = 0.5$ N-m-s.

Using $F = -c_1 q_3 - c_2 q_4$ in the equations of motion it can be seen that q^* is still an equilibrium point for the system. The response of this model starting from the initial condition $q(0) = [0, \pi, 0, 0.1]^T$ is shown in Fig. 6.13. From these plots it is observed that the system oscillates about the equilibrium

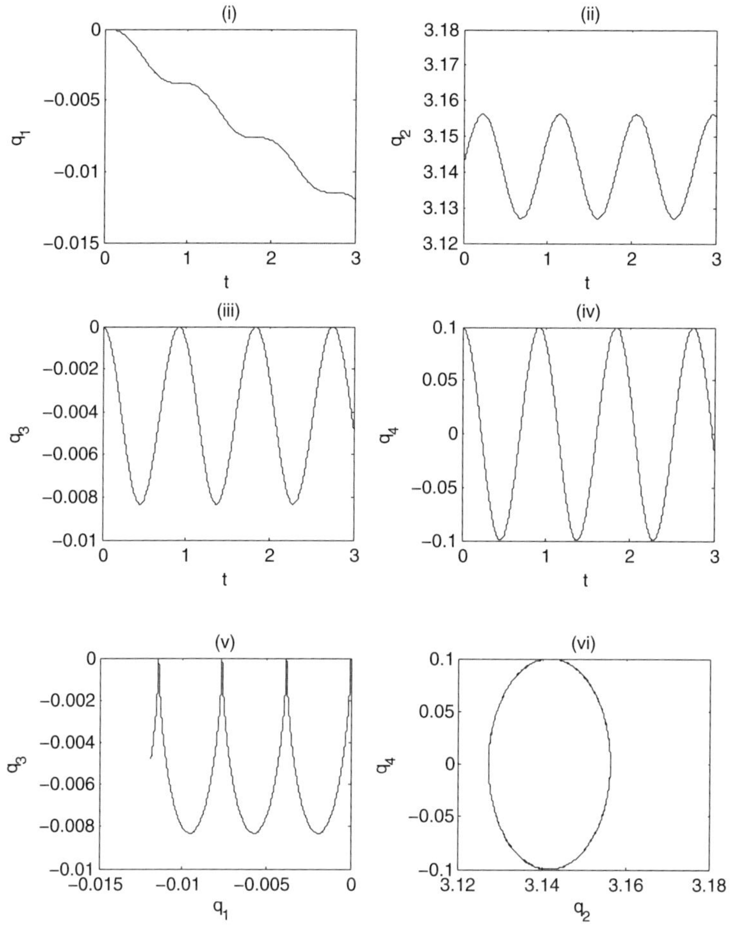

Fig. 6.12 Example 6.19: Case 1

with decreasing amplitude. In fact $\lim_{t\to\infty} q(t) = q^*$. That is, the perturbed system eventually returns to the equilibrium q^*.

The state portraits of this response are shown in plots (v) and (vi), which clearly shows that trajectory of the system is from the initial condition, $q(0)$, to the equilibrium q^* along a spiral path. Since the motion is such that $q(t) \to q^*$, the point q^* is called an *asymptotically stable equilibrium*. Thus, when compared to Case 1, adding damping to the system has made the equilibrium q^* asymptotically stable.

Case 3: $\theta(0) = 0$ and $F = 0$.
The point $\bar{q} = [0, 0, 0, 0]^T$ is also an equilibrium of the system, since the

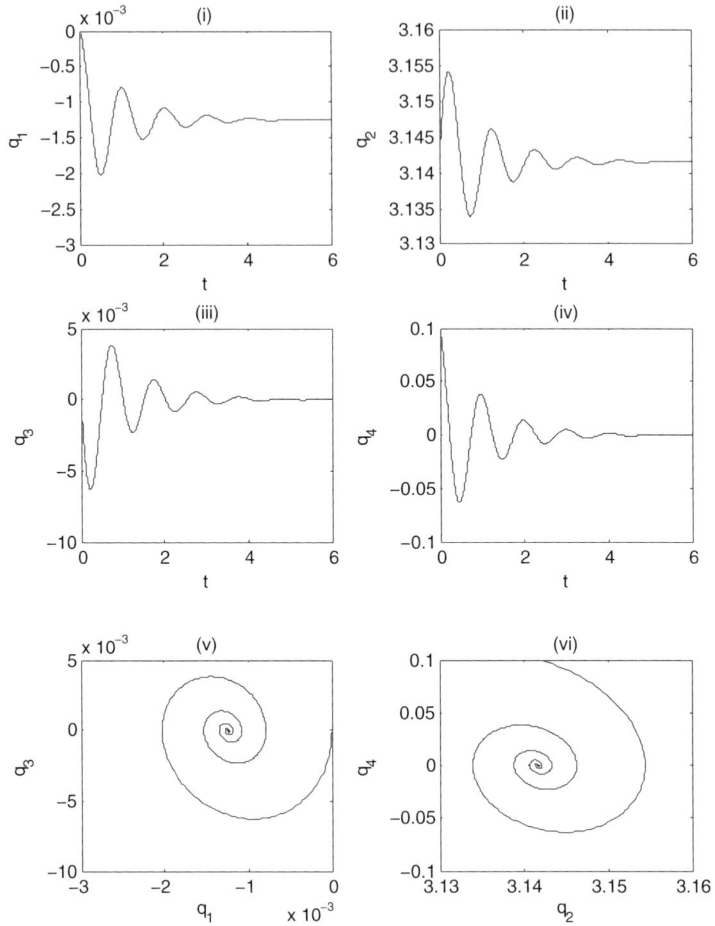

Fig. 6.13 Example 6.19: Case 2

equations of motion are satisfied identically with $q(t) = \bar{q}$, $\dot{q}(t) = 0$ and $F = 0$. Now, consider the initial condition $q(0) = [0, 0, 0, 0.1]^T$ which is a perturbation from the equilibrium $\bar{q}$. The results of the system simulation, in this case, are shown in Fig. 6.14. These plots show that the system trajectory is such that the state $q(t)$ moves 'far' away from the equilibrium $\bar{q}$. As a result, the point $\bar{q}$ is called an *unstable equilibrium* in this case.

Case 4: $\theta(0) = 0$ and $F = -k_1 q_1 - k_2 q_2 - k_3 q_3 - k_4 q_4$.
This simulation shows that the unstable equilibrium $\bar{q}$ can be made stable by using a force input

$$F = -k_1 q_1 - k_2 q_2 - k_3 q_3 - k_4 q_4,$$

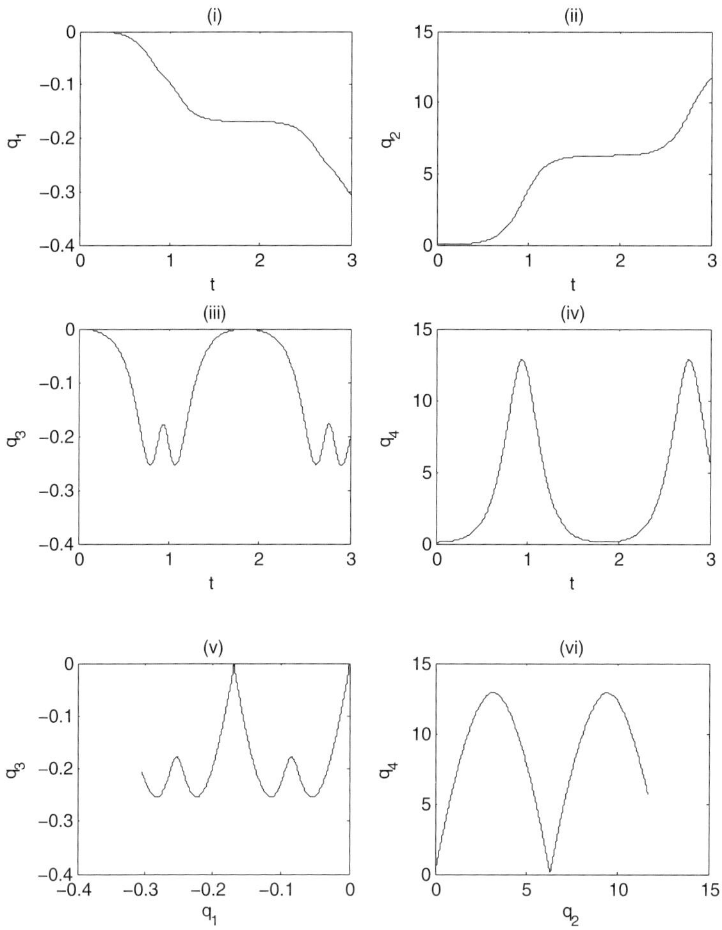

Fig. 6.14 Example 6.19: Case 3

where k_1, k_2, k_3 and k_4 are constants. In control system terminology this is called *linear state feedback control.*

To develop this result, first linearize the equations of motion about the equilibrium point $\bar{q}$. Let $q = \bar{q} + w$, $\dot{q} = \dot{\bar{q}} + \dot{w}$, and $F = \bar{F} + u$, where x, $\dot{x}$ and u are small variations in the state, state derivative and force input from the equilibrium, respectively. Note that, for this system, at the equilibrium point $\dot{\bar{q}} = 0$ and $\bar{F} = 0$. Now, the equations of motion, (a), can be written as $\Phi(q, \dot{q}, F) = \Phi(\bar{q} + w, \dot{\bar{q}} + \dot{w}, \bar{F} + u) = 0$. Writing this equation as a Taylor series centered at $\bar{q}$, $\dot{\bar{q}}$, $\bar{F}$ gives

$$\Phi(\bar{q} + w, \dot{\bar{q}} + \dot{w}, \bar{F} + u) = \Phi(\bar{q}, \dot{\bar{q}}, \bar{F}) + \Phi_{\dot{q}}\dot{w} + \Phi_q w + \Phi_F u$$

$$+ \text{higher order terms.}$$
$$= \Phi_{\dot{q}}\dot{w} + \Phi_q w + \Phi_F u = 0.$$

To obtain the last expression, the higher order terms in w, $\dot{w}$ and u are neglected from the Taylor series expansion, and we use the fact that at the equilibrium $\Phi(\bar{q}, \dot{\bar{q}}, \bar{F}) = 0$. In addition,

$$\Phi_{\dot{q}} = \frac{\partial \Phi}{\partial \dot{q}}(\bar{q}, \dot{\bar{q}}, \bar{F}) = \begin{bmatrix} 1 & 0 & 0 & 0 \\ 0 & 1 & 0 & 0 \\ 0 & 0 & m_1 + m_2 & m_2 l \\ 0 & 0 & m_2 l & m_2 l^2 \end{bmatrix}$$

$$\Phi_q = \frac{\partial \Phi}{\partial q}(\bar{q}, \dot{\bar{q}}, \bar{F}) = - \begin{bmatrix} 0 & 0 & 1 & 0 \\ 0 & 0 & 0 & 1 \\ 0 & 0 & 0 & 0 \\ 0 & m_2 g l & 0 & 0 \end{bmatrix}$$

$$\Phi_F = \frac{\partial \Phi}{\partial F}(\bar{q}, \dot{\bar{q}}, \bar{F}) = - \begin{bmatrix} 0 \\ 0 \\ 1 \\ 0 \end{bmatrix}$$

Since $\Phi_{\dot{q}}$ has an inverse the equation $\Phi_{\dot{q}}\dot{w} + \Phi_q w + \Phi_F u = 0$ can be written as

$$\begin{aligned} \dot{w} &= -\Phi_{\dot{q}}^{-1}(\Phi_q w + \Phi_F u) \\ &= Aw + Bu, \end{aligned}$$

where

$$A = -\Phi_{\dot{q}}^{-1}\Phi_q = \begin{bmatrix} 0 & 0 & 1 & 0 \\ 0 & 0 & 0 & 1 \\ 0 & -(m_2 g)/m_1 & 0 & 0 \\ 0 & (1 + m_2/m_1)\frac{g}{l} & 0 & 0 \end{bmatrix}, \text{ and}$$

$$B = -\Phi_{\dot{q}}^{-1}\Phi_F = \begin{bmatrix} 0 \\ 0 \\ 1/m_1 \\ -1/(m_1 l) \end{bmatrix}.$$

Using the input (feedback control)

$$F = u = -k_1 w_1 - k_2 w_2 - k_3 w_3 - k_4 w_4 = -Kw,$$

where $K = [k_1\ k_2\ k_3\ k_4]$, the *closed-loop* system becomes

$$\begin{aligned} \dot{w} &= Aw + Bu = Aw - BKw = (A - BK)w \\ &= A_{\text{cl}} w \end{aligned}$$

$$= \begin{bmatrix} 0 & 0 & 1 & 0 \\ 0 & 0 & 0 & 1 \\ -k_1/m_1 & -(k_2+m_2 g)/m_1 & -k_3/m_1 & -k_4/m_1 \\ k_1/(m_1 l) & (m_1+m_2+\frac{k_2}{g})\dfrac{g}{m_1 l} & k_3/(m_1 l) & k_4/(m_1 l) \end{bmatrix} \begin{bmatrix} w_1 \\ w_2 \\ w_3 \\ w_4 \end{bmatrix}.$$

The control system design problem becomes one of finding the *gains*, K, so that the eigenvalues of $A_{\text{cl}} = A - BK$ are all, strictly, in the left-hand complex plane. The eigenvalues of the matrix A_{cl} are the roots of the characteristic equation

$$\begin{aligned} 0 &= \det(A_{\text{cl}} - \lambda I) \\ 0 &= \lambda^4 + \rho_3\lambda^3 + \rho_2\lambda^2 + \rho_1\lambda + \rho_0, \end{aligned}$$

where

$$\rho_3 = \frac{k_3}{m_1} - \frac{k_4}{m_1 l}, \ \rho_2 = \frac{k_1}{m_1} - (m_1 + m_2 + \frac{k_2}{g})\frac{g}{m_1 l}, \ \rho_1 = -\frac{g k_3}{m_1 l}, \ \rho_0 = -\frac{k_1 g}{m_1 l}.$$

To place the roots of the characteristic equation at $\lambda = -10 + i, -10 - i, -10 + 2i, -10 - 2i$, requires the gains

$$K = [-133.980 \quad -115.000 \quad -52.296 \quad -18.074].$$

Figure 6.15 shows the simulation results obtained using the force input $F = -Kq$. These plots show that the system returns quickly to the equilibrium position $\bar{q}$. Thus, by using the feedback control $F = -Kq$ we have been able to stability the equilibrium point $\bar{q}$. Note however that this control scheme is only valid for small perturbations from the equilibrium $\bar{q}$. (See Problem 13.)

References

1. W. E. Boyce and R. C. DiPrima, *Elementary Differential Equations and Boundary Value Problems*, Third Edition, John Wiley and Sons, 1977.
2. J. J. Craig, *Introduction to robotics: Mechanics and Control*, Addison-Wesley, 1986.
3. A. F. D'Souza and V. K. Garg, *Advanced Dynamics: Modeling and Analysis*, Prentice-Hall, 1984.
4. B. C. Fabien and R. A. Layton, "Modeling and simulation of physical systems II: An approach to solving Lagrangian DAEs," *4th IASTED International Conference in Robotics and Manufacturing*, 200–204, 1996.
5. T. E. Fortmann and K. Hitz, *An Introduction to Linear Control Systems*, Marcel Dekker, 1977.

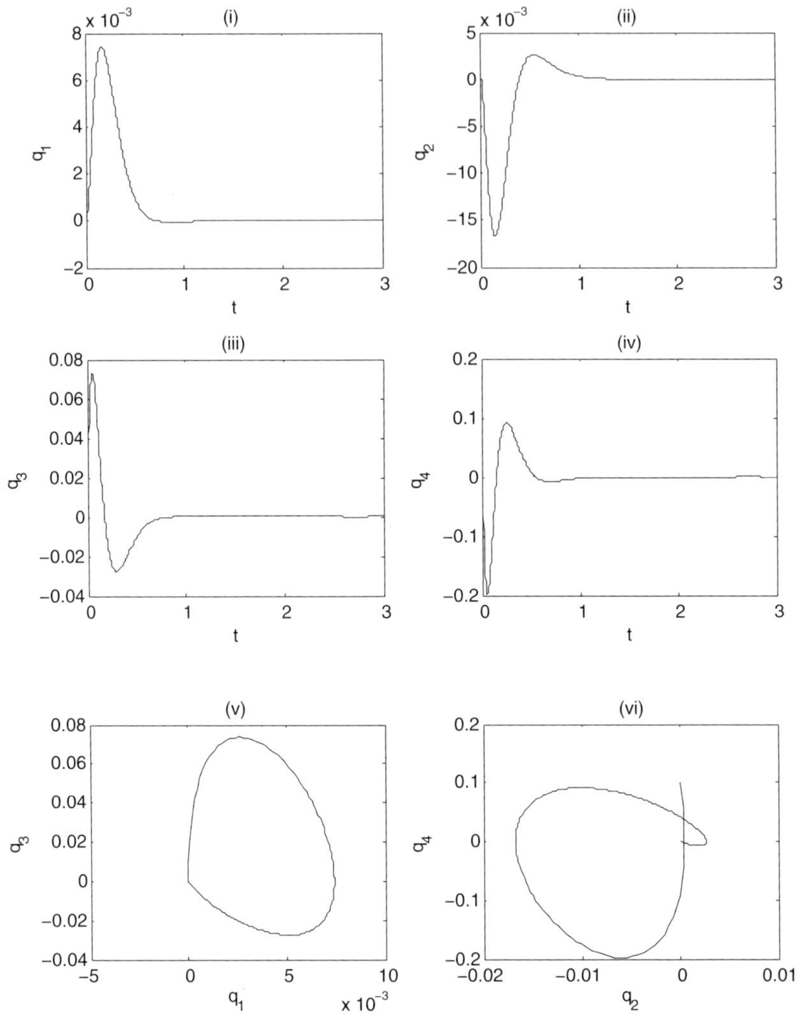

Fig. 6.15 Example 6.19: Case 4

6. R. Lamey, *The Illustrated Guide to PSpice*, Delmar Publishers Inc., 1995.
7. R. Lozano, B. Brogliato, O. Egeland and B. Maschke *Dissipative Systems Analysis and Control: Theory and Applications*, Springer, 2000.
8. A. P. Malvino, *Transistor Circuit Approximation*, 3rd ed., McGraw-Hill, 1980.
9. L. Meirovitch, *Methods of Analytical Dynamics*, McGraw-Hill, 1970.
10. D. R. Merkin, *Introduction to the Theory of Stability*, Springer, 1997.
11. A. Möschwitzer, *Semiconductor Devices, Circuits, and Systems*, Oxford University Press, 1991.

12. B. Nobel and J. W. Daniel, *Applied Linear Algebra*, Prentice Hall, 1977.
13. B. Paul, *Kinematics and Dynamics of Planar Machinery*, Prentice-Hall, 1979.
14. G. Sandor and A. Erdman, *Advanced Mechanism Design: Analysis and Synthesis, Volume 2*, Prentice-Hall, 1984.
15. R. J. Smith, *Circuits Devices and Systems*, John Wiley and Sons, 1976.
16. J. E. Slotine and W. Li, *Applied Nonlinear Control*, Prentice-Hall, 1991.
17. G. Strang, *Linear Algebra and it Applications*, Harcourt Brace Jovanovich, 1980.
18. M. Vidyasagar, *Nonlinear Systems Analysis*, Prentice-Hall, 1978.
19. D. M. Wiberg, *Theory and Problems of State Space and linear Systems*, Schaum's Outline Series, McGraw-Hill, 1971.

Problems

1. Writing the matrix exponential, e^{At}, as a power series, verify the properties given in section 6.1.2.
2. Use the Lyapunov direct and indirect methods to determine the stability of the equilibrium points for the systems shown below.

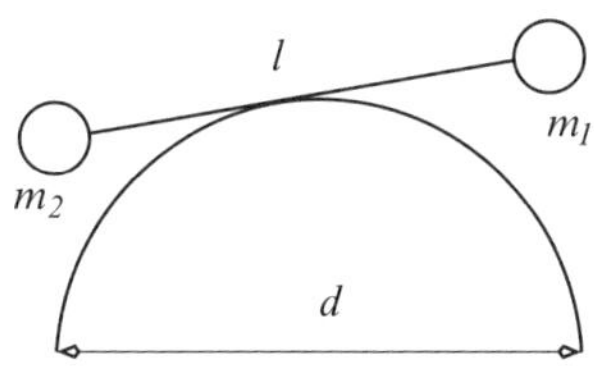

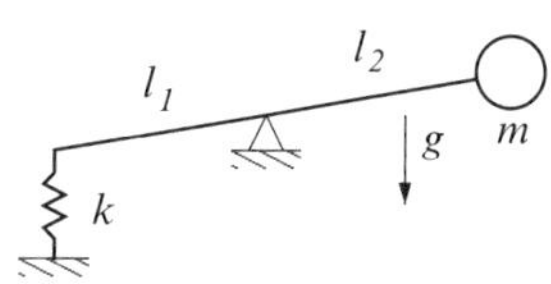

3. Repeat the simulation in Example 6.11 assuming the voltage input is of the form

$$v(t) = \begin{cases} 3, & \text{if } 0 \le t < 2 \\ 0, & \text{if } 2 \le t < 4 \\ 3, & \text{if } 4 \le t \end{cases}$$

Hint: Use the option `T_EVENT` *to ensure that the solver does not integrate past the critical time points,* $t = 2$ *and* $t = 4$.

4. 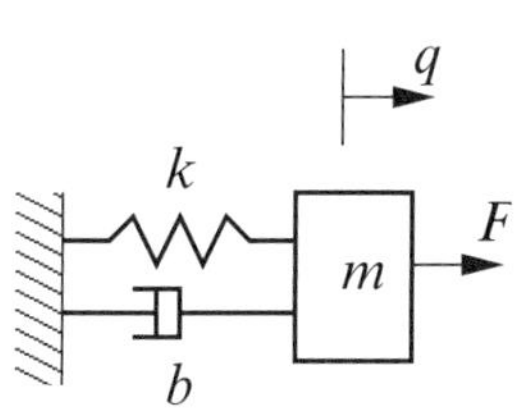

For the linear system shown here; (i) Derive Lagrange's equations of motion. (ii) Assume that the force input is a constant, and obtain an analytical solution to the equations of motion. (iii) Using the parameters, $m = 0.3$ kg, $k = 45$ N/m, $b = 0.75$ N-s/m, and $F = 0.5$ N. Solve the equations of motion using the function `ride` in the interval $0 \le t \le 5$ seconds, with initial conditions $q(0) = 0$ and $\dot{q}(0) = 0$. (iv) Compare the numerical and analytical solutions for this problem.

5. Derive Lagrange's equation of motion for the the system shown below. Using the parameters, $m = 0.3$ kg, $k = 45$ N/m, $b = 0.75$ N-s/m, and $F = 0.5$ N. Solve the equations of motion using the function `ride` in the interval $0 \leq t \leq 5$ seconds. Assume that the system is initially at rest. Compare the results obtained for this problem with the results obtained for the previous problem.

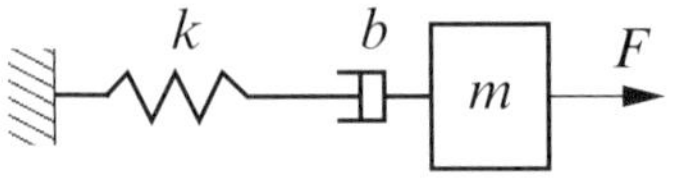

6. The double pendulum shown here has the following parameters; $m_1 = 2$, $m_2 = 1$, $l_1 = 1$, $l_2 = 2$, and $g = 9.81$. (i) Derive Lagrange's equation of motion for the system using θ_1 and θ_2 as the generalized coordinates. (ii) Determine the equilibrium points for the system. (iii) Use the Lyapunov direct and indirect methods to evaluate the stability of the equilibrium points. (iv) Solve the equations of motion using the function `ride`. (v) Plot the trajectory of m_1, i.e., x_1, y_1, and the trajectory of m_2, i.e., x_2, y_2. Apply the following initial conditions;
(a) $\theta_1 = \pi/8$, $\theta_2 = \pi/8$, $\dot{\theta}_1 = 0$, and $\dot{\theta}_2 = 0$. (b) $\theta_1 = 0$, $\theta_2 = \pi/2$, $\dot{\theta}_1 = 0$, and $\dot{\theta}_2 = 0$. (c) $\theta_1 = \pi/2$, $\theta_2 = 0$, $\dot{\theta}_1 = 0$, and $\dot{\theta}_2 = 0$. (d) $\theta_1 = 3\pi/4$, $\theta_2 = \pi/2$, $\dot{\theta}_1 = 0$, and $\dot{\theta}_2 = 0$.

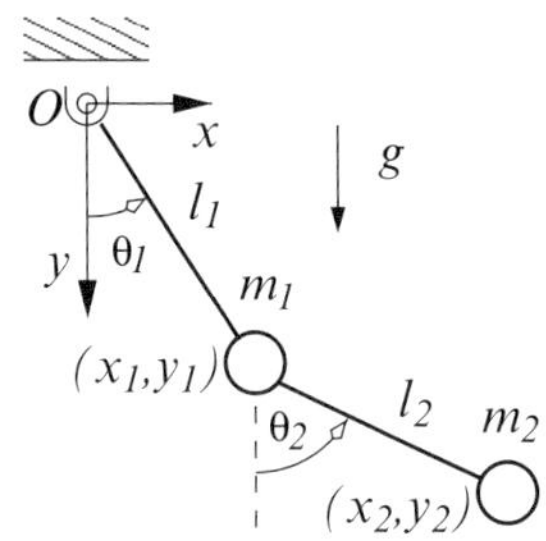

7. For the double pendulum shown in Problem 6; (i) Derive Lagrange's equation of motion using x_1, y_1, x_2 and y_2 as the displacement variables for the system. (ii) Solve these equations of motion using the function `ride`, with the parameters and initial conditions given in Problem 6. (iii) Compare the results obtained here with those obtained in Problem 6. In particular, compute the errors in the trajectories of x_1, y_1, x_2, and y_2. Also, evaluate the errors in the displacement constraints.
8. Derive Lagrange's equation of motion for the full-wave bridge rectifier shown below.

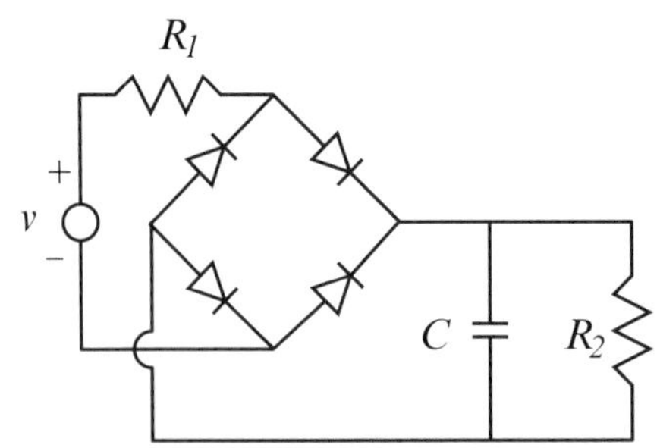

Using $R_1 = 8$ ohm, $R_2 = 48$ ohm, $C = 3900 \times 10^{-6}$ farad, $v(t) = 12\sin 120\pi t$, and $\alpha = 40$ for the diode, simulate the behavior of the system. Compare the steady-state voltage across R_2 obtained here, with that obtained in Example 6.14

9. Simulate the behavior of the voltage clipping circuit shown here by deriving Lagrange's equation of motion for the system, and using the function `ride`.

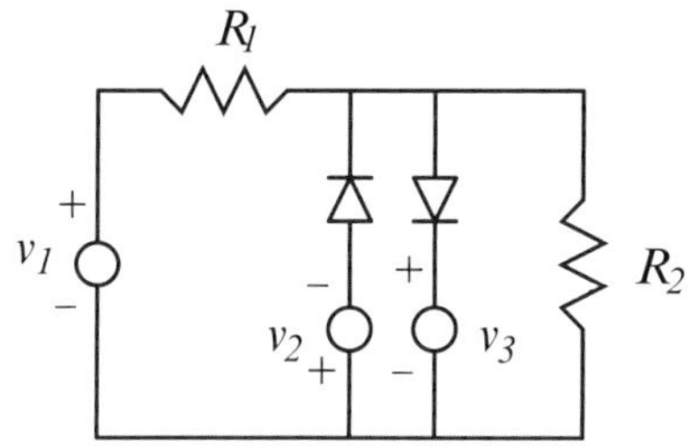

The model parameters are; $v_1(t) = 24\sin 2\pi t$ volt, $v_2 = 6$ volt, $v_3 = 12$, volt, $R_1 = 10$ ohm, and $R_2 = 1000$ ohm. Also, the current through the diodes satisfy $\dot{q}_d = I_s(e^{\alpha v_d} - 1)$, where the reverse saturation current is $I_s = 10^{-12}$ amp, the thermal voltage is $(1/\alpha) = 25 \times 10^{-3}$ volt, and v_d is the voltage across the diode. Compare the input voltage v_1 and the voltage across the resistor R_2. Find the numerical solution in the interval $0 \leq t \leq 4$ second.

10. Repeat the simulation in Example 6.15 using the inputs; (i) $v_2(t) = 0.002\sin(2\pi t)$, (ii) $v_2(t) = 0.001\sin(20\pi t)$, (iii) $v_2(t) = 0.001\sin(200\pi t)$, (iv) $v_2(t) = 0.001\sin(2000\pi t)$, and (v) $v_2(t) = 0.001\sin(20000\pi t)$. Compare the peak to peak steady state result obtained for v_{R_3} in each case.
11. Simulate the behavior of the fourbar mechanism given in Example 2.16. Use the following proportions for the model; $QA = 3$ cm, $AB = 1.5$ cm, $BC = 5$ cm, $BE = 2.5$ cm, $QD = 4.5$ cm, $DC = 3$ cm. Assume that all links are uniform bars with the following mass properties; $m_{AB} = 0.037$ kg, $I_{AB} = 1.621 \times 10^{-6}$ kg-m^2, $m_{BC} = 0.123$ kg, $I_{BC} = 2.881 \times 10^{-5}$ kg-m^2, $m_{DC} = 0.074$ kg, $I_{DC} = 7.409 \times 10^{-6}$ kg-m^2. In each case, the moment of inertia is about the center of mass. Take the torque input to be $\tau = 10(\dot{\theta}_1 - 2\pi)$ N-m. Here, τ is the torque applied to the crank AB, and θ_1 is the angle the crank makes with the horizontal axis. Derive the Lagrangian differential-algebraic equations of motion for the mechanism (see Example 4.3), and put these equations in GGL stabilized index-2 form, Note you must also find consistent initial conditions for this model.
12. Derive Lagrange's equation of motion for the slider-crank mechanism shown in Example 2.17. (see also Example 4.2.) Assume that the mechanism has the following proportions; $QA = 0.25$ cm, $AB = 1.5$ cm, $BE = 2.5$ cm, $BC = 5$ cm. The mass properties of the uniform links are; $m_{AB} = 0.037$ kg, $I_{AB} = 1.621 \times 10^{-6}$ kg-m^2, $m_{BC} = 0.123$ kg, $I_{BC} = 2.881 \times 10^{-5}$ kg-m^2, The slider D has mass $m_D = 0.2$ kg. The crank AB has a torque input $\tau = 10(\dot{\theta}_1 - 2\pi)$ N-m, where θ_1 is the crank

angle with respect to the horizontal. Simulate the behavior of this system by writing the LDAEs in GGL stabilized index-2 form.

13. Repeat Example 6.15, Case 4 with initial condition $q = [\pi/4, 0, 0, 0]^T$. Discuss the result obtained.
14. Derive Lagrange's equations of motion for the system shown below.

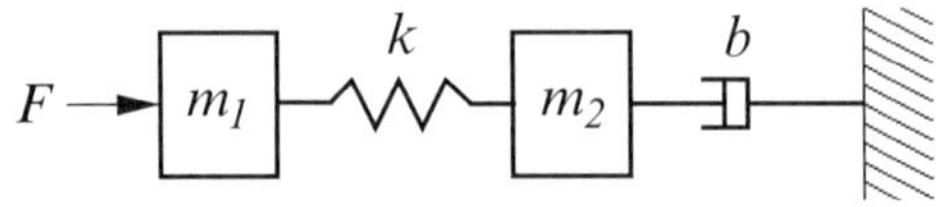

Simulate the behavior of the system using the parameters; $m_1 = 1$ kg, $m_2 = 0.1$ kg, $k = 25$ N/m, $b = 0.1$ N-s/m, and $F = 0.1 \sin \omega t$. Solve the problem for $\omega = 0.5$, 5 and 50 radians/s.

15. Derive the Lagrangian differential-algebraic equations for the 2-stage amplifier shown below. Simulate the behavior of the system using the function `ride`. The model parameters are defined in Example 6.15.

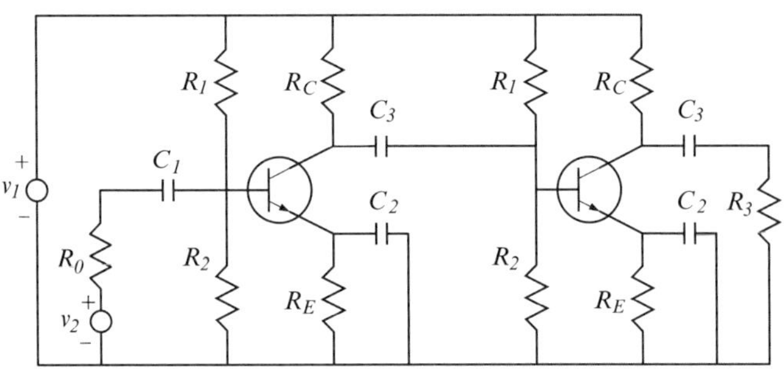

Index